T0190278

INTRODUCTION TO OPTICAL ENGINEERING

INTRODUCTION TO OPTICAL ENGINEERING

FRANCIS T. S. YU
Pennsylvania State University

XIANGYANG YANG
University of New Orleans

CAMBRIDGE
UNIVERSITY PRESS

CAMBRIDGE UNIVERSITY PRESS
Cambridge, New York, Melbourne, Madrid, Cape Town, Singapore,
São Paulo, Delhi, Dubai, Tokyo

Cambridge University Press
The Edinburgh Building, Cambridge CB2 8RU, UK

Published in the United States of America by Cambridge University Press, New York

www.cambridge.org
Information on this title: www.cambridge.org/9780521574938

First published 1997

A catalogue record for this publication is available from the British Library

Library of Congress Cataloguing in Publication data

Yu, Francis T. S., 1934–
    Introduction to optical engineering/Francis T. S. Yu, Xiangyang Yang
        p. cm.
    Includes bibliographical references.
    ISBN 0-521-57366-1 (hc). – ISBN 0-521-57493-5 (pbk).
    1. Electro-optical devices. 2. Optical instruments. 3. Optics.
I. Yang, Xiangyang II. Title
TA1750.Y8 1997
621.36 – dc20   96-2864
                                                                  CIP

ISBN 978-0-521-57366-5 Hardback
ISBN 978-0-521-57493-8 Paperback

Transferred to digital printing 2010

To our undergraduates,
without them we would not have the graduates

TABLE OF CONTENTS

PREFACE

Since the first edition of this book (previously entitled, *Principles of Optical Engineering*) was published, significant technical advances in optical engineering have been made; especially in the areas of telecommunication, medicine, and entertainment. Optical fiber communication has become a key to new developments such as the compact disc used in the entertainment industry and optical scanners used in the local supermarkets. With this new technology, knowledge of the optical field has become a requirement for the electrical engineering student. The purpose of this book is to introduce engineering students to some of the basic concepts in modern optical engineering and bridge the gap for those interested in continuing on with the higher-level electro-optics courses.

We have not only changed the title of the book to give a better representation of the content, but also made substantial revisions in an effort to update the sections on optical detectors, spatial light modulators, as well as fiber optics. Chapters 5 and 6, as well as various others, have been rewritten. New chapters such as optical instruments and interferences have been added. Most of the new material has been used in the classroom at the Electrical Engineering Departments at Penn State and New Orleans.

This book is not intended to cover the vast domain of optical engineering; rather, it comprises fundamental material that we believe to be important and useful to the undergraduate electrical engineering student. The book is suitable as a junior- and senior-level course text, and we have found that it is possible to teach the whole book, without significant omission, in one semester.

Finally, we would like to express our appreciation to our undergraduate students for their active participation, feedback, and criticism. In view of the enormous amount of contributions in the field, we apologize in advance for those omissions. We are indebted to Mrs Debby Pruger for her patience in typing our manuscript.

Francis T. S. Yu
University Park, PA
Xiangyang Yang
New Orleans, LA

PRINCIPLES OF REFLECTION AND REFRACTION

We shall begin our discussion with the basic phenomena of *reflection* and *refraction* at a boundary surface that has been formed by the meeting of two different media. Although light is electromagnetic in nature, its wavelength is much shorter than that of a radio wave. The velocity of light wave propagation in a dielectric medium is given by

$$v = \frac{1}{\sqrt{\mu\varepsilon}},$$
(1.1)

where μ and ε are the permeability and the permittivity (i.e., the dielectric constant) of the medium.

Let us consider a monochromatic (i.e., a single wavelength) light wave that impinges on a plane boundary surface between two media as depicted in Figure 1.1. If we assume that the two media have different permeabilities and permittivities, the velocities of the light waves within these two media will be different. For simplicity, we assume that these two media are transparent; hence, the incident light wave would penetrate the second medium continuously. From past experience, we know that penetrating light rays will bend away from the direction of incidence. This bending causes optical illusions such as the broken appearance of a tablespoon when it is immersed in clear water. Thus, we see that when a reflected light ray and a refracted light ray originate at the boundary surface, only a part of the incident ray is transmitted (i.e., refracted), while the rest is reflected.

1.1 Snell's Law of Refraction

Let us now discuss one of the most important theories of refraction, Snell's law of refraction. The angle of the incident light ray θ_1 is called the *incident angle*; the angle of the refracted light ray θ_2 is called the *refraction angle*; and the angle of the reflected light ray ϕ, is called the *reflection angle*.

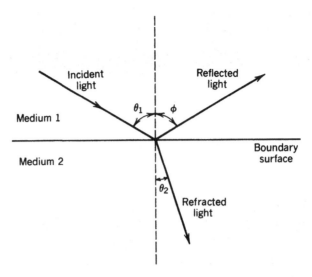

FIGURE 1.1 Reflection and refraction of light.

Since the boundary surface between the two media is assumed to be a planar surface, it can be shown, by the law of refraction, that the reflection angle is equivalent to the incident angle. It can also be shown that the ratio of the sine of the incident angle to the sine of the refraction angle is equal to the ratio of their velocities of wave propagation in the two media; that is,

$$\frac{\sin \theta_1}{\sin \theta_2} = \frac{v_1}{v_2},\tag{1.2}$$

where v_1 and v_2 are the velocities of the wave propagation in medium 1 and medium 2, respectively.

To show that Eq. 1.2 holds, we assume a monochromatic plane wave illumination, as depicted in Figure 1.2; here the thick lines represent the constant propagation of the phase train. Since the two media are assumed to be different, the velocities of wave propagation are therefore not the same. Thus the reflected and refracted wavefronts can be represented as

$$\overline{AP_2} = v_1 t, \qquad \overline{P_1B} = v_1 t, \quad \text{and} \quad \overline{P_1C} = v_2 t,\tag{1.3}$$

where t denotes the time variable.

From the triangles of P_1AP_2, P_1BP_2, and P_1CP_2 and the law of sines, we have

$$\overline{P_1P_2} = \frac{\overline{AP_2}}{\sin \theta_1} = \frac{\overline{P_1B}}{\sin \phi} = \frac{\overline{P_1C}}{\sin \theta_2}.\tag{1.4}$$

With substitutions from Eq. 1.3, we reduce Eq. 1.4 to

$$\frac{v_1}{\sin \theta_1} = \frac{v_1}{\sin \phi} = \frac{v_2}{\sin \theta_2},\tag{1.5}$$

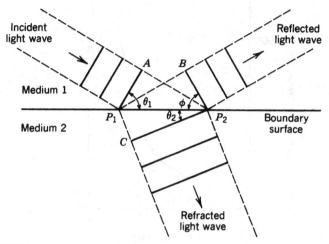

FIGURE 1.2 Reflection and refraction of monochromatic wave trains.

from which we conclude that

$$\phi = \theta_1, \tag{1.6}$$

that is, the angle of reflection is equal to the angle of incidence.

We now define the *indices of refraction* of the two media by the following equations,

$$\eta_1 \triangleq \frac{c}{v_1}, \tag{1.7}$$

$$\eta_2 \triangleq \frac{c}{v_2}, \tag{1.8}$$

where c is the velocity of light propagation in a vacuum ($c = 3 \times 10^8$ m/s). Thus, Eq. 1.2 can be written in terms of the refractive index ratio

$$\frac{\sin \theta_2}{\sin \theta_1} = \frac{\eta_1}{\eta_2}, \tag{1.9}$$

or, alternatively,

$$\eta_1 \sin \theta_1 = \eta_2 \sin \theta_2, \tag{1.10}$$

where Eq. 1.9 or Eq. 1.10 is best known as *Snell's law of refraction*.

Strictly speaking, to find out how an incident light wave is reflected and refracted from a boundary surface, we should start with the boundary conditions from the electromagnetic standpoint. It can be shown that the reflected and refracted light waves are affected by the polarization of the incident light wave. In Figure 1.3, however, we illustrate a reflected light wave coming from an unpolarized incident light source. Thus, we see that,

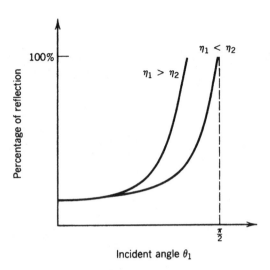

FIGURE 1.3   Reflection as a function of the angle of incidence.

in order to obtain a total reflection, the angle of incidence should be smaller for $\eta_1 > \eta_2$ than for $\eta_1 < \eta_2$.

EXAMPLE 1.1

In Figure 1.1 we assume that the refractive index for medium 1 is $\eta_1 = 1.3$, and that for medium 2 it is $\eta_2 = 1.5$. If a light ray is traveling in the upper medium and impinging on the boundary surface at an angle of $45°$ to the normal axis, determine the corresponding angles of reflection and refraction.
We note that

$$\eta_1 = 1.3, \qquad \theta_1 = 45°, \qquad \eta_2 = 1.5.$$

From Eq. 1.6 the angle of reflection is

$$\phi = \theta_1 = 45°.$$

By Snell's law of refraction, as expressed in Eq. 1.10, the angle of refraction can be evaluated as follows:

$$1.3 \sin 45° = 1.5 \sin \theta_2$$
$$\sin \theta_1 = 0.613$$
$$\theta_2 = 37.8°.$$

EXAMPLE 1.2

Given the two media described in Example 1.1 and the light ray impinging from the lower medium on the boundary surface, at an angle of incidence equal to 35°, calculate the angle of refraction in the upper medium.
    We note that

$$\eta_1 = 1.5, \qquad \eta_2 = 1.3, \quad \text{and} \quad \theta_1 = 35°.$$

By applying Snell's law, we can calculate the angle of refraction in the upper medium:

$$1.5 \sin 35° = 1.3 \sin \theta_2$$
$$\sin \theta_2 = 0.66$$
$$\theta_2 = 41.4°.$$

Again we see that the angle of reflection is equal to the angle of incidence:

$$\phi = \theta_1 = 35°.$$

1.2 Huygens' Principle

We now discuss *Huygens' principle*, one of the most significant principles in diffraction optics. Through Huygens' principle, it is possible to obtain, by using graphs, the shape of a wavefront at any instant when given a wavefront of an earlier instant. The principle essentially states that every point of a wavefront can be regarded as a secondary point source from which a small spherical wavelet is generated and spreads in all directions at a velocity of wave propagation. Thus, a new wavefront can be constructed graphically by drawing a surface tangent to all the secondary spherical wavelets. If the velocity of propagation is not constant at all parts of the wavefront, each spherical wavelet should be drawn to an appropriate wave velocity.
    Figure 1.4 gives an illustrated example for the use of Huygens' principle. The known wavefront is shown as the surface Σ, called the *primary wavefront*, and the directions of wave propagation are indicated by small arrows. To determine the wavefront after an interval of time Δt with a wave velocity v, we construct a series of spheres (i.e., the secondary spherical wavelets) of radius $r = v\Delta t$ from each point of the surface wavefront Σ. These spheres represent the secondary wavelets. Now if we draw a surface tangent to all the surfaces of the spheres, we get the shape of a new wavefront, called the

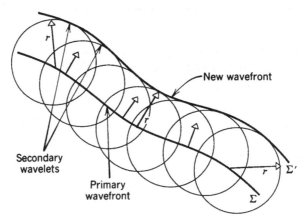

FIGURE 1.4 Huygens' principle.

secondary wavefront. We note that the wave velocities at every point on the surface are assumed to be equal. This is a typical example of light wave propagation in a homogeneous isotropic medium.

Furthermore, according to Huygens' principle, a backward wavefront should form from the secondary spherical wavelets. This phenomenon has never been observed, however. The discrepancy can be explained by examining the wave propagation within the surrounding closed surface. None the less, the application of Huygens' principle lies in predicting a new wavefront. During the period immediately after Huygens' principle was developed, little physical significance was attached to the secondary wavelets, but as the wave nature of light propagation became more fully understood, Huygens' principle took on a deeper physical meaning.

Huygens' principle is a very useful technique for demonstrating the *diffraction phenomenon*, which will be discussed in Chapter 8.

EXAMPLE 1.3

Let Σ be the primary wavefront at time t. At each point on Σ draw a spherical wavelet of radius $r = v\Delta t$, where $v = c/\eta$, and c is the velocity of light. Although η varies continuously in the medium, the radii of the wavelets vary according to η, as shown in Figure 1.5. As a result, the surface Σ', which is tangent to all the secondary wavelets, is the new wavefront. By similar procedures, we can obtain another new wavefront Σ'' from Σ', and so on, as shown in the figure. Accuracy in predicting the new wavefronts varies inversely with time Δt. Thus, smaller values of Δt correspond to more accurate predictions of the new wavefront.

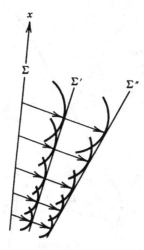

FIGURE 1.5 Wavefront propagation in the nonhomogeneous medium.

This phenomenon in wave propagation in a nonhomogeneous medium may explain several interesting optical illusions in the Earth's atmosphere, namely the mirage and looming effects of distant objects.

1.3   Refraction and Reflection

In Section 1.1 we derived Snell's law of refraction, which is very useful in geometrical optics and in physical optics. In Section 1.2 we discussed the applications of Huygens' principle for predicting new wavefronts in either homogeneous or nonhomogeneous media. We now show that the laws of refraction and reflection can also be derived from Huygens' principle.

Let us consider a plane monochromatic light wave that impinges on a plane boundary surface Σ between two homogeneous media. The velocities of light wave propagation in each of the media are different. When applying Huygens' principle, we see that all the points on surface Σ act as centers of propagation for the secondary spherical wavelets. The secondary wavelets, originating at the boundary surface Σ, give rise to both forward and backward propagations. This results in both refracted and reflected wavefronts, as shown in Figure 1.6.

We note that the wavelets propagating backward have a velocity equal to that of the incident light wave, since these wavelets are in the same medium as the incident wave. The radii of these reflected wavelets can be determined by the following equation,

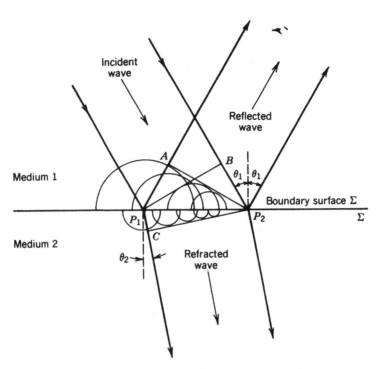

FIGURE 1.6 Reflection and refraction by Huygens' principle.

$$r_n = \frac{c\Delta t_n}{\eta_1}, \qquad n = 1, 2, 3, \ldots, \tag{1.11}$$

where c is the velocity of light, η_1 is the refractive index of the upper medium, and Δt_n is the time interval after the incident wavefront touches the boundary surface. The oblique light ray of line $\overline{P_1B}$ is shown in Figure 1.6. The radius of the secondary wavelet, with its center at point P_1, is

$$r_1 = \overline{P_1A} = \frac{c\Delta t_1}{\eta_1}. \tag{1.12}$$

The radii of the secondary wavelets with centers at other points on the boundary surface in line $\overline{P_1P_2}$ can be determined by

$$r_n = \overline{OP_2} \sin\theta_1, \tag{1.13}$$

where $\overline{OP_2}$ is the distance between the center of a secondary wavelet and point P_2, as shown in the figure, and θ_1 is the angle of reflection.

Similarly, we can also determine the radii of the refracted secondary wavelets by

$$r'_n = \frac{c\Delta t_n}{\eta_2}, \qquad n = 1, 2, 3, \ldots, \tag{1.14}$$

where η_2 is the refractive index of the lower medium. Again if the radius of the refracted wavelet at point P, is

$$r_1' = \overline{P_1 C} = \frac{c \Delta t_1}{\eta_2}, \tag{1.15}$$

the radii of the other refracted wavelets, whose centers are on line $\overline{P_1 P_2}$, must be

$$r_n' = \overline{OP_2} \sin \theta_2, \tag{1.16}$$

where $\overline{OP_2}$ is the distance between the center of a refracted wavelet and the point P_2, and θ_2 is the angle of refraction.

To maintain plane-reflected and plane-refracted wavefronts (from Figure 1.6), we see that

$$\overline{BP_2} = \overline{P_1 A} = \frac{c}{\eta_1} \Delta t_1 = v_1 \Delta t_1, \tag{1.17}$$

where v_1 is the velocity of wave propagation in the upper medium. By the law of sines, we have

$$\overline{P_1 P_2} = \frac{\overline{BP_2}}{\sin \theta_1} = \frac{\overline{P_1 A}}{\sin \theta_1} = \frac{\overline{P_1 C}}{\sin \theta_2}. \tag{1.18}$$

Since the ratio of the wave propagations in the two media is

$$\frac{\sin \theta_1}{\sin \theta_2} = \frac{v_1}{v_2}, \tag{1.19}$$

where v_2 is the velocity of wave propagation in the lower medium, and

$$v_1 = \frac{c}{\eta_1}, \qquad v_2 = \frac{c}{\eta_2}, \tag{1.20}$$

we can combine Eqs. 1.19 and 1.20 to obtain the following:

$$\frac{\sin \theta_1}{\sin \theta_2} = \frac{\eta_2}{\eta_1}. \tag{1.21}$$

This equation gives Snell's law of refraction as derived from Huygens' principle.

1.4 Spherical Wave and Image Formation

In order to gain a deeper appreciation of Snell's law of refraction, we begin with the following situation. A spherical wavefront is reflected and refracted at a plane boundary surface.

Let us consider a monochromatic point source S located in the upper medium of a boundary surface, as shown in Figure 1.7. By applying

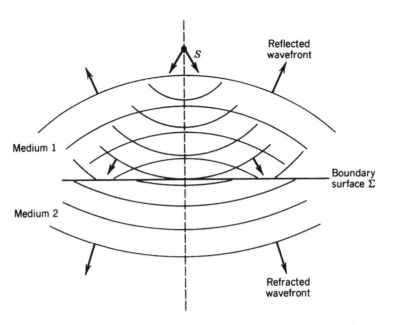

FIGURE 1.7 The effects of reflection and refraction by a spherical wave.

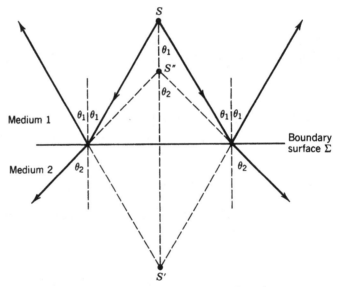

FIGURE 1.8 Light ray representation of a point source.

Huygens' principle, we can show that a divergent spherical wavefront, with the same radius of curvature, is reflected from the boundary surface, and a refracted spherical wavefront, with a different radius of curvature, penetrates the lower medium.

If we use a light ray of representation, Figure 1.7 can be replaced by Figure 1.8. In this figure we see that when an incident light ray hits the boundary surface, it is reflected with the same angle of incidence.

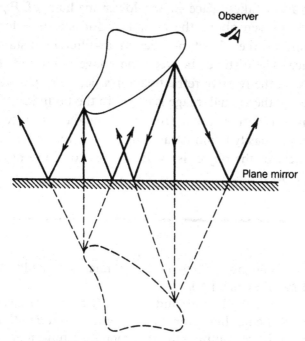

FIGURE 1.9 Image formation by reflection.

If we extend the reflected light rays below the boundary surface and through the lower medium, they converge at point S', which lies opposite S at an equal distance from the boundary surface. Similarly, if we extend the refracted light rays into the upper medium, the extended light rays converge at point S'', which lies on line SS'.

We have shown that although the wavefronts originate from point S, they appear to diverge from point S'. This phenomenon is caused by reflection from a plane boundary surface. Standing in the upper medium, we see that image point S' lies below the boundary surface, directly opposite object point S. Point S' is called the *virtual image* because the light rays appear to converge below the boundary surface. Images formed above the boundary surface, on the other hand, are called *real images*.

Since we can think of an extended, diffuse (i.e., scattered) object as composed of many infinitesimal points, the image of the object can therefore be found by the ray-tracing technique. For example, a finite extended object lies above a plane mirror. Every point of the object results in an image point behind the mirror, as illustrated in Figure 1.9. Looking from above the mirror, an observer sees a virtual image of the object. Notice that a small cone of reflected light rays carries the entire virtual image. Since these reflected light rays are identical to the light rays that emanate from the object, the observer perceives them as they actually originated from the object.

If the observation takes place in the lower medium of Figure 1.8, the virtual image S'' is seen above the boundary surface. It is located on the same normal line as the object but lies at a different distance from the boundary surface. The distance between the image point and the boundary surface depends on the relative refractive index between the two media. For example, if $\eta_1 > \eta_2$, the virtual image is closer to the boundary surface, but if $\eta_1 < \eta_2$, the image is farther away from the boundary surface. The second phenomenon can be easily demonstrated by placing an object in clear water. The "lifting" effect of the object is evidently due to the refractive phenomenon of the light waves.

EXAMPLE 1.4

Using the ray-tracing method, locate the virtual images formed by the two plane mirrors M_1 and M_2 of Figure 1.10.

Let us extend mirror M_1 by line m_1 and mirror M_2 by line m_2, as shown in the figure. Draw perpendicular lines from object points A and B with respect to the mirror plane of M_1. If we continue the ray tracings behind mirror plane M_1 at distances equal to the distances from object points A and B, we can find the position of the virtual image $A'B'$ created by M_1. Similarly, by using image $A'B'$ and the mirror plane of M_2, we can find another virtual image $A''B''$.

We stress that the locations of these virtual images can be easily verified by employing the ray-tracing method. For example, two light rays originating from point A of the object strike mirror plane M_1. Using the law of reflection, we can show that the light rays from the object are reflected from mirror M_1 toward mirror M_2, and are then reflected from mirror M_2, as shown by the arrowheads

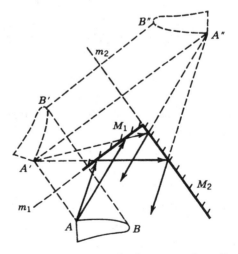

FIGURE 1.10 Image formation by successive reflections.

in the figure. If the reflected rays from mirrors M_1 and M_2 are projected backward behind the mirrors, two virtual images can be found. The resulting images coincide perfectly with the images formed by the extended-mirrors method that we described first.

EXAMPLE 1.5

Calculate the distance from image point S'' in Figure 1.8 to the boundary surface.

Let us call this distance y_2, and the distance between the object point S and the boundary surface y_1. From Snell's law of refraction (Eq. 1.10), we have

$$\eta_1 \sin \theta_1 = \eta_2 \sin \theta_2.$$

From the right triangle in Figure 1.8, we see that

$$y_1 \tan \theta_1 = y_2' \tan \theta_2. \tag{1.22}$$

It is apparent that

$$\frac{y_1 \tan \theta_1}{\eta_1 \sin \theta_1} = \frac{y_2 \tan \theta_2}{\eta_2 \sin \theta_2}. \tag{1.23}$$

Furthermore, since

$$\tan \theta_1 = \frac{\sin \theta_1}{\cos \theta_1}$$

and

$$\tan \theta_2 = \frac{\sin \theta_2}{\cos \theta_2},$$

Eq. 1.23 can be written as

$$y_2 = y_1 \frac{\eta_2}{\eta_1} \frac{\cos \theta_2}{\cos \theta_1}. \tag{1.24}$$

We note further that the ratio $\cos \theta_2 / \cos \theta_1$ varies with respect to the angle of incidence, θ_1, and that y_2 is not the same for the light rays that diverge from S. Thus, if the refracted wavefront is not spherical, it will not appear to diverge from a point. However, if the observer is looking vertically upward from a point near the normal line below the boundary surface, the angles of refraction and incidence are very small:

$$\cos \theta_2 \simeq \cos \theta_1. \tag{1.25}$$

Equation 1.24 can be reduced to

$$y_2 = y_1 \frac{\eta_2}{\eta_1}. \tag{1.26}$$

Thus, if $\eta_1 > \eta_2$, the image appears to be closer to the boundary surface, and if $\eta_1 < \eta_2$, the image appears to be farther away from the boundary surface.

1.5 Total Reflection and Dispersion

Total reflection is a phenomenon that we frequently encounter in everyday life. To illustrate the phenomenon, we sketch a number of light rays emanating from a point source S, located in the lower medium. These light rays strike the plane boundary surface Σ of the upper medium shown in Figure 1.11. From Snell's law of refraction we can obtain the sine of the refraction angle,

$$\sin \theta_r = \frac{\eta_i}{\eta_r} \sin \theta_i, \tag{1.27}$$

where θ_i is the angle of incidence.

If the ratio of η_i and η_r is greater than 1 (i.e., $\eta_i > \eta_r$), then $\sin \theta_r$ is greater than $\sin \theta_i$. Thus, when θ_r is 90°, θ_i is still smaller than 90° (i.e., $\theta_i < 90°$). But when $\theta_r = 90°$, the refracted light ray will be propagated on the boundary surface in the upper medium, as shown in the figure. This angle of incidence, which produces a refracted light ray tangent to the boundary surface, is called the *critical angle of incidence* and is represented by θ_c in the figure. If, however, the angle of incidence is greater than the critical angle of incidence (i.e., $\theta_i > \theta_c$), all the light rays will be reflected back into the lower medium. This second phenomenon is known as *total internal reflection*.

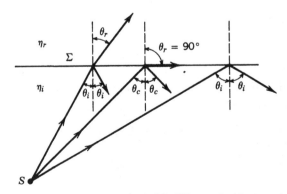

FIGURE 1.11 Effects on refraction with different incident angles $\eta_i > \eta_r$.

By using Snell's law of refraction, we see that total internal reflection can occur only when a light ray from one medium strikes the boundary surface of another medium that has a *smaller* index of refraction. Thus, the critical angle of incidence is

$$\theta_c = \arcsin\left(\frac{\eta_r}{\eta_i}\right), \tag{1.28}$$

where $\theta_r = 90°$.

EXAMPLE 1.6

A light ray travels under the surface of a clear-water lake. The index of refraction of the water is $\eta = 1.33$. Compute the critical angle of incidence.

Since it is a water-to-air surface boundary, we have

$$\theta_c = \arcsin\left(\frac{1}{1.33}\right) = 48.75°.$$

Thus, if the angle of incidence is greater than the critical angle (i.e., $\theta_i > \theta_c$), the light ray will be totally reflected back under the surface. In practice, however, the boundary surface is usually not uniform. Therefore, when the incident angle exceeds the critical angle, we expect only a small portion of light rays to be refracted.

EXAMPLE 1.7

Suppose a light ray is incident at an angle θ_i on the upper surface of a transparent glass plate, as shown in Figure 1.12. Assuming that both surfaces of the glass plate are optically smooth and parallel to one another, compute the

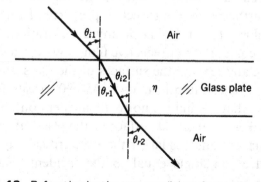

FIGURE 1.12 Refraction by the two parallel surfaces of a glass plate.

corresponding angles of refraction below both the first and second surfaces of the glass plate.

Let η be the index of refraction of the glass plate and η_a be the index of refraction of air. Applying Snell's law to the refraction on the first boundary surface we have

$$\eta_a \sin \theta_{i1} = \eta \sin \theta_{r1}.$$

and on the second boundary surface we have

$$\eta \sin \theta_{i2} = \eta_a \sin \theta_{r2}.$$

Since

$$\theta_{i2} = \theta_{r1},$$

we obtain

$$\theta_{r2} = \theta_{i1}.$$

Thus, we see that the first angle of incidence (where the light is traveling from air to glass) is equal to the second angle of refraction (where the light is traveling from glass to air). The final emergent light is parallel to but not in line with the incident ray. In fact, after passing through any number of transparent parallel plates, a ray of light is displaced only from its original path. The refracted ray will therefore still be parallel to the incident light.

We emphasize that total internal reflection inside a glass prism is of considerable importance when dealing with optical instruments. The critical angle for light passing from air to a glass surface can be shown to be

$$\sin \theta_c = \frac{1}{1.5} = 0.67,$$
$$\theta_c = 42°,$$

where we assume that the refractive index for glass is $\eta = 1.5$.

Total reflection is one of the major advantages of the glass prism over the standard metallic-coated reflector, since no metallic surface coating offers 100% reflection. Furthermore, the reflective property of glass prisms is permanent and not subject to tarnishing or any deterioration. When the glass prism has a surface coating of nonreflective film, however, a small amount of light is lost entering and leaving the surfaces of the glass prism.

A commonly used reflecting prism, a 45°–45°–90° prism, is shown in Figure 1.13. We assume that the light is normally incident on one of the vertical faces and that it strikes the inclined surface of the prism at an incident angle of 45°. Since the incident angle is larger than the critical angle, the light will be totally reflected at an angle equal to the incident. Thus, the light is reflected 90° by the prism.

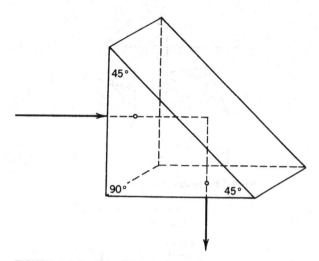

FIGURE 1.13 A 45°–45°–90° totally reflection prism.

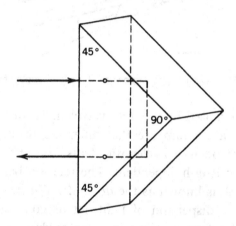

FIGURE 1.14 A totally reflection Porro prism.

We will, however, obtain different results when we project the light on the same prism, but at a different angle of incidence, as shown in Figure 1.14. After entering the prism, the light will be totally reflected back opposite to the incident light. In geometric optics, a 45°–45°–90° prism arranged in this manner is often called a *Porro* prism. Two or more *Porro* prisms can also be used to change the angle and direction of incident light, as shown in Figure 1.15.

Let us now consider the *dispersion* of light waves caused by a prism. When a beam of sunlight strikes a triangular prism, the light rays emerges on the other side as a rainbow of colors, as shown in Figure 1.16. In previous sections, we have dealt only with monochromatic light (i.e., light with a single wavelength) when discussing the phenomenon of refraction. Now we shall

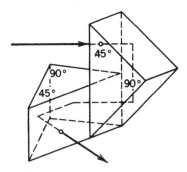

FIGURE 1.15 Reflection by two Porro prisms.

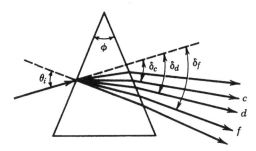

FIGURE 1.16 Effect of light dispersion. δ_d, Mean angle of deviation.

discuss the refraction caused by different wavelengths of light. Although the velocities of light in a vacuum are the same, the wave propagation in a material is dependent on its wavelength. Thus, the refractive index of the material must be wavelength dependent. The relation between wave propagation and wavelength is known as the *dispersion effect*.

To demonstrate the dispersion of light, let us consider a light ray that strikes the surface of a triangular prism at an angle θ_i, as shown in Figure 1.16. The prism's apex angle is denoted by ϕ, and its refractive index is denoted by η. The medium on either side of the prism is air, and the *angle of deviation* δ can be determined by using Snell's law of refraction. If the angle of incidence is decreasing, the angle of deviation δ can be shown first to decrease, and then to increase. There is a minimum angle of deviation δ_m if the light ray passes the triangular prism symmetrically. We can also show that the refractive index η of the prism satisfies the following equation:

$$\eta = \frac{\sin[(\phi + \delta_m)/2]}{\sin(\phi/2)}. \tag{1.29}$$

Furthermore, for small angles, $\sin\theta \approx \theta$, the preceding equation can be approximated by

$$\eta \simeq \frac{\phi + \delta_m}{\phi}, \tag{1.30}$$

in which we have

$$\delta_m \simeq (\eta - 1)\phi. \tag{1.31}$$

This is a very useful approximation.

Assume that a ray of white light strikes the triangular prism shown in Figure 1.16. Since a larger angle of deviation corresponds to a larger refractive index (see Figure 1.17), violet light would emerge with the largest angle of deviation, and red light would emerge with the smallest angle. Hence, the light emerging from the prism would disperse into a spectrum of rainbow colors wherein each color corresponds to a different angle of deviation. The difference between the angles of deviation of *any* two rays is called the *angle of dispersion*.

The *dispersive power* of a given material is defined as

$$W \triangleq \frac{\eta_f - \eta_c}{\eta_d - 1}, \tag{1.32}$$

where η_c, η_d, and η_f are the refractive indices of the material with respect to the red, yellow, and blue wavelengths. Notice that these wavelengths are arbitrarily selected for reference.

We further note that this dispersive power is the *measure* of the refractive property of a given material to bend spectral lines of light in different angles. A prism (composed of a given material) has this ability to separate polychromatic light into spectral lines of colors.

Consider a narrow beam of polychromatic visible light that is incident to a triangular prism with a narrow apex angle. The minimum angles of deviation for red, yellow, and blue light are denoted by δ_c, δ_d, and δ_f, respectively. From Eq. 1.31 we have

$$\delta_c = (\eta_c - 1)\phi, \tag{1.33}$$
$$\delta_d = (\eta_d - 1)\phi, \tag{1.34}$$
$$\delta_f = (\eta_f - 1)\phi, \tag{1.35}$$

where δ_d is the *mean angular deviation* of the spectrum. From these equations we see that

$$\delta_c - \delta_f = (\eta_c - \eta_f)\phi.$$

Dividing by Eq. 1.34, we have

$$\frac{\delta_f - \delta_c}{\delta_d} = \frac{\eta_f - \eta_c}{\eta_d - 1} = W, \tag{1.36}$$

which is the dispersive power defined by Eq. 1.32. From this we see that the dispersive power is also equal to the ratio of the dispersion between the blue and red angles of deviation (i.e., $\delta_f - \delta_c$), to the mean deviation δ_d of the entire spectrum.

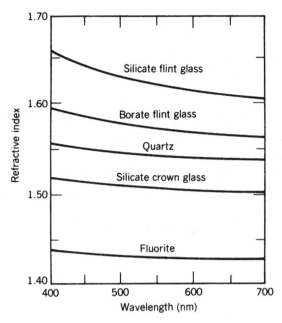

FIGURE 1.17 Refractive indices as a function of wavelength for different media.

EXAMPLE 1.8

Figure 1.17 shows the refractive indices of silicate flint glass and borate flint glass as functions of wavelength. We use Eq. 1.32 to determine their corresponding dispersive powers, and we assume that the spectral lines of 475, 550, and 650 nm can be used to determine the reference indices of the materials. From Figure 1.17 we obtain the following refractive indices for silicate flint glass: $\eta_f = 1.635$, $\eta_d = 1.625$, and $\eta_c = 1.618$. Substituting these into Eq. 1.36, we obtain the dispersive power for silicate flint glass:

$$W = \frac{1.635 - 1.618}{1.625 - 1} = 0.0272.$$

Using this same method, we can obtain the refractive indices for borate flint glass: $\eta_f = 1.58$, $\eta_d = 1.57$, and $\eta_c = 1.565$. Thus, the dispersive power for silicate crown glass is

$$W = \frac{1.58 - 1.565}{1.57 - 1} = 0.0175.$$

EXAMPLE 1.9

Given an apex of $9°$, calculate the mean angular deviation and the power of dispersion produced by a prism composed of silicate flint glass and one composed of borate flint glass. Since the apex angle is small, Eq. 1.31 can be applied.

For silicate flint glass:

Mean deviation $= \delta_d = (\eta_d - 1)\phi = (1.62 - 1)9° = 5.58°$.
Dispersion $= W\delta_d = 0.0272 \times 5.58° = 0.15°$.

For borate flint glass:

Mean deviation $= \delta_d = (\eta_d - 1)\phi = (1.57 - 1)9° = 5.13°$.
Dispersion $= W\delta_d = 0.0175 \times 5.13° = 0.089°$.

1.6 Polarization

Let us assume a monochromatic plane wave is propagating along the z axis, as illustrated in Figure 1.18, in which the electric and magnetic fields are

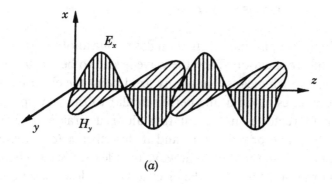

(a)

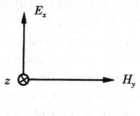

(b)

FIGURE 1.18 A linearly polarized light. *(a)* Propagation in the z-direction. *(b)* The electric and magnetic vectors.

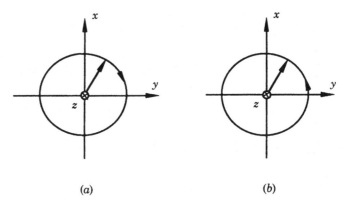

(a)   (b)

FIGURE 1.19 (a) A clockwise circularly polarized light propagation in the z-direction. (b) A counterclockwise circularly polarized light.

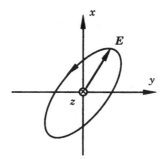

FIGURE 1.20 An elliptically polarized light.

oscillating along straight lines at right angles to one another, perpendicular to the direction of propagation. Then the wave is said to be *linearly polarized* (in the x direction), which means that at any fixed point the electric (or magnetic) field vector oscillates along a line perpendicular to the wave propagation. If the electric field vector is not oscillating along a straight line but changes direction as the wave propagation, and it describes a (clockwise) circle, as shown in Figure 1.19a, then the light wave is called *clockwise circularly polarized* light. On the other hand, if the electric vector describes a counterclockwise circle, as shown in Figure 1.19b, it is termed *counterclockwise circularly polarized* light. Furthermore, any electric field vector E can be decomposed into components E_x and E_y. If $E_x \neq E_y$, and assuming a constant phase shift between them, the resultant electric field vector E may describe an ellipse, as shown in Figure 1.20, in which the light wave is *elliptically polarized* light. Nevertheless, if the electric field vectors are arranged in a manner symmetrically distributed in all directions perpendicular to the propagation, as shown in Figure 1.21, the light field is known as *unpolarized* or *natural* light.

If linearly polarized light is incident on a plane of a boundary surface, as shown in Figure 1.22, then two distinct types of polarized reflection and refraction are observed: namely, s-polarized light and p-polarized light.

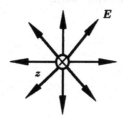

FIGURE 1.21 An unpolarized or natural light.

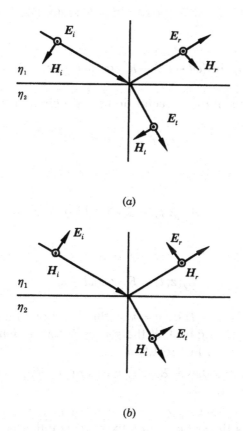

(a)

(b)

FIGURE 1.22 Reflection and refraction of s-polarized light.

EXAMPLE 1.10

An electric field vector of a polarized light is given by

$$E = iE_x \sin(\omega t - kz) + jE_y \sin(\omega t - kz + \vartheta),$$

where i and j are unit vectors.

Describe the following polarization status:

(a) $E_x = E_y$ and $\vartheta = \dfrac{\pi}{2}$.

(b) $\vartheta = 0$ or π.

(c) $E_x \neq E_y$ and $\vartheta = \dfrac{\pi}{4}$.

Answers

(a) Since two constituent fields have equal amplitude:

$E_x = E_y = E_0$, while $\vartheta = \pi/2$, the resulting E vector has a fixed amplitude as given by

$$|E|^2 = E_x^2 \sin^2(\omega t - kz) + E_y^2 \cos^2(\omega t - kz) = E_0^2 = \text{const.}$$

However, the locus of the E vector describes a circular locus as given by

$$E_x^2(z, t) + E_y^2(z, t) = E_0^2,$$

where

$$E_x(z, t) = E_x \sin(\omega t - kz),$$

and

$$E_y(z, t) = E_y \cos(\omega t - kz).$$

Since $E_y(z, t)$ leads $E_x(z, t)$ by $\vartheta = \pi/2$, the direction of the E vector rotates clockwise when looking back at the source. This is a *left circularly polarized* light, as shown in Figure 1.23a.

(b) For $\vartheta = 0$ or π, we have $E_x(z, t) = E_x \sin(\omega t - kz)$, and $E_y(z, t) = \pm E_y \cos(\omega t - kz)$, respectively.

It can be shown that $E_y(z, t) = E_x(z, t) \tan \alpha$, in which the resulting E vector describes a linear locus, at a fixed angle either at $+\alpha$ or $-\alpha$, shown in Figure 1.23b, as given by

$$\alpha = \tan^{-1} \frac{E_x}{\pm E_y},$$

and

$$|E|^2 = (E_x^2 + E_y^2) \sin^2(\omega t - kz).$$

Thus we see that for $\vartheta = 0$ or π, the light wave is linearly polarized.

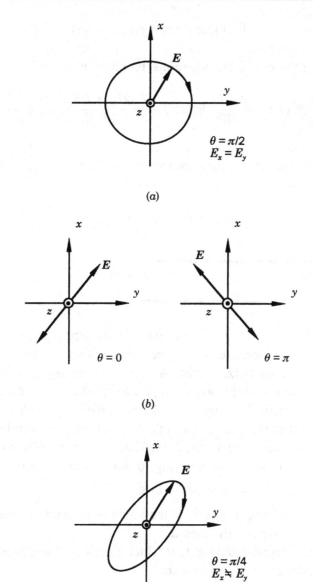

FIGURE 1.23 Reflection and refraction of p-polarized light.

(c) When $E_x \neq E_y$ and $\vartheta = \pi/4$, the electric field is elliptically polarized light. This can be shown as follows:

$$E_x(z, t) = E_x \sin(\omega t - kz)$$

$$E_y(z, t) = E_y \sin\left(\omega t - kz + \frac{\pi}{4}\right)$$

$$\frac{E_Y}{\sqrt{2}}[\sin(\omega t - kz) + \cos(\omega t - kz)]$$

By combining the preceding equations, one can show that

$$\frac{2}{E_y^2}E_y^2(z,t) + \frac{2 + E_y^2}{2E_x^2}E_x^2(z,t) - \frac{2\sqrt{2}E_y(z,t)E_x(z,t)}{E_yE_x} = \frac{E_y^2}{2},$$

which represents elliptically polarized light. Since $E_y(z,t)$ leads $E_x(x,t)$ by $\pi/4$, it is a left elliptically polarized wave, as shown in Figure 1.23c.

1.7 Polarizers

By referring to the reflection and refraction of polarized light discussed in the preceding section, the phenomena can be used to filter out the linear polarized light from an unpolarized light. At a particular angle of incidence at which the electric field vector is assumed parallel to the plane of incidence, no light would be reflected. This angle is called the *polarizing angle* (between the two media), such that the sum of the reflected and refracted angles is 90°. In other words, at the polarizing angle, the reflected and the refracted rays are at right angles to one another. By referring to Snell's law, we have

$$\eta_1 \sin \vartheta_p = \eta_2 \sin \vartheta_r, \tag{1.37}$$

where ϑ_p is the polarizing angle, ϑ_r is the angle of refraction, and η_1 and η_2 are the refractive indices of the media.

In order to have no reflected light, at which the electric vector is parallel to the plane of incidence, it is required that

$$\vartheta_p + \vartheta_r = 90° \tag{1.38}$$

Therefore, it can be shown that

$$\tan \vartheta_p = \frac{\eta_2}{\eta_1}, \tag{1.39}$$

which is known as *Brewster's law*.

Suppose a beam of unpolarized light is incident at the polarizing angle on a reflecting surface. Each of the linearly polarized waves of the incident beam can be decomposed into electric vectors that are parallel or perpendicular to the plane of the incidence. At the polarizing angle, none of the electric vector components parallel to the plane of incident would be reflected. In other words, all the reflected light rays are linearly polarized with electric field vectors

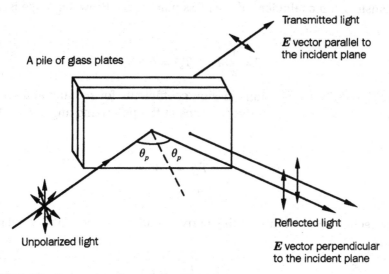

FIGURE 1.24 Separation of unpolarized light into two beams of linearly polarized light.

perpendicular to the plane of incidence. However, in practice, it is rather difficult to achieve a situation where all the perpendicular components are reflected or all the transmitted light rays are parallel components. However, by illuminating a beam of unpolarized light at the polarizing angle on a pile of glass plates, for example, more and more of the perpendicular components may be filtered out by the layers of reflections, as illustrated in Figure 1.24. Thus, we see that unpolarized light can be decomposed into two linearly polarized beams. The device that is capable of transmitting (or reflecting) only waves in which all the electric vectors are in the same direction is called a *polarizer*.

EXAMPLE 1.11

Assume that the refractive index of a silicate glass is $\eta = 1.5$,

(a) Calculate the polarizing angle of the glass.

(b) Assuming that the transmission coefficient of the glass at the polarizing angle is about 85%, what would be the percentage of the overall reflected linearly polarized light?

(c) Repeat (b) for two tightly sandwiched glass plates.

Answers

(a) Since the refractive index of air is $\eta_2 = 1$, substitution in Eq. 1.39 gives the polarizing angle as

$$\vartheta_p = \tan^{-1}\left(\frac{1.5}{1}\right) = 56.3°.$$

(b) The transmission coefficient of the glass plate at the Brewster angle is given by

$$T = \frac{1}{2}(T_p + T_s) = 0.85,$$

where T_p and T_s are the transmission coefficients for parallel and perpendicular components, respectively. Since at the polarizing angle $T_p = 1$, we have

$$\frac{1}{2}(1 + T_s) = 0.85$$

$$T_s = 0.76.$$

The reflection coefficient for the perpendicular components (s-polarized light) is

$$R_s = 1 - T_s = 0.30.$$

Thus the overall percentage of reflected light from the unpolarized light is

$$R = \frac{1}{2}R_s = 15\%.$$

(c) For two sandwiched glass plates, the overall reflected light would be

$$R = \frac{1}{2}R_s(1 + T_s^2)$$

$$= \frac{0.3}{2}[1 + (0.7)^2] = 22.35\%.$$

Notice that the second term is due to the contribution from the second glass plate.

1.8 Fresnel Equations of Reflection

When a plane wave is projected onto a boundary of two transparent media, part of the energy is reflected back into the first medium and the rest is refracted into the second medium. The directions of the reflected and refracted beams are determined by the reflection law and Snell's law of refraction, which were discussed in Section 1.1. In this section, the ratio between the reflected and refracted energy will be studied.

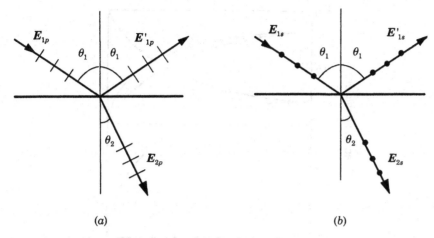

FIGURE 1.25 Light rays at a boundary surface. The direction of the electric vector is shown (a) parallel to the incident plane and (b) normal to the incident plane.

It has been observed in the past that the ratio between the reflected and refracted light intensities depends not only on the incident angle and refractive indices of the media, but also on the polarization state of the incident beam. To model the polarization dependency, we can decompose the incident beam into two orthogonal polarization components. The vibration of the electric field of one component is parallel to the plane of incidence and is called p-polarization. The vibration of the other component is perpendicular to the incidence and is called s-polarization (s stands for the German *senkrecht*, meaning perpendicular). The polarization of the p and s components is illustrated as bars and dots, respectively, in Figure 1.25.

Assume that the refraction indices of the two media are η_1 and η_2, respectively. The reflection and refraction coefficients of both p and s-polarizations at the boundary surface, as derived by Fresnel using the elastic-solid theory, are given by:

$$r_p = \frac{E'_{1p}}{E_{1p}} = \frac{\tan(\theta_2 - \theta_1)}{\tan(\theta_2 + \theta_1)}, \tag{1.40}$$

$$t_p = \frac{E_{2p}}{E_{1p}} = \frac{2\sin\theta_2\cos\theta_1}{\sin(\theta_1 + \theta_2)\cos(\theta_1 - \theta_2)}, \tag{1.41}$$

$$r_s = \frac{E'_{1s}}{E_{1s}} = \frac{\sin(\theta_2 - \theta_1)}{\sin(\theta_2 + \theta_1)}, \tag{1.42}$$

$$t_s = \frac{E_{2s}}{E_{1s}} = \frac{2\sin\theta_2\cos\theta_1}{\sin(\theta_1 + \theta_2)}. \tag{1.43}$$

These equations are known as the *Fresnel equations* and the coefficients are called the *Fresnel coefficients*. The symbols E_{1s}, E'_{1s}, E_{2s}, E_{1p}, E'_{1p}, and

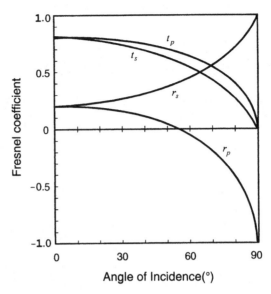

FIGURE 1.26 Fresnel coefficients for an air–glass boundary surface.

E_{2p} represent the incident, reflected, and refracted electric vectors as shown in Figure 1.25. We note that the Fresnel equations can also be derived using electromagnetic theory. The detailed derivation is beyond the scope of this chapter, and we refer interested readers to the book by Hecht and Zajac [4].

Plots of the Fresnel coefficients for an air–glass boundary are given in Figure 1.26, in which we see that r_p goes to zero when $\theta_1 + \theta_2$ approaches to $\pi/2$. At this condition, the angle of incidence is known as the Brewster angle, as described in Section 1.7. We note that light reflected at the Brewster angle becomes completely linearly polarized with the electric vector normal to the plane of incidence (i.e., s-polarization). Beyond the Brewster angle, the reflectance coefficient r_p changes sign, which indicates that there is a phase reversal (or a phase shift of π) of the reflected electric vector E'_{1p}.

EXAMPLE 1.12

A light beam is incident on the surface of an air–glass boundary at the Brewster angle. Assume that the refractive index of the glass is 1.5. Calculate the Fresnel coefficients of the p-polarization components.

The Brewster angle is given by

$$\theta_1 = \tan^{-1}\left(\frac{\eta_2}{\eta_1}\right) = \tan^{-1}\left(\frac{1.5}{1.0}\right) = 56.31°.$$

The reflection angle is determined by Snell's law, as given by

$$\theta_2 = \frac{\eta_1 \sin \theta_1}{\eta_2} = 33.69°.$$

The Fresnel coefficients are calculated as:

$$r_p = \frac{\tan(\theta_2 - \theta_1)}{\tan(\theta_2 + \theta_1)} = 0,$$

$$t_p = \frac{2 \sin \theta_2 \cos \theta_1}{\sin(\theta_1 + \theta_2) \cos(\theta_1 - \theta_2)} = 0.6667.$$

It is noted that

$$(r_p)^2 + (t_p)^2 \neq 1.$$

It seems that the conclusion reached in the preceding example is in contradiction to the law of energy conservation. However, this is not the case, as the optical intensity is defined as the energy crossing a unit area per second by which the cross-sectional area of the reflected beam is differently from that of the incident and the reflected beam, as illustrated in Figure 1.27. If we take this difference into consideration, the reflectance and the transmittance at the boundary surface can be written as:

$$R_p = (r_p)^2 = \frac{\tan^2(\theta_2 - \theta_1)}{\tan^2(\theta_2 + \theta_1)}, \tag{1.44}$$

$$T_p = \frac{\eta_2 \cos \theta_2}{\eta_1 \cos \theta_1} (t_p)^2 = \frac{\eta_2 \cos \theta_2}{\eta_1 \cos \theta_1} \cdot \frac{4 \sin^2 \theta_2 \cos^2 \theta_1}{\sin^2(\theta_1 + \theta_2) \cos^2(\theta_1 - \theta_2)}, \tag{1.45}$$

$$R_s = (r_s)^2 = \frac{\sin^2(\theta_2 - \theta_1)}{\sin^2(\theta_2 + \theta_1)}, \tag{1.46}$$

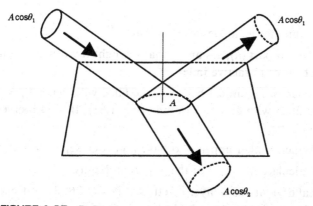

FIGURE 1.27 Reflection and refraction of an incident beam.

$$T_s = \frac{\eta_2 \cos \theta_2}{\eta_1 \cos \theta_1}(t_s)^2 = \frac{\eta_2 \cos \theta_2}{\eta_1 \cos \theta_1} \cdot \frac{4 \sin^2 \theta_2 \cos^2 \theta_1}{\sin^2(\theta_1 + \theta_2)}. \tag{1.47}$$

It can be seen that the energy is conserved, such that

$$R_p + T_p = 1, \tag{1.48}$$

$$R_s + T_s = 1. \tag{1.49}$$

The value of reflectance at normal incidence deserves special mention. It cannot be directly obtained from Fresnel equations by simply setting $\theta_1 = 0°$, since the substitution gives rise to an indeterminate result. However, as both θ_1 and θ_2 become very small when the incident beam approaches normal incidence, the tangential factors of small angles can be replaced by sinusoidal approximations. This gives rise to the following result

$$R_p = R_s = \left(\frac{\eta_2 - \eta_1}{\eta_2 + \eta_1}\right)^2 \tag{1.50}$$

for a small incident angle. For example, if the reflective index of glass is $\eta_2 = 1.5$, then the reflectance at the air–glass boundary surface with normally incident light would be about 4%.

REFERENCES

1. F. W. SEARS, *Optics*, Addison-Wesley, Reading, MA, 1949.
2. B. ROSSI, *Optics*, Addison-Wesley, Reading, MA, 1957.
3. F. A. JENKINS and H. E. WHITE, *Fundamentals of Optics*, fourth edition, McGraw-Hill, New York, 1976.
4. E. HECHT and A. ZAJAC, *Optics*, Addison-Wesley, Reading, MA, 1974.

PROBLEMS

1.1 A light ray is incident on a plane surface separating two transparent media of refractive indices 1.65 and 1.35. If the angle of incidence is 35° and the light ray originates in the medium of higher refractive index.

(a) Compute the angle of refraction.

(b) Repeat part (a), assuming that the light ray originates in the medium of lower refractive index.

1.2 Consider two identical beakers, one filled with water ($\eta = 1.33$) and the other filled with carbon disulfide ($\eta = 1.63$). If the observer views them from above.

(a) Which beaker appears to have a greater depth of liquid?

(b) Calculate the ratio of the apparent depths.

1.3 A beaker contains a layer of ethanol ($\eta = 1.36$) 3 cm deep, which floats on a layer of water ($\eta = 1.33$) 4 cm deep. If an observer views the beaker from above, calculate the apparent distance from the surface of the ethanol to the bottom of the water layer.

1.4 A glass plate about 4 cm thick is held about 5 cm above a printed page. If its view is from above, calculate the location of the printed page. We assume that the refractive index of the glass is $\eta = 1.5$ and that for air is $\eta = 1$.

1.5 Consider that a light ray is obliquely incident at an angle of 45° on the surface of a 4 cm thick glass plate. We assume that air ($\eta = 1$) is on both sides of the plate and that the refractive index of the glass is $\eta = 1.5$. Calculate the transverse displacement as the incident ray emerges through the plate.

1.6 Assume that a layer of oil ($\eta = 1.63$) about 0.5 cm thick is floating on a body of water ($\eta = 1.33$). A light ray originating from the body of water impinges on the boundary surface between the water and the oil at an incident angle of 25°.

(a) Calculate the angle of refraction in the oil.

(b) If the medium above the oil layer is air ($\eta = 1$), will the light ray be totally reflected?

(c) What is the critical angle of incidence between the oil and the air?

1.7 A cone of spherical wavefronts, which originates from a point source S, is incident on a boundary surface between two transparent media, as shown in Figure 1.28. Use Huygens' principle to construct the reflected and the refracted wavefronts.

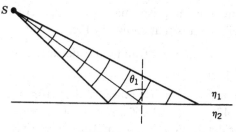

FIGURE 1.28

1.8 An observer is looking perpendicularly through a pond of water. If the bottom of the pond appears to be at a depth of 5 ft, calculate its actual depth. Assume that the refractive index of water is $\eta = 1.33$.

1.9 Suppose a light ray is incident on a rectangular glass plate, which is submerged in clear water, as shown in Figure 1.29. Calculate the light ray's maximum angle of incidence on the left vertical surface of the glass that will make total internal reflection occur at the top surface of the glass. We assume that the refractive indices of the glass and the water are $\eta = 1.5$ and $\eta = 1.33$, respectively.

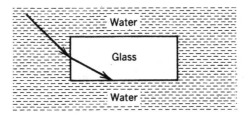

FIGURE 1.29

1.10 Repeat Problem 1.9 for the same glass plate when it is surrounded by air ($\eta = 1$).

1.11 If a 45°–45°–90° prism is submerged in a body of water ($\eta = 1.33$), calculate the refractive index for the prism that would be required for the normally incident light ray (as shown in Figure 1.13) to be totally reflected.

1.12 Assume that a light ray is normally incident on the shorted face of a 30°–60°–90° prism ($\eta = 1.6$), as depicted in Figure 1.30. If a drop of liquid is placed on the top of the prism, calculate the refractive index of the liquid necessary to produce a total internal reflection.

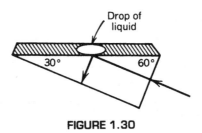

FIGURE 1.30

1.13 Given an equiangular prism made of silicate crown glass, the index of refraction of which is that shown in Figure 1.17.

 (a) Calculate the angles of minimum deviation for wavelengths of 700 and 400 mm.

 (b) Determine the dispersive power of the prism. We assume that spectral lines of 475, 550, and 650 nm (i.e., blue, yellow, and red colors) are used for determining η_f, η_d, and η_c, respectively.

1.14 Given a prism made of silicate crown glass that has an apex angle of 10°, calculate the mean angular deviation and the dispersive power of the prism.

1.15 Assume a natural light of intensity I_0 is incident on two cascaded polarizers P_1 and P_2, which are perpendicularly polarized to each other.

 (a) Calculate the transmitted light intensity.

 (b) A third polarizer P_3 of 45° polarization direction, with respect to the two polarizers, is inserted between P_1 and P_2. Evaluate the transmitted light intensity.

1.16 (a) Since natural light or circularly polarized light can be regarded as composed of two linearly, perpendicularly polarized lights of equal amplitude, what is the major distinction between them?

(b) What are the major distinctions between partially linearly polarized light and elliptically polarized light?

1.17 If the critical angle of a quartz (assuming it is surrounded by air) is 40°, what is the Brewster angle for the quartz?

1.18 By referring to Eqs 1.45 and 1.47, show that the following equations can be derived:

$$T_p = \frac{\eta_2 \cos\theta_2}{\eta_1 \cos\theta_1} \cdot \frac{4\sin^2\theta_2 \cos^2\theta_1}{\sin^2(\theta_1+\theta_2)\cos^2(\theta_1-\theta_2)}$$

and

$$T_s = \frac{\eta_2 \cos\theta_2}{\eta_1 \cos\theta_1} \cdot \frac{4\sin^2\theta_2 \cos^2\theta_1}{\sin^2(\theta_1+\theta_2)}.$$

1.19 Plot the reflectances of p- and s-polarized incident light at an air–glass boundary surface. Assume that the refractive index of the glass is 1.5.

1.20 Calculate the reflectances for a normally incident beam at the air–material boundary surface for the following materials:

(a) Crown glass, $\eta = 1.526$.

(b) Quartz, $\eta = 1.547$.

(c) Diamond, $\eta = 2.426$.

(d) Rutile, $\eta = 2.946$.

2 LENSES AND ABERRATIONS

A *lens* is defined as an optical element consisting of two or more refractive surfaces that share a common optical axis. A lens consisting of two refractive surfaces is called a *simple lens*, and one that has more than two refractive surfaces is a *compound lens*. Hence, a compound lens is made by combining two or more simple lenses. Generally, the space between the refractive surfaces of the lenses is filled either with air or with the appropriate refractive fluid. High-quality lenses are generally compound lenses, although there are some exceptions. Most compound lenses are aberration free and are called *diffraction limited.*

When a light ray passes through a simple lens, it is refracted by both surfaces of the lens. If a lens, simple or compound, causes a negligible deviation of the refracted ray, it is regarded as a *thin lens*. On the other hand, if the deviation cannot be ignored, the lens is called a *thick lens*.

2.1 Image Formation

Again with some exceptions, the surfaces of all practical lenses are spherical in shape. We now discuss the reflection and refraction of the spherical surface shown in Figure 2.1.

In the figure, a spherical surface of radius R separates two transparent media. Medium 1 is represented by its index of refraction η_1, and medium 2 is represented by η_2. We assume that all light rays scattered from the object point S will ultimately converge to point P, called the *image point*. The distance d_2 from P to the vertex of the spherical surface is called the *image distance*, and distance d_1 from S to the vertex is called the *object distance*. Thus, by the law of sines, we have

$$\frac{\sin(\pi - \theta_1)}{\sin \phi_1} = \frac{R + d_1}{R}.$$

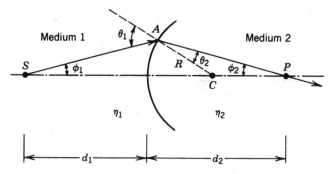

FIGURE 2.1 Refraction at a spherical surface.

Since $\sin(\pi - \theta_1) = \sin\theta_1$, the preceding equation reduces to

$$\sin\theta_1 = \frac{R + d_1}{R}\sin\phi_1. \tag{2.1}$$

From triangle SAP we see that

$$\phi_1 + \phi_2 + \theta_2 + (\pi - \theta_1) = \pi,$$

which reduces to

$$\phi_2 = \theta_1 - \theta_2 - \phi_1. \tag{2.2}$$

Using triangle ACP and utilizing the law of sines, we have

$$\frac{d_2 - R}{\sin\theta_2} = \frac{R}{\sin\phi_2}.$$

Thus, the image distance can be written as

$$d_2 = R + R\frac{\sin\theta_2}{\sin\phi_2} = R\left[1 + \frac{\sin\theta_2}{\sin(\theta_1 - \theta_2 - \phi_1)}\right]. \tag{2.3}$$

From this equation we see that the image distance d_2 increases as the emission angle ϕ_1 decreases. This means that the light rays originating from the object point S do not in actuality all intersect at a common point after the refraction. In other words, the refracted wavefront is no longer spherical in shape, so it creates an image aberration. This type of image aberration is commonly called *spherical aberration*, as will be discussed in Section 2.4.

In addition, we note that if the emission angle ϕ_1 is sufficiently small, the incident light rays can be considered parallel to the optical axis and hence are called *paraxial rays*. Thus, if ϕ_1 is small, the incident and refraction angles of θ_1 and θ_2 will also be small. As a result, the sines of these angles can be approximated by $\sin\phi_1 \simeq \phi_1$, $\sin\phi_2 \simeq \phi_2$, $\sin\theta_1 \simeq \theta_1$, and $\sin\theta_2 \simeq \theta_2$. Thus, from Eqs 2.1, 2.2, and 2.3, we obtain the following useful relation,

$$\frac{\eta_1}{d_1} + \frac{\eta_2}{d_2} = \frac{\eta_2 - \eta_1}{R}, \tag{2.4}$$

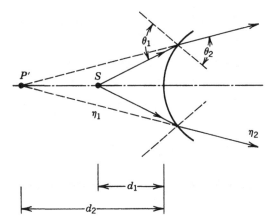

FIGURE 2.2 Divergent refraction at a spherical surface.

which is known as the *lens maker's equation* or the *Gaussian formula* for a single spherical surface. The lens maker's equation is derived under the paraxial-ray condition, in which it is independent of the emission angle ϕ_1. Thus, the light rays originating from the object point, under the *paraxial approximation*, will ultimately converge to a common image point after refraction.

It is apparent that if the object is closer to the boundary surface, a divergent refracted wavefront will be obtained, as shown in Figure 2.2. Thus, we see that the image point P located behind the spherical surface, as shown in Figure 2.1, is the *real image*, and the image point P' appearing in front of the spherical surface, as shown in Figure 2.2, is the *virtual image*.

Since a *focal point* is defined as the point through which all the incoming paraxial rays eventually intersect, the *focal length* must be the distance from the vertex of the spherical surface to the focal point. If we assume that the surface is illuminated by a plane wavefront (i.e., parallel rays), the object distance is $d_1 = \infty$ and the image distance is $d_2 = f_2$. By applying the lens maker's equation, Eq. 2.4, we have

$$f_2 = \frac{\eta_2}{\eta_2 - \eta_1} R, \qquad (2.5)$$

which is called the *second focal length* of the spherical surface. On the other hand, if we assume that the object point is located at $d_1 = f_1$, the image distance is $d_2 = \infty$. Thus, by Eq. 2.4 we obtain

$$f_1 = \frac{\eta_1}{\eta_2 - \eta_1} R, \qquad (2.6)$$

which is known as the *first focal length* of the spherical surface.

The *lateral magnification* of a refractive image can be obtained by using the paraxial ray-tracing technique, as depicted in Figure 2.3. Lateral magnification is defined as

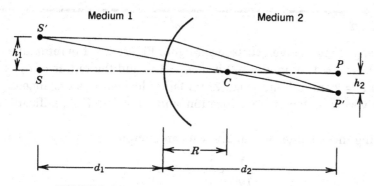

FIGURE 2.3 Lateral magnification by refraction.

$$M = \frac{h_2}{h_1}, \tag{2.7}$$

where h_1 and h_2 are the separation of the object and image points, respectively.

We now derive an expression for lateral magnification in terms of the object and image distances and the refractive indices of the media. From the right triangles of SCS' and PCP' in Figure 2.3, we see that

$$M = \frac{h_2}{h_1} = \frac{CP}{CS} = \frac{d_2 - R}{d_1 + R}. \tag{2.8}$$

Substituting Eqs 2.1 and 2.3 into Eq. 2.8, we have

$$M = \frac{h_2}{h_1} = \frac{\sin \theta_2 \sin \phi_1}{\sin \theta_1 \sin \phi_2}. \tag{2.9}$$

Using Snell's law of refraction, we can reduce this expression to

$$M = \frac{h_2}{h_1} = \frac{\eta_1 \sin \phi_1}{\eta_2 \sin \phi_2}, \tag{2.10}$$

which can be written as

$$h_1 \eta_1 \sin \phi_1 = h_2 \eta_2 \sin \phi_2. \tag{2.11}$$

This is *Abbe's sine condition*, a well-known relation that must hold to prevent a lens from producing a coma, or pear-shaped spot (see Section 2.4).

Since the emission angle ϕ_1 and the receiving angle ϕ_2 are relatively small, that is, $\sin \phi_1 \simeq h_1/d_1$ and $\sin \phi_2 \simeq h_2/d_2$, Eq. 2.10 can be further reduced to the following form:

$$M = -\frac{\eta_1 d_2}{\eta_2 d_1}. \tag{2.12}$$

EXAMPLE 2.1

Given the spherical convex surface shown in Figure 2.1, the refractive index of medium 1 is $\eta_1 = 1$, of medium 2 it is $\eta_2 = 1.5$, and the radius of the spherical surface is $R = 20\,\text{mm}$. If an object 2 mm tall is located at a distance of 100 mm from the vertex, calculate the location and the lateral magnification of the image.

By applying these values to Eq. 2.4, we can compute the image distance as

$$\frac{1}{100} + \frac{1.5}{d_2} = \frac{1.5 - 1}{20}$$
$$d_2 = 100\,\text{mm}.$$

Since the image distance d_2 is a positive quantity, a *real* image is formed behind the spherical surface, in medium 2, at a distance of 100 mm from the vertex.

Using Eq. 2.12, we can calculate the corresponding lateral magnification,

$$M = \frac{h_2}{h_1} = -\frac{1 \times 100}{1.5 \times 100} = -0.667.$$

Thus, we see that the image is inverted and appears to be smaller (i.e., demagnified) than the object.

EXAMPLE 2.2

Let us continue the preceding problem. If the index of refraction of medium 1 is filled with water, which has a refractive index of $\eta = 1.33$, calculate the location of the refractive image and the corresponding lateral magnification.

Again we use Eq. 2.4 to find the image distance:

$$\frac{1.33}{100} + \frac{1.5}{d_2} = \frac{1.5 - 1.33}{20}$$
$$d_2 = -312.5\,\text{mm}.$$

Since the image distance is a negative quantity, a virtual image is formed in front of the spherical surface in medium 1, at a distance of 312.5 mm from the vertex.

Again from Eq. 2.12, the lateral magnification can be evaluated:

$$M = \frac{h_2}{h_1} = -\frac{(1.33)(-312.5)}{1.5 \times 100} = 2.77.$$

Thus, we see that the image is erected vertically and is about 2.77 times larger (i.e., magnified) than the actual object.

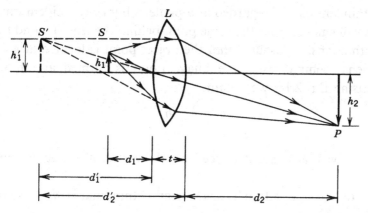

FIGURE 2.4 Image formation by a simple lens.

2.2 Simple Lenses

We now consider the refraction produced by a simple lens. We assume that the light rays scattered from an object point S are incident on a simple lens L, as shown in Figure 2.4. The refractive index of the lens is η, and R_1 and R_2 are the radii of curvature of the first and second surfaces of the lens, respectively. Using the principles of reflection and refraction, we see that the first surface of the lens produces a virtual image at point S' behind object point S, and the lens produces a real image at point P behind the lens. By applying the lens maker's equation, Eq. 2.4, we can determine the distance of the virtual image,

$$\frac{1}{d_1} + \frac{\eta}{d_1'} = \frac{\eta - 1}{R_1}, \tag{2.13}$$

where R_1 is the radius of curvature of the first surface, and d_1 and d_1' are the object and the virtual-image distances, as they are measured from the vertex of the first surface.

Similarly, the distance of the real image can be determined for the second surface,

$$\frac{\eta}{d_2'} + \frac{1}{d_2} = \frac{1 - \eta}{R_2}, \tag{2.14}$$

where R_2 is the radius of curvature of the second surface, and d_2' and d_2 are the virtual-image and the real-image distances, as they are measured from the vertex of the second surface. Notice that Eqs 2.13 and 2.14 can be used to calculate the image positions and the lateral magnifications for all types of lens, including compound lenses.

Since a thin lens can be regarded as a plane, a light ray incident on a point on the lens will emerge from the same point behind the lens. To find the focal point of a thin lens, we assume that the object point is located at a distance from the lens, behind which the image point is formed at infinity (i.e., $d_2 = \infty$). Using Eq. 2.14, we obtain

$$\frac{\eta}{d_2'} = \frac{1 - \eta}{R_2},\tag{2.15}$$

where d_2' is the virtual-image distance from the vertex of the second surface of the lens.

Since the thickness of a thin lens is negligibly thin, that is, $d_1' \simeq d_2'$, Eq. 2.15 can also be written as

$$\frac{\eta}{d_1'} = \frac{1 - \eta}{R_2}.\tag{2.16}$$

By substituting Eq. 2.16 in Eq. 2.13, we have

$$\frac{1}{d_1} = (\eta - 1)\left(\frac{1}{R_1} + \frac{1}{R_2}\right),\tag{2.17}$$

where R_1 and R_2 are the radii of curvature of the first and second surfaces of the lens, η is the refractive index of the lens, and d_1 is the distance from the object to the center plane of the lens. By the definition of the first focal length, that is, $f_1 - d_1$, we see that

$$\frac{1}{f_1} = (\eta - 1)\left(\frac{1}{R_1} + \frac{1}{R_2}\right).\tag{2.18}$$

Similarly, the second focal length can be found from

$$\frac{1}{f_2} = (\eta - 1)\left(\frac{1}{R_1} + \frac{1}{R_2}\right),\tag{2.19}$$

which is identical to Eq. 2.18. Thus, when the refractive indices on both sides of the lens are equal, the front and the back focal lengths are also equal. For a *thin lens in air*, we therefore simply write

$$\frac{1}{f} = (\eta - 1)\left(\frac{1}{R_1} + \frac{1}{R_2}\right),\tag{2.20}$$

which is also known as the *spherical lens maker's equation*, derived from Eq. 2.4.

Let us now derive several useful expressions for thin lenses. We assume that an extended object is located in front of a thin lens, as shown in Figure 2.5. The height of the object is denoted by h_1, and h_2 is the height of the image. We see that the light rays from object point S converge at image point P. We also know that the intersection of any two rays from S is adequate for

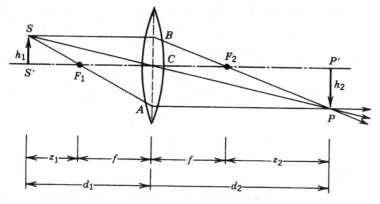

FIGURE 2.5 Image formation by a thin lens.

determining the position of image point P. From triangles $SS'F_1$, F_1CA, BCF_2, and $F_2P'P$, we see that

$$\frac{h_1}{z_1} = \frac{h_2}{f}, \qquad \frac{h_1}{f} = \frac{h_2}{z_2}.$$

(2.21)

By dividing these equations, we have

$$f^2 = z_1 z_2,$$

(2.22)

which is known as the *Newtonian form lens equation*. Similarly, from triangles SBA, F_1CA, BAP, and BCF_2, we have

$$\frac{h_1 + h_2}{d_1} = \frac{h_2}{f}, \qquad \frac{h_1 + h_2}{d_2} = \frac{h_1}{f}.$$

(2.23)

Thus, the lateral magnification is

$$M = \frac{h_2}{h_1} = -\frac{d_2}{d_1},$$

(2.24)

where the negative sign represents an inverted image. Then, by adding Eqs 2.23, we obtain

$$\frac{1}{d_1} + \frac{1}{d_2} = \frac{1}{f},$$

(2.25)

which is the famous *Gaussian lens equation*. We further note that in the Gaussian lens equation the object distances are measured from the lens, whereas in the Newtonian form lens equation they are measured from the focal points.

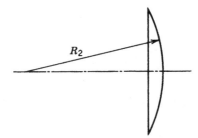

FIGURE 2.6

EXAMPLE 2.3

Consider a planoconvex thin lens having a focal length of 20 cm (see Figure 2.6). If the refractive index of the lens is 1.52, calculate the radius of curvature of the convex surface.

Since the first surface of the lens is a plane, its radius of curvature is $R_1 = \infty$. By applying the spherical lens maker's equation, Eq. 2.20, we obtain

$$\frac{1}{20} = (1.52 - 1)\left(\frac{1}{\infty} + \frac{1}{R_2}\right)$$

$$R_2 = 10.4 \, \text{cm}.$$

EXAMPLE 2.4

Consider again the planoconvex thin lens shown in Figure 2.6 as described in Example 2.3. If the radius of curvature of the second surface on the planoconvex thin lens is 30 cm, calculate the focal length of the lens.

From the previous example we have $\eta = 1.52$. By applying Eq. 2.20, we obtain

$$\frac{1}{f} = (1.52 - 1)\left(\frac{1}{\infty} + \frac{1}{30}\right)$$

or

$$f = \frac{30}{0.52} = 57.69 \, \text{cm}.$$

Notice that if we reverse the planoconvex lens, we still obtain the same focal length. This result shows that, when the indices of refraction between two sites on the lens are the same, the first and second focal lengths are equal.

EXAMPLE 2.5

An object is located 35 cm to the left of a thin convex lens. If the focal length is 20 cm, calculate the position and the lateral magnification of the image.

In this problem we have $d = 35$ cm and $f = 20$ cm. By applying Eq. 2.25, the Gaussian lens equation, we obtain

$$\frac{1}{35} + \frac{1}{d_2} = \frac{1}{20}$$
$$d_2 = 46.66 \text{ cm}.$$

We see that the image is real and that it is located 46.66 cm behind the lens.

Moreover, from the Newtonian lens equation, Eq. 2.22, we have

$$(20)^2 = (35 - 20)z_2$$
$$z_2 = 26.66 \text{ cm},$$

which is in agreement with the result obtained with the Gaussian lens equation.

In computing the lateral magnification, we use Eq. 2.24,

$$M = \frac{46.66}{35} = -1.33.$$

Thus, we see that the image is inverted and is about 1.33 times larger than the object.

2.3 Phase Retardation by Thin Lenses

An important property of thin lenses is *phase retardation,* or *delay.* As we have illustrated in the preceding chapter, light rays that propagate in a dielectric medium can be described by *wave theory.* Thus, a wavefront that propagates through a transparent dielectric medium (e.g., a lens) exhibits severe wavefront deformation.

Let us consider a light ray entering at a point on one side of a thin lens and emerging from the same point on the other side of the lens. Since the lens is assumed to be thin, the transversal displacement of the light ray inside the lens is negligible. Therefore, only the phase is retarded when a wavefront passes through a thin lens. We stress that the amount of phase retardation is proportional to the variation in thickness of the lens.

In order to calculate the phase retardation of a lens, let us first consider the variation in thickness of a convergent (i.e., positive) lens, as shown in

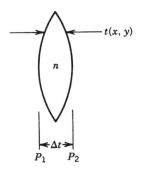

FIGURE 2.7 Variation in the thickness of a convex lens.

Figure 2.7. The phase retardation $\phi(x, y)$ on the wavefront as it passes through (i.e., is refracted by) the lens can be written as

$$\phi(x, y) = k[\Delta t + (\eta - 1)t(x, y)], \qquad (2.26)$$

where $t(x, y)$ is the variation in thickness of the lens, Δt is its maximum thickness, η is its refractive index, $k = 2\pi/\lambda$ is the wave number, λ is the wavelength, and x and y are the spatial coordinates of the lens. We note that the quantity

$$\phi_1(x, y) = k\eta t(x, y)$$

represents the phase retardation caused by the lens alone, and

$$k[\Delta t - t(x, y)]$$

represents the phase retardation caused by free space (or air) alone. Notice that the *phase transform* function of the lens can be represented by the following:

$$T(x, y) = \exp[i\phi(x, y)] = \exp\{ik[\Delta t + (\eta - 1)t(x, y)]\}. \qquad (2.27)$$

If the thin lens is illuminated by a monochromatic wavefront $u_i(x, y)$ at plant P_1 (see Figure 2.7), the wavefront that emerges through the lens can be expressed as

$$u_o(x, y) = u_i(x, y)T(x, y), \qquad (2.28)$$

where $T(x, y)$ is the phase transform of the thin lens. Thus, we see that the output wavefront is the product of the input wavefront multiplied by the phase transform function of the lens. For convenience, an analog system representation of Figure 2.7 is given in Figure 2.8.

To derive the phase transform of a lens, we first consider the variation in thickness of a positive lens. If we divide the lens into left and right halves, as shown in Figure 2.9, the variation in thickness of the left half can be written as

$$t_1(x, y) = \Delta t_1 - [R_1 - (R_1^2 - \rho^2)^{1/2}]$$

$$= \Delta t_1 - R_1\left\{1 - \left[1 - \left(\frac{\rho}{R_1}\right)^2\right]^{1/2}\right\}, \qquad (2.29)$$

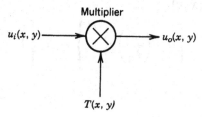

FIGURE 2.8 System analog diagram of a lens transformation.

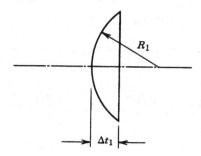

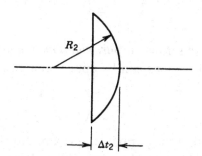

FIGURE 2.9 Left and right halves of a convex lens.

where Δt_1 is the maximum thickness variation of the left half of the lens, R_1 is the radius of curvature, and

$$\rho^2 = x^2 + y^2.$$

Similarly, the variation in thickness in the right half is

$$t_2(x, y) = \Delta t_2 - [R_2 - (R_2^2 - \rho^2)^{1/2}]$$

$$= \Delta t_2 - R_2 \left\{ 1 - \left[1 - \left(\frac{\rho}{R_2} \right)^2 \right]^{1/2} \right\}. \qquad (2.30)$$

Thus, the overall thickness variation of the lens is

$$t(x,y) = t_1(x,y) + t_2(x,y)$$

$$= \Delta t - R_1 \left\{ 1 - \left[1 - \left(\frac{\rho}{R_1} \right)^2 \right]^{1/2} \right\} - R_2 \left\{ 1 - \left[1 - \left(\frac{\rho}{R_2} \right)^2 \right]^{1/2} \right\},$$

$$(2.31)$$

where $\Delta t = \Delta t_1 + \Delta t_2$.

To simplify Eq. 2.31 we shall consider only the relatively small region of the lens that lies near the optical axis. In this case, we have

$$R_1 \gg \rho, \quad R_2 \gg \rho.$$

Using the binomial series expansion, we can make the following approximations:

$$\left[1 - \left(\frac{\rho}{R_1} \right)^2 \right]^{1/2} \simeq 1 - \frac{1}{2} \left(\frac{\rho}{R_1} \right)^2 \qquad (2.32)$$

and

$$\left[1 - \left(\frac{\rho}{R_2} \right)^2 \right]^{1/2} \simeq 1 - \frac{1}{2} \left(\frac{\rho}{R_2} \right)^2. \qquad (2.33)$$

These approximations are known as *paraxial approximations*. Thus, the overall thickness of the positive lens can be approximated as

$$t(x,y) \simeq \Delta t - \frac{\rho^2}{2} \left(\frac{1}{R_1} + \frac{1}{R_2} \right). \qquad (2.34)$$

It should also be noted that the paraxial approximations of Eqs 2.32 and 2.33 provide the same result by replacing the spherical surfaces of the lens with *parabolic* surfaces.

Thus, the phase transform function of a positive lens can be written as

$$T(x,y) = e^{ik\eta\Delta t} \exp \left[-ik(\eta - 1) \frac{\rho^2}{2} \left(\frac{1}{R_1} + \frac{1}{R_2} \right) \right], \qquad (2.35)$$

and its focal length can be identified as

$$f = \frac{R_1 R_2}{(\eta - 1)(R_1 + R_2)}. \qquad (2.36)$$

By substituting Eq. 2.36 into Eq. 2.35, we have

$$T(x,y) = C_1 \exp \left(-i \frac{k}{2f} \rho^2 \right), \qquad (2.37)$$

where $C_1 = e^{ik\eta\Delta t}$ is a complex constant, f is the focal length of the lens, and $\rho = \sqrt{x^2 + y^2}$.

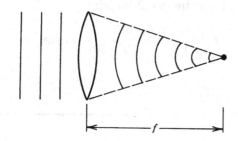

FIGURE 2.10 Convergent effect of a positive lens.

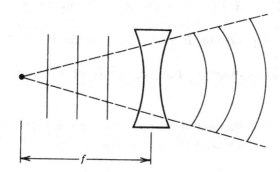

FIGURE 2.11 Divergent effect of a negative lens.

We stress that Eq. 2.37 is very useful in the processing of optical signals. The negative exponent expresses the convergent effect of a wavefront after it is refracted by the lens. In other words, if parallel light rays (in the form of a plane wavefront) are normally incident on the lens, the wavefront emerges as a spherical wavefront after being refracted by the lens, as illustrated in Figure 2.10.

Furthermore, the phase transform function of a divergent (or negative) lens can also be derived by the same approximation. Thus, we obtain

$$T(x, y) = C_2 \exp\left(i\frac{k}{2f}\rho^2\right), \qquad (2.38)$$

where C_2 is a complex constant. Notice that Eq. 2.38 is essentially identical to Eq. 2.37, except that it contains a *positive* quadratic phase factor rather than a negative one. Thus, a plane wavefront will emerge as a divergent wavefront after passing through a negative lens, as shown in Figure 2.11.

EXAMPLE 2.6

Consider a positive thin lens for which the refractive index is $\eta = 1.5$ and the radii of curvature of the front and back surfaces are $R_1 = 10\,\text{cm}$ and $R_2 = 30\,\text{cm}$, respectively. Calculate the focal length of the lens.

Using Eq. 2.36, we find that the focal length is

$$f = \frac{(10)(30)}{(1.5 - 1)(10 + 30)} = 15 \, \text{cm}.$$

EXAMPLE 2.7

Consider two positive lenses of focal lengths f_1 and f_2 that are cascaded, as shown in Figure 2.12a. If this set of lenses is illuminated by monochromatic wavefront $u_i(x, y)$:

(a) Draw an analog system diagram to represent its input–output optical setup.

(b) Calculate the resultant focal length of this set of lenses.

Answers

(a) The analog system diagram of the optical system is shown in Figure 2.12b, where the phase transforms of the lenses are

$$T_1(x, y) = C \exp\left[-i\frac{k}{2f_1}(x^2 + y^2)\right]$$

and

$$T_2(x, y) = C \exp\left[-i\frac{k}{2f_2}(x^2 + y^2)\right].$$

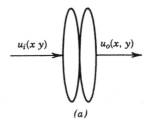

(a)

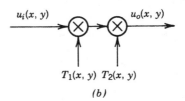

(b)

FIGURE 2.12

(b) The overall phase transform of this setup is

$$T_1(x,y)T_2(x,y) = C\exp\left[-i\frac{k}{2}(x^2+y^2)\left(\frac{1}{f_1}+\frac{1}{f_2}\right)\right].$$

Thus, the resultant focal length is

$$f = \frac{f_1 f_2}{f_1 + f_2}.$$

2.4 Primary Aberrations

We shall now discuss the primary aberrations of lenses. In the preceding sections we have shown several useful relations—of object and image distances, of focal lengths, of refractive indices, and of radii of curvatures. However, these relations were primarily derived under the assumption of paraxial approximation, which means that they can be applied only to the light rays that have originated from an axial object and are making small incident angles.

However, the expressions derived in the preceding sections may not hold for all light rays that reach a lens either from axial object points or from object points located away from the optical axis. Because of the finite size of the lens, the cone of light rays that forms the image point should not be treated as a very small cone. Thus, the nonparaxial rays from an object point will not, in general, intersect at the same image point after refraction by the lens.

In addition, from Eq. 2.36, the focal-length equation, we see that the focal length depends on the refractive index of the lens, and that the refractive index varies with the wavelength of the incident light. Strictly speaking, a broad spectrum of color images of different sizes will be in different locations in the image space of the lens. Thus, any deformation or distortion of the image that is caused by the lens and that makes it different from the *first-order approximation* is generally called a *lens aberration*. Aberrations caused primarily by the wavelength of the light source are called *chromatic aberrations*. If the light source is monochromatic, however, these aberrations are called *monochromatic aberrations*. Notice that the lens aberration has nothing to do with any physical imperfection of the lens, but is due to the refraction of light rays at the spherical surfaces.

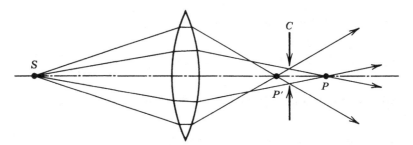

FIGURE 2.13 Spherical aberration of a lens.

One of the most common aberrations in lenses is spherical aberration. Picture a bundle of light rays originating from an axial object point S and impinging on a positive lens, as shown in Figure 2.13. The light rays that are incident near the rim of the lens are imaged at point P', which is closer to the lens than point P where the paraxial rays are imaged. The light rays incident in the intermediate zone of the lens are imaged between point P and P'. Thus, there exists no clear, sharp image behind the lens. If we insert an observation screen between image points P and P', we find that the image consists of a circular disc image because the cones of light rays from various axial image points overlap. However, there is a location, known as the *circle of least confusion* or *best focus*, where the cross-sectional area of the circular disc image is smallest. The phenomenon of confusion and imprecision in image formation is generally called *spherical aberration* of the lens.

Spherical aberration of a lens can be reduced by inserting a circular diaphragm to block the light rays near the rim and allow only the center position of the light rays to pass through the lens. However, the decrease in light transmission reduces image resolution, a subject we will discuss in the next section. We stress that spherical aberration of lenses can be minimized by designing the surface curvatures of the lenses properly; for example, by grinding the surfaces to appropriate *aspherical* shapes.

Another primary aberration of the lens is known as *coma*. Coma is similar to spherical aberration in being produced by light rays that are not incident near the optical axis of the lens. However, coma differs from spherical aberration in that the object point images as a pear-shaped spot rather than as a disc, as illustrated in Figure 2.14. The aberration is named after the coma of a comet, the gaseous envelope surrounding its nucleus.

Let us discuss qualitatively how a coma is formed. For simplicity, we assume that coma is the only lens aberration present, because if other aberrations are considered, the coma will be further enhanced. Consider an object point S that is located slightly away from the optical axis of the lens, as shown in Figure 2.14. A very small pencil cone of light rays from the object point that passes through the center zone of the lens converges to image point

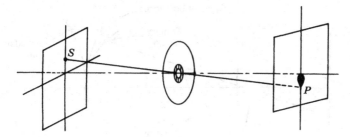

FIGURE 2.14 Coma.

P. However, a narrow hollow cone of light rays from *S*, the shaded zone, will be imaged as a circle above the image point *P*. The hollow cones of light rays from within the shaded zone will image as smaller circles below the circle from the shaded zone. Hollow cones that are larger than the shaded zone will image as larger circles above this circle. The total effect of these superpositions of circular images takes the blurred shape of a comet.

Like spherical aberration, coma may be corrected by the proper design of the surface curvature of the lens. Unfortunately, the curvature necessary to correct coma is generally not the curvature that corrects spherical aberration. Therefore, a lens designed for minimum spherical aberration cannot be totally free from coma.

We shall now consider the aberrations caused by *astigmatism* and *curvature of field*. Since these two aberrations arise from basically the same effect, we shall discuss them simultaneously.

Basically, astigmatism is similar to coma except for the spread of the image in relation to the optical axis. Unlike coma, the spread of which is along the optical axis, that of astigmatism is over a plane perpendicular to the optical axis. As in coma, however, the image is distorted because the object point is not located on the optical axis of the lens.

Referring to Figure 2.15, we see that, after being refracted by the lens, the light rays from the object point *S* converge to a horizontal line image called the *primary image*, after which they converge to a vertical line image called

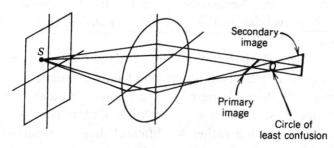

FIGURE 2.15 Astigmatism.

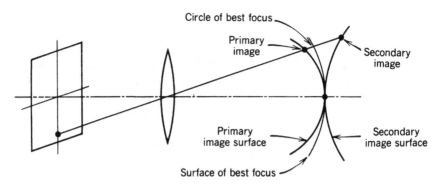

FIGURE 2.16 Curvature of field.

the *secondary image*. The cross-section of the refracted light beam is generally elliptical in shape, except at the primary and secondary images. Somewhere between the two line images it becomes a circle. The circular image at this location is called the *least-confused* image or the *best-focused* image.

Furthermore, if the object point is moving on a plane that is perpendicular to the optical axis of the lens, the loci of the primary and secondary images describe the *primary image surface* and the *secondary image surface*, respectively, as illustrated in Figure 2.16. The *surface of best focus* is the locus of least confusion, since all three of these surfaces are tangential at the same point on the optical axis. Thus, when the object point is located on the optical axis, the primary image, the secondary image, and the point of least confusion all converge to a point on the optical axis. Since the surface of best focus is generally a curved surface, the aberration caused by this curve is known as *curvature of field*, and the aberration caused by the noncoincident primary and secondary image surfaces is called *astigmatism* of the lens.

The curvatures of the primary image surface, the secondary image surface, and the surface of best focus are all affected by the curvatures of the lens surfaces. For both curvature of field and astigmatism to be eliminated, the primary and secondary image surfaces must be in the same plane. In practice, however, it is not possible to make them coincident and planar. In other words, it is not possible to eliminate simultaneously both curvature of field and astigmatism of a lens. Nevertheless, it is possible to eliminate either one or the other of these two aberrations by properly locating aperture stops near the lens surface.

Apparently, spherical aberration, coma, astigmatism, and curvature of field are aberrations in which an object point fails to image into an image point. We shall now discuss one of the last monochromatic aberrations, namely, *distortion*. Distortion is an aberration that is caused not by the best focus of the image, but rather by different magnifications at different distances of the object point from the axis. For example, if the magnification

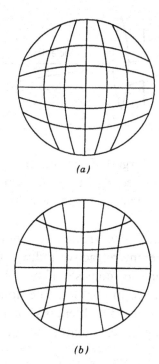

(a)

(b)

FIGURE 2.17 Distortion. (a) Barrel distortion. (b) Pincushion distortion.

decreases with increasing distance of the object point from the axis, the outer portion of the image becomes smaller. Say a circular screen is imaged by the lens; the image has laterally bulging sides, as shown in Figure 2.17a. This distortion is *barrel shaped*. On the other hand, if the magnification increases as the distance of the object point from the axis increases, the image has concave sides or a *pincushion* distortion, as shown in Figure 2.17b. Again, these distortions can be corrected by properly locating stops near the surface of the lens.

Spherical aberration, coma, astigmatism, curvature of field, and distortion are generally known as the *five primary aberrations* of the lens. All five are primarily due to curvature of the refractive surfaces of the lens.

We now briefly discuss the *chromatic aberration* of a lens. Referring to Eq. 2.36, we notice that the focal length varies with the refractive index of the lens. Since the refractive index of the lens varies with the wavelength (see Figure 1.17), a lens will image a chromatic object into a series of images of different colors, at different locations and magnifications. The variation of image distances in relation to the refractive index of the lens is called *axial* or *longitudinal chromatic aberration*. The variation in image size caused by the refractive index of the lens is called *lateral chromatic aberration*. Examples of these chromatic aberrations are shown in Figure 2.18. The violet image is the closest to the lens and the smallest in size, whereas the red image is the

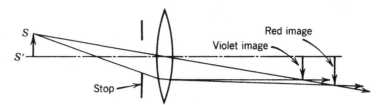

FIGURE 2.18 Chromatic aberrations of a lens.

farthest from the lens and the largest. The wavelengths of the other colors are imaged between the violet and red, and their sizes are correspondingly larger and smaller, depending on which of these two colors images they are closer to.

Chromatic aberration may be corrected to some degree by combining several thin lenses with different refractive indices. One such combination, called an *achromatic doublet lens*, consists of two thin lenses that have different refractive indices and are in contact. The doublet lens is designed so that the two lenses have identical focal lengths for two colors of light (e.g., red and violet), thus offsetting the chromatic aberrations of one lens by those of the other.

EXAMPLE 2.8

Consider a positive lens made of silicate flint glass and whose surfaces have radii of curvature of $R_1 = R_2 = 30$ cm. Compute the focal lengths of the lens if it is illuminated by $\lambda_1 = 650$ mm (red light) and $\lambda_2 = 450$ mm (blue light), respectively.

Since the refractive index of the lens is a function of wavelength, that is, $\eta(\lambda)$, the focal length varies as the wavelength changes. Using Eq. 2.36, we have

$$f(\lambda) = \frac{R}{2[\eta(\lambda) - 1]}.$$

From Figure 1.17 we see that $\eta(\lambda_1) = 1.618$ and $\eta(\lambda_2) = 1.64$ for silicate flint glass. Thus, the focal lengths for λ_1 and λ_2 are

$$f(\lambda_1) = \frac{30}{2(1.618 - 1)} = 24.27 \text{ cm}$$

and

$$f(\lambda_2) = \frac{30}{2(1.64 - 1)} = 23.44 \text{ cm}.$$

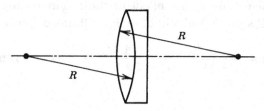

FIGURE 2.19

EXAMPLE 2.9

Consider a doublet lens, as shown in Figure 2.19. We assume that the positive lens is made of silicate flint glass and that the planoconcave lens, the negative lens, is made of silicate crown glass. If the radius of curvature of the surfaces is $R = 30$ cm, calculate the focal lengths of this doublet lens under red light ($\lambda_1 = 650$ mm) and blue light ($\lambda_2 = 450$ nm) illuminations.

From Eq. 2.20 we see that the focal lengths of the positive lens and the planoconcave lens can be computed, respectively, as

$$f_1 = \frac{R}{2[\eta(\lambda) - 1]}$$

and

$$f_2 = \frac{R}{\eta(\lambda) - 1}.$$

Again from Figure 1.17, we see that $\eta(\lambda_1) = 1.618$ and $\eta(\lambda_2) = 1.64$ for silicate flint glass, and $\eta(\lambda_1) = 1.505$ and $\eta(\lambda_2) = 1.515$ for silicate crown glass. Thus, the corresponding focal lengths are

$$f_1(\lambda_1) = \frac{30}{2(1.618 - 1)} = 24.27 \text{ cm},$$

$$f_1(\lambda_2) = \frac{30}{2(1.64 - 1)} = 23.44 \text{ cm},$$

$$f_2(\lambda_1) = \frac{30}{1.505 - 1} = 59.4 \text{ cm},$$

and

$$f_2(\lambda_2) = \frac{30}{1.515 - 1} = 58.25 \text{ cm}.$$

Since the two lenses are in close contact, the resultant focal lengths can be calculated using the following equation,

$$\frac{1}{f} = \frac{1}{f_1} - \frac{1}{f_2} \quad \text{or} \quad f = \frac{f_1 f_2}{f_2 - f_1},$$

where the minus sign reflects the effect of the negative lens. Therefore, the resultant focal lengths under red and blue light illuminations are, respectively,

$$f(\lambda_1) = \frac{f_1(\lambda_1)f_2(\lambda_1)}{f_2(\lambda_1) - f_1(\lambda_1)} = \frac{(24.27)(59.4)}{59.4 - 24.27} = 41.03 \, \text{cm}$$

and

$$f(\lambda_2) = \frac{f_1(\lambda_2)f_2(\lambda_2)}{f_2(\lambda_2) - f_1(\lambda_2)} = \frac{(23.44)(58.25)}{58.25 - 23.44} = 39.22 \, \text{cm}.$$

2.5 Resolution Limit

In this section we discuss the resolution limit of lenses from the standpoint of *diffraction*. The phenomenon of diffraction is discussed in greater detail in Chapter 8. Lens aberrations aside, the image of an object point is not simply the intersection of the light rays that are refracted by the lens; rather, it is a *diffraction pattern*. Thus, there is an ultimate limit in the resolution of a lens. A lens or an optical system is said to be able to *resolve* two object points if the corresponding diffraction patterns are sufficiently separated. In the following, we discuss Baron Rayleigh's widely adopted resolution criterion, which he discovered in 1888.

Consider an optical imaging system or a simple lens, one that is used to image two closely spaced objects, as shown in Figure 2.20. We assume that the lens if *stigmatic*—that is, any light ray from an object point S impinging on the lens will eventually pass through image point P—and that the *optical lengths* (lengths measured in wavelengths of light) of all paths from S to P are

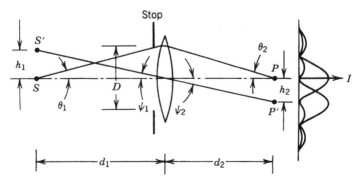

FIGURE 2.20 Resolution limit of a lens.

equal. Although the geometrical length of the light ray passing through the center of the lens is shorter than the lengths of other rays from S to P, the number of wavelengths is the same, since the lens is thicker at the center. Because of the lens's higher index of refraction, wave propagation in the lens is slower than it is in the air.

In addition to the geometrical ray tracing of the refracted light rays, diffraction also takes place. Thus, the lens can be regarded as an open, circular aperture in which there is diffraction at the *far-field condition*. The result is that the images at points P and P' are not very sharp, each having a far-field circular diffraction pattern around it. Therefore, the images of the object points tend to be confused when they are close together. A question is thereby posed: How close can objects be and still be *resolved?*

The *Rayleigh criterion* states that two equally bright object points can be resolved by the lens of an optical system if the center maximum irradiance of the diffraction pattern of one image falls exactly on the first minimum, that is, the first dark fringe, of the diffraction pattern of the other. In other words, the distance between the centers of the two circular diffraction patterns should equal the radius of the central bright disc.

Looking again at Figure 2.20, we see that the irradiance curves for image points P and P' are shown at the right. The *minimum resolvable angular separation* can be written as

$$\psi_{\min} = \frac{1.22\lambda}{D}, \tag{2.39}$$

where D is the diameter of the open aperture, and λ is the wavelength of the light source. Therefore, according to Rayleigh's criterion, if the angular separation ψ is smaller than $\psi_{\min}$, the images cannot be resolved.

Moreover, Figure 2.20 shows that the angular separation ψ_1 of the object points equals that of the image points, ψ_2. If the lens is the objective lens of an astronomical telescope, the distance d_1 will be infinitely large, and the minimum resolvable angular separation of Eq. 2.39 will best describe the lens's resolving power. Conversely, for closer observation of the object, as with a microscope, it is better to express Eq. 2.39 in terms of *minimum resolvable separation*; that is,

$$h_{1,\min} = \frac{1.22\lambda d_1}{D}, \tag{2.40}$$

where $\psi = h/d_1$.

When Eq. 2.40 is applied to an optical instrument, it is customary to use Abbe's sine condition, Eq. 2.11:

$$h_1\eta_1 \sin\theta_1 = h_2\eta_2 \sin\theta_2. \tag{2.41}$$

To find the resolving power of the objective of a microscope, we first assume that air is on both sides of the optical system, so that $\eta_1 = \eta_2$. From Figure 2.20 we see that $h_1/h_2 = d_1/d_2$, so that Eq. 2.41 reduces to

$$d_1 \sin \theta_1 = d_2 \sin \theta_2. \tag{2.42}$$

Since the angle θ_2 is very small in microscopic application,

$$\sin \theta_2 \simeq \theta_2 = \frac{D}{2d_2}. \tag{2.43}$$

Substituting Eq. 2.43 into Eq. 2.42, we have

$$\frac{D}{d_1} = 2 \sin \theta_1. \tag{2.44}$$

Therefore, from Eq. 2.40, we obtain

$$h_{1,\min} = \frac{1.22\lambda_1}{2 \sin \theta_1} = \frac{0.61\lambda}{\text{NA}}, \tag{2.45}$$

where $\text{NA} = \sin \theta_1$ is the value called the *numerical aperture* of the objective lens. The NA represents the angular radius of the shaft of rays from an object when focused by the lens. Manufacturers usually include the NA for purposes of rating the objective of the microscope.

When we consider an oil immersion microscope, $\eta_1 \neq \eta_2$ (i.e., $\eta_2 = 1$), Abbe's sine condition becomes

$$\eta_1 h_1 \sin \theta_1 = h_2 \sin \theta_2. \tag{2.46}$$

Since ψ_1 and ψ_2 are generally not identical, in the present case we have $\eta_1 \psi_1 = \psi_2$. Equation 2.39 still holds, provided we express it in terms of ψ_2 and λ_2, that is,

$$\psi_{2,\min} = \frac{1.22\lambda_2}{D} \tag{2.47}$$

or

$$\psi_{1,\min} = \frac{1.22\lambda_2}{\eta_1 D}. \tag{2.48}$$

Thus, the minimum separation of the point object is

$$h_{1,\min} = \frac{1.22\lambda_2 d_1}{\eta_1 D}. \tag{2.49}$$

Since $\eta_1 \psi_1 = \psi_2$, we have

$$\frac{\eta_1 h_1}{d_1} = \frac{h_2}{d_2}. \tag{2.50}$$

Substituting Eq. 2.50 into Eq. 2.46, we obtain

$$d_1 \sin \theta_1 = d_2 \sin \theta_2, \tag{2.51}$$

which is the same as Eq. 2.42. If we use Eqs 2.42 and 2.43, then Eq. 2.49 can be written as

$$h_{1,\min} = \frac{0.61\lambda_2}{\eta_1 \sin\theta_1} = \frac{0.61\lambda_2}{\text{NA}}, \tag{2.52}$$

where $\text{NA} = \eta_1 \sin\theta_1$. Thus, it is possible for an oil immersion microscope to have a numerical aperture greater than unity.

EXAMPLE 2.10

A lateral object is located at the near point of 250 mm from an adult human eye. The minimum distance of the object for which the eye can accommodate is called the distance of distinct vision, and the point corresponding to the minimum distance is called the *near point*. The near point is the point nearest the eye at which an object is properly focused on the retina when the maximum degree of accommodation is employed.

Assume that the diameter of the pupil of the eye is 2 mm, the diameter of the eyeball is 25 mm, the refractive index of the vitreous humor is $\eta_2 = 1.33$, and that the object is illuminated by a light source of $\lambda_1 = 550 \times 10^{-6}$ mm. Calculate both the numerical aperture of the eye and the minimum resolution limit of the eye.

Referring to Figure 2.21, we see that the sine of the half-angle θ_1 is

$$\text{NA} = \sin\theta_1 = \frac{1}{250} = 0.004.$$

From the Rayleigh resolution criterion of Eq. 2.45, we see that the minimum resolvable separation is

$$h_{1,\min} = \frac{0.61 \times 550 \times 10^{-6}}{0.004} = 0.084 \text{ mm},$$

which is the resolution limit of a normal adult eye.

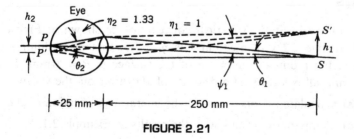

FIGURE 2.21

EXAMPLE 2.11

Using the data in the preceding example, calculate both the minimum resolution on the retina in the eye and the minimum angular separation of the lateral object points.

Referring to the image space (i.e., in the eye), we can write the resolution limit on the retina as

$$h_{2,\text{min}} = \frac{0.61\lambda_2}{\sin\theta_2}.$$

Since $\eta_2\lambda_2 = \lambda_1$ (i.e., $\eta_1 = 1$), the minimum resolution on the retina can be calculated as follows:

$$\begin{aligned} h_{2\,\text{min}} &= \frac{0.61\lambda_1}{\eta_2\sin\theta_2} \\ &= \frac{0.61 \times 550 \times 10^{-6}}{1.33 \times 1/25} = 6.3 \times 10^{-3}\,\text{mm}. \end{aligned}$$

Referring to Figure 2.21, we see that the minimum angular separation can be determined by Eq. 2.48:

$$\begin{aligned} \psi_{1,\text{min}} &= \frac{1.22\lambda_1}{\eta_1 D} \\ &= \frac{1.22 \times 550 \times 10^{-6}}{2 \times 10^{-1}} = 3.35 \times 10^{-3}\,\text{rad}. \end{aligned}$$

REFERENCES

1. F. W. SEARS, *Optics*, Addison-Wesley, Reading, MA, 1949.
2. B. ROSSI, *Optics*, Addison-Wesley, Reading, MA, 1957.
3. F. A. JENKINS and H. E. WHITE, *Fundamentals of Optics*, fourth edition, McGraw-Hill, New York, 1976.
4. E. HECHT and A. ZAJAC, *Optics*, Addison-Wesley, Reading, MA, 1974.
5. F. T. S. YU, *Optical Information Processing*, Wiley-Interscience, New York, 1983, Sections 3.6 and 6.1.

PROBLEMS

2.1   Look again at the spherical convex surface shown in Figure 2.1. The refractive index of medium 1 is $\eta_1 = 1.5$, that of medium 2 is $\eta_2 = 1$, and the radius of curvature of the surface is $R = 20\,\text{mm}$. If an object 2 mm tall is located at a distance of 100 mm from the vertex,

(a) Calculate the position and the lateral magnification of the image.

(b) Compare the results with the results in Example 2.1.

2.2   Refer to the spherical convex surface shown in Figure 2.1 and described in Example 2.1.

     (a) Calculate the *first* and *second* focal lengths of the spherical surface.

     (b) Where are the focal points located?

2.3   Find the first and second focal lengths of the spherical surface in Problem 2.1. Compare your results with those of Problem 2.2.

2.4   If the object in Example 2.1 is located at a distance $d_1 = 20\,\text{mm}$ from the vertex, calculate where the image will be located.

2.5   A small tropical fish is at the center of a spherical fishbowl 30 cm in diameter. Assuming that the indices of refraction of the glass and the water are the same, that is, $\eta = 1.33$, calculate the position and lateral magnification of the image of the fish when the observer views the fish from outside the bowl.

2.6   Using the lens maker's equation, derive the focal point for a concave mirror.

2.7   If an object is located at a distance two times the focal length of a concave mirror.

     (a) Compute the location of the image from the vertex of the concave mirror.

     (b) Calculate the lateral magnification of the image.

2.8   Find the focal length of a thin lens if both sides of the lens are filled with water. We assume that R_1 and R_2 are the radii of curvature of the first and second surfaces of the lens, and that η and η' are the refractive indices of the lens and the water, respectively.

2.9   If one side of a thin lens is filled with ethanol and the other side is filled with water, find the first and the second focal lengths of the lens. Here R_1 and R_2 are the radii of curvature of the lens's surfaces, and η_1, η, and η_2 are the refractive indices of the ethanol, the lens, and the water, respectively.

2.10   Repeat Problem 2.9 for the conditions that one side of the lens is filled with air and the other side is filled with water.

2.11   Given a positive thin lens with a refractive index of $\eta = 1.5$, assume that one side of the lens is filled with air ($\eta = 1$) and that the other side is filled with water ($\eta = 1.33$). The radii of curvature of the lens's surfaces are $R_1 = 10\,\text{cm}$ and $R_2 = 15\,\text{cm}$. Calculate the first and the second focal lengths of the lens.

2.12   If an object 3 mm tall is located at a distance 1.5 times the first focal length in Problem 2.11, calculate both the image location and the lateral magnification of the image.

2.13 The object is located at a distance 1.5 times the second focal length (i.e., in the water side) of the lens described in Problem 2.11. Compute both the location and the lateral magnification of the image.

2.14 If an object is located at 0.5 times the focal length of a thin lens, calculate the location and the lateral magnification of the image. Assume that both sides of the lens are filled with air, that the index of refraction of the lens is $\eta = 1.5$, and that the radius of curvature of the surfaces is $R = 15$ cm.

2.15 Find the phase transform of a planoconcave thin lens, as shown in Figure 2.22.

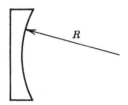

FIGURE 2.22

2.16 Picture an achromatic doublet lens as shown in Figure 2.23; the doublet lens is composed of a positive lens with a refractive index η_1 and a negative lens with a refractive index η_2.

(a) Calculate the phase transform of the doublet lens.

(b) Compute the focal length of the doublet lens.

(c) Draw an analog system diagram of the doublet lens. Assume that the complex light, $u_i(x,y)$, is normally incident at the lens.

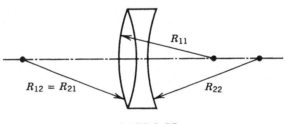

FIGURE 2.23

2.17 Consider two cascaded lenses as given in Example 2.7. If an object is located at a distance $1.2f$, where f is the resultant focal length, calculate the location and the lateral magnification of the image.

2.18 An optical system consists of both a planoconvex and a planoconcave thin lens, as shown in Figure 2.24.

(a) Draw an analog system diagram to represent this optical system. In this diagram $u_i(x,y)$ will represent the complex light illumination.

(b) Compute the overall focal length of the optical system.

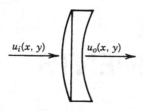

$u_i(x, y)$ $u_o(x, y)$

FIGURE 2.24

2.19 A positive lens is made of silicate flint glass and has a radius of curvature of $R = 20$ cm, for both surfaces. If a white-light plane wave is normally incident on the lens,

(a) Calculate the smeared focal length. Assume that the spectral line of the white light varies from 400 to 700 nm. (Note that the refractive indices of the lens can be found in Figure 1.17.)

(b) If an object is located at a distance $d_1 = 30$ cm from the lens, calculate the positions and the lateral magnifications of the images for $\lambda = 400$ and 700 nm.

2.20 An optical imaging system is shown in Figure 2.20. Two lateral object points, separated by $h_1 = 0.001$ mm, are located at a distance $d_1 = 50$ cm from the lens.

(a) If these objects are illuminated by a red light of $\lambda = 650$ nm, calculate the minimum size of lens required to resolve the object.

(b) Repeat part (a) for violet light illumination of 400 nm. Compare the result with your result for part (a).

2.21 An object is located at a distance $d_1 = 100$ cm from an imaging lens. Assume that the diameter of the imaging lens is $D = 1$ cm. If the object is illuminated by a green light of $\lambda = 550$ nm, calculate the minimum resolvable separation of the image (i.e., $h_{2,\mathrm{min}}$). Assume $f = 40$ cm.

2.22 If the object space in Problem 2.21 – that is, the left-hand side of the lens in Figure 2.20 – is filled with mineral oil of $\eta_1 = 1.4$,

(a) Calculate the minimum angular separation $\psi_{1,\mathrm{min}}$.

(b) Compare the results with those for Problem 2.21.

2.23 If the image space referred to in Problem 2.22 is filled with mineral oil of $\eta_2 = 1.4$, and the object space is filled with air,

(a) Calculate the minimum angular separation $\psi_{1,\mathrm{min}}$.

(b) Compute the numerical aperture of the lens.

(c) Compare the results with those for Problem 2.22.

2.24 If both the object and the image spaces in Figure 2.20 are filled with transparent fluids of η_1 and η_2,

(a) Calculate the resolution limit of the optical system.

(b) Determine the numerical aperture of the lens.

3 OPTICAL INSTRUMENTS

In the last chapter we discussed image-formation properties and aberrations of optical lenses. One of the objectives of optical engineering is to design high-performance optical instruments using optical lenses, as well as mirrors and prisms, as fundamental building blocks. The subject of optical instruments covers such a large scope that it is impossible to discuss all its aspects in a book devoted to an introduction to optical engineering. We will thus limit our discussion to a few standard and commonly used optical instruments, such as photographic cameras, microscopes, telescopes, and projectors. The emphasis will be placed on the optical architecture and working principles. These discussions will serve to illustrate some applications of the theories given in previous chapters, and will be of interest to the students who have used, or expect to use, some of these instruments.

3.1 The Human Eye

The human eye is perhaps the simplest and yet the most powerful optical instrument we have ever had. This marvelous gift of nature consists of only a single convergent lens and is able to form the sharp image of an object at a wide range of distances and under a broad range of illumination conditions.

A cross-sectional diagram of the human eye is shown in Figure 3.1. An object is imaged by the lens onto the back inner surface of the eye, called the retina. As shown in Figure 3.1, what is formed on the retina is a real but inverted image of the object. It is amazing that, while all the images are inverted, our brain interprets them as being upright. The iris diaphragm in front of the lens changes its aperture diameter by unconscious muscular action to adapt to the brightness of the environment; it opens wider for faint light and closes down for bright sunlight. The diameter of the iris can vary between approximately 1.5 and 7 mm.

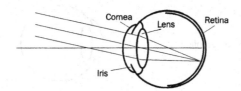

FIGURE 3.1 The human eye.

The retina of the eye contains two types of light-sensitive cells: *rods* and *cones*. These cells receive light pulses and convert them into electric currents. How these electric currents are produced by the cones and the rods, and how they are perceived by the brain is not yet fully understood. It is known, however, that the cones respond to bright light only and are responsible for our distinction of colors. Rods, on the other hand, are sensitive to faint light, to the motion of an object, and to slight variation in intensity.

The resolving power of the eye (i.e., its ability to detect fine details of the object being observed) is determined by the density of cone cells. The cones are closely packed in the retina; each cone subtends slightly less than 0.15 mrad (0.5′) to the center of the lens. The eye can distinguish two points only if their images are separated by one cone; otherwise, the two points will appear as only one. Thus, the eye can, at best, distinguish points that subtend about 0.3 mrad (1′) to the center of the lens. For an object located at the normal distance (25 cm) this corresponds to a resolution limit of about 0.075 mm. This figure is often used in determining the required magnification of an optical instrument.

A feature of the human eye is its capability to focus automatically on objects at different distances. This is called *accommodation* and is achieved by changing the focal length of the lens through muscular contraction. With a twitch of the muscle holding it, the lens can either flatten to decrease its focal length or fatten to increase its focal length. The ability of accommodation varies from a person to another. A normal eye can see clearly objects located at infinity to about 25 cm. Excessive departure from normality can be corrected by the use of optical lenses, or glasses as they are called in our daily life. There are three common abnormalities: myopia, hyperopia, and astigmatism. These are discussed in the following.

Myopia or *nearsightedness*, is the condition in which the optical power of the lens is too big, such that light rays emanating from an infinitely distant object, essentially parallel rays, are focused in front of the retina, as shown in Figure 3.2a. The person therefore sees a blurred image of distant objects. As the object draws closer, it is brought slowly into focus on the retina, as illustrated in Figure 3.2b. The distance at which this happens is called the

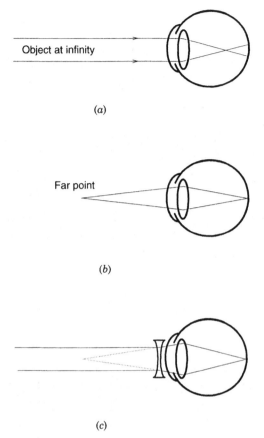

FIGURE 3.2 (a) A myopic eye focusing on a distant object. (b) The far point for a myopic eye. (c) Correction of a myopic eye using a negative lens.

far point. The most commonly employed and the most effective way of correcting for myopia is to use a negative lens, as shown in Figure 3.2c. The function of the negative lens is simply to refract the light rays from distant objects so that they appear to emanate from the far point. Rays from objects that are nearer will appear to come from points at distances nearer than the far point, and can therefore be accommodated by the eye.

EXAMPLE 3.1

Assume a myopic eye cannot see clearly objects located at distances greater than 1.5 m. What type of corrective lens is needed?

As just explained, a lens is required to make rays from infinitely distant objects appear to come from the far point (i.e., 1.5 m). So the object distance is infinity and the image distance is −1.5 m. The minus sign on the image distance is by convention, since the object and the image are both on the same side of the lens. The focal length f of the lens is therefore given by

$$\frac{1}{f} = \frac{1}{\infty} + \frac{1}{-1.5}.$$

Hence, $f = -1.5\,\text{m}$. A negative lens with a focal length of 1.5 m is required.

Hyperopia, or *farsightedness*, is almost the opposite of the nearsightedness. The optical power of the lens is too small, such that the true focus of a relatively close object point lies behind the retina, as illustrated in Figure 3.3a. The nearest point at which an object can be imaged sharply onto the retina is called the *near point* (Figure 3.3b). A hyperopic eye cannot accommodate for objects nearer than the near point. When a positive corrective lens is used, an object placed at the normal reading distance (about 25 cm) is

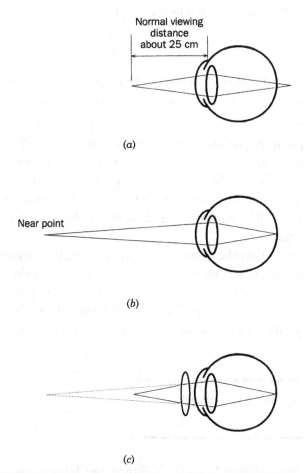

(a)

(b)

(c)

FIGURE 3.3 (a) A hyperopic eye focusing on a distant object. (b) A hyperopic eye focusing at a normal viewing distance. (c) Correction of a hyperopic eye using a positive lens.

slightly focused and appears to be located at the near point, as shown in Figure 3.3c. Objects located further than the normal reading distance will appear to come from points at distances further than the near point, and can therefore be accommodated by the eye.

EXAMPLE 3.2

A hyperopic person can clearly see objects no closer than 150 cm. What type of corrective lens is needed for such a person to read this page clearly at the normal distance of 25 cm?

The light rays from an object located 25 cm from the eye should appear to come from a distance of 150 cm. Therefore, we have

$$\text{Object distance} = 25\,\text{cm}$$
$$\text{Image distance} = -150\,\text{cm}.$$

The focal length of the lens is therefore given by

$$\frac{1}{f} = \frac{1}{25} + \frac{1}{-150} = \frac{1}{30}$$

Hence, the focal length required is 30 cm. It is a positive or convergent lens.

In *astigmatism* the lens has different focal lengths in two orthogonal directions, for example in horizontal and vertical directions. As discussed in Section 2.4, an object point will be focused into a horizontal line and a vertical line at different positions. Therefore, it is impossible to form a sharp image on the retina. This defect is corrected by means of a cylindrical lens, which compensates for the difference in focal lengths of the lens in the horizontal and vertical directions.

It should be noted that the above description of the eye is very incomplete and lacking in detail. The response of the eye to varying light levels, to fine detail, and to different wavelengths is very complicated, and by no means fully understood. The description given here is intended merely as a sketch to show how the eye works as an optical instrument to form images of objects.

3.2 Camera and Photographic Film

The optical instrument that is most similar to the human eye is the photographic camera. It is also the most popular optical instrument used in our

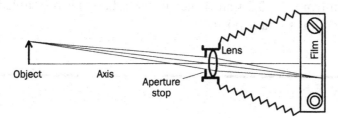

FIGURE 3.4 A camera.

daily life. In this section, we discuss the basic principles of the camera and the characteristics of photographic film.

3.2.1 CAMERA

Like the eye, the camera consists of a lens, an "iris" (the aperture stop), and a "retina" (the light-sensitive photographic film). Figure 3.4 schematically illustrates the configuration of a camera. A real, inverted image is formed by a lens, or combination of lenses, on the surface of a photographic film, which is later developed and printed to obtain the final picture. The aperture stop is an iris diaphragm, the diameter of which can be varied to control the exposure of the film. The ratio of the focal length of the lens and the diameter of the stop, f/d, is named the F-number and is an important parameter for a photographer to choose in order to obtain high-quality pictures. As an example, if the focal length of a camera lens is 50 mm and the diameter of the aperture stop is 6.25 mm, the F-number is $f/d = 8$ and is often denoted as $f/8$. It should be noted that the F-number is inversely proportional to the diameter of the aperture. The smaller the F-number, the larger the aperture.

The object of which an image is being taken may be stationary or moving relative to the camera. If the object is moving, a short exposure is necessary in order to obtain an unblurred image on the film. This requires a large aperture to collect sufficient light. The light entering a camera is proportional to the brightness of the object, is proportional to the area of the aperture stop, and is inversely proportional to the square of the focal length. This relation is given by

$$E \propto \frac{BA}{f^2} = \frac{B\pi d^2}{4f^2} \tag{3.1}$$

where E is the energy collected by the camera lens, B denotes the brightness of the object, A is the area of the aperture, and d is the diameter of the aperture stop. For any given object, we can write

$$E \propto \frac{d^2}{f^2} = \frac{1}{(F\text{-number})^2} \tag{3.2}$$

It can be seen from Eq. 3.2 that if the aperture diameter is doubled the light falling on the film is quadrupled.

EXAMPLE 3.3

A student takes pictures of a school bus. While the school bus is stopped, he takes a picture with an F-number of $f/8$ and an exposure time of $1/30$ s. Then the bus starts moving, he wants to take another picture with an exposure time of $1/120$ s. What F-number should be used?

By reducing the shutter speed from $1/30$ to $1/120$ s, the exposure time is decreased fourfold. In order to keep the energy entered into the camera identical, the student needs to increase the area of the aperture four times. Since the area of the aperture is proportional to the square of its diameter, the student should double the diameter of the aperture stop. Because he does not change the focal length, the F-number should be adjusted from $f/8$ to $f/4$.

In order to take pictures of faintly illuminated subjects, or of objects in rapid motion, which require a very short exposure, a lens of large aperture (i.e., small F-number) is required. As we discussed in Chapter 2, a lens of large aperture is subject to many aberrations. In order to correct the aberration caused by a single lens, in a real camera a combination of lenses is generally used. The design of such camera lenses is beyond the scope of this book, but as an example Figure 3.5 shows a schematic illustration of the cross-section of a so-called "Dagor lens". The lens consists of two triplets with the aperture stop located in the middle.

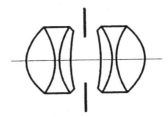

FIGURE 3.5 A compound camera lens (i.e., Dagor lens).

3.2.2 PHOTOGRAPHIC FILM

Photographic film is generally composed of a substrate, made of a transparent glass plate or acetate film, and a layer of photographic emulsion, as shown in Figure 3.6. The emulsion consists of a large number of tiny photosensitive silver halide particles suspended uniformly in a supporting gelatin.

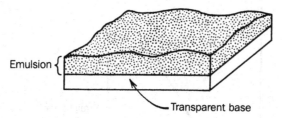

FIGURE 3.6 Section of a photographic film. The emulsion is composed of silver halide particles suspended in gelatin.

When the photographic emulsion is exposed to light, some of the silver halide grains absorb optical energy and undergo a series of complex physical and chemical changes to form tiny metallic silver particles. These particles are called *development centers*. The grains that are not exposed or that do not absorb sufficient light energy remain unchanged. If the developed film is then subject to a chemical fixing process, the unexposed silver halide grains are removed, leaving only the metallic silver particles in the gelatin. These remaining grains are opaque, so the transmittance of the developed film depends on their density. The relation between the intensity of the transmittance and the density of the developed grains was first demonstrated in 1890 by F. Hurter and V. C. Driffield, who showed that the photographic density, the density of the metallic silver particles per unit area, could be expressed as

$$D = -\log T_i \tag{3.3}$$

where T_i is the intensity transmittance and is defined as

$$T_i(x, y) = \left\langle \frac{I_o(x, y)}{I_i(x, y)} \right\rangle \tag{3.4}$$

The angle bracket represents the localized ensemble average, and $I_i(x, y)$ and $I_o(x, y)$ are the input and output irradiances, respectively, at point (x, y).

One of the most commonly used descriptions of the photosensitivity of a given photographic film is the Hurter–Driffield curve, or the H-and-D curve, as shown in Figure 3.7. It is the plot of the density D of the developed grains versus the logarithm of the exposure E. The plot shows that if the exposure is below a certain level, the photographic density is quite independent of the exposure; this minimum density is usually referred as *gross fog*. As the exposure increases beyond the toe of the curve, the density begins to increase in direct proportion to log E. The slope of the straight-line portion of the H-and-D curve is usually referred as the *film gamma*, γ. If the exposure is increased further, the density saturates after an intermediate region called the *shoulder*. In the saturated region there is no further increase in the density of the developed grains as the exposure increases.

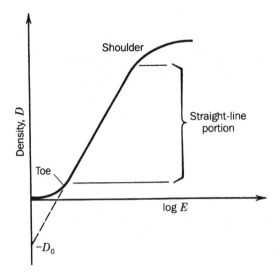

FIGURE 3.7 The Hurter–Driffield (H-and-D) curve.

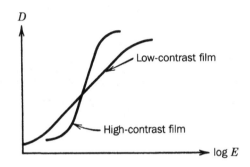

FIGURE 3.8 High- and low-γ films.

Conventional photography is usually carried out within the linear region of the H-and-D curve. A film with a high γ value is called a high-contrast film, whereas one with a low γ value is referred as a low-contrast film, as illustrated in Figure 3.8. The value of γ is, however, affected not only by the type of photographic emulsion used, but also by the chemical of the developer and the time taken by the developing process. In practice, it is possible to achieve a prescribed value of γ, with a fair degree of accuracy, by using a suitable film, developer, and developing time.

If a given film is recorded in the straight-line region of the H-and-D curve, the photographic density can be written as

$$D = \gamma_n \log E - D_o = \gamma_n \log(It) - D_o \qquad (3.5)$$

where the subscript n means that a negative film is being used, $-D_o$ is the point where the projection of the straight-line portion of the H-and-D curve intercepts the density coordinate (as shown in Figure 3.7), I is the incident irradiance, and t is the exposure time. By substituting Eq. 3.3 into Eq. 3.5, we have

$$T_{in} = K_n I^{-\gamma_n} \qquad (3.6)$$

where $K_n = 10^{D_o} t^{-\gamma_n}$ is a positive constant.

It is apparent that the intensity transmittance is highly nonlinear with respect to the irradiance. It is, however, possible to obtain a positive linear relation between intensity transmittance and the incident irradiance by a two-step process. This is normally used in photography to obtain the final positive picture. In the first step, a negative film is loaded into the camera and is exposed. The film is then developed and another negative film (or a photographic printing paper) is laid under it. An incoherent light is then transmitted through the first film to expose the second one. If the negative image recorded on the first film needs to be enlarged or reduced it is not placed in contact with the second film or the printing paper, but is imaged with suitable magnification onto the printing paper. By developing the exposed second film (or the printing paper) to obtain the prescribed γ value, we can get a transparency (or a positive picture) with a linear relation between the intensity of transmittance (or reflectance) and the incident irradiance.

To illustrate this two-step process, assume the resultant intensity transmittance of the second film be

$$T_{ip} = K_{n2} I_2^{-\gamma_{n2}} \qquad (3.7)$$

where K_{n2} is a positive constant, and the subscript $n2$ denotes the second-step negative. The irradiance I_2 incident on the second film can be written as

$$I_2 = I_1 T_{in} \qquad (3.8)$$

where

$$T_{in} = K_{n1} I^{-\gamma_{n1}}$$

where the subscript $n1$ denotes the first-step negative. The illuminating irradiance of the first film during the second exposure is I_1, and I is the irradiance originally incident on the first film. By substituting Eq. 3.8 into Eq. 3.7 we obtain

$$T_{ip} = K I^{\gamma_{n1} \gamma_{n2}} \qquad (3.9)$$

where

$$K = K_{n2} K_{n1}^{-\gamma_{n2}} I_1^{-\gamma_{n2}}$$

is a positive constant. Hence, we see that a linear relation between the intensity transmittance (or reflectance) of the final picture can be obtained if the overall γ is made to be unity; that is, if $\gamma_{n1} \gamma_{n2} = 1$.

EXAMPLE 3.4

Using the two-step process model, show that it is possible to encode two image transparencies into a single photographic printing paper such that the intensity reflectance of the final picture is the sum of the two image irradiances.

First, we select a low γ film (e.g., $\gamma_{n1} < 1$) for the sequential recording of the two image irradiances, that is, $I_1(x, y)$ and $I_2(x, y)$. Assume the recording is in the straight-line portion of the H-and-D curve, so that the intensity transmittance of the recorded film would be (according to Eq. 3.6)

$$T_{in}(x, y) = K[I_1(x, y) + I_2(x, y)]^{-\gamma_{n1}}.$$

By contact printing T_{in} onto a high-contrast (e.g., $\gamma_{n2} > 1$) printing paper, the intensity reflectance of the developed printing paper is

$$R_p(X, Y) = K_1[I_1(x, y) + I_2(x, y)]^{\gamma_{n1}\gamma_{n2}}$$

which is a positive image. If the product of the two γ values is unity (i.e., $\gamma_{n1}\gamma_{n2} = 1$), then

$$R_p(X, Y) = K_1[I_1(x, y) + I_2(x, y)].$$

It is seen that the reflectance, that is, the brightness, of the positive picture is the sum of two image irradiances.

3.3 Telescope

When the eye attempts to distinguish two points on a very distant object, it is attempting to separate two bundles of parallel rays which are inclined at small angles to each other. If this angle is smaller than the resolving power of the human eye (about 0.15 mrad), an optical instrument must be used to increase the angular separation between the two bundles of parallel rays. Such an instrument is called a telescope.

A diagram of the astronomical telescope is shown in Figure 3.9. The telescope consists of two convergent lenses: a long-focus lens called the *objective lens*, and a short-focus lens called the *eyepiece*. Rays from one point on the distant object are brought to a focus and form an inverted point image on the back focal plane of the objective. The eyepiece is suitably placed such that its front focal plane coincides with the back focal plane of the objective. Thus the point image is transformed into a bundle of parallel rays by the eyepiece, with their angle to the optical axis being enlarged. These parallel rays, when viewed with the human eye, form a point image on the retina.

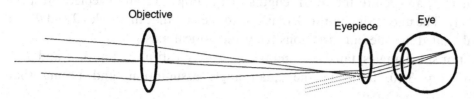

FIGURE 3.9 Schematic diagram of an astronomical telescope.

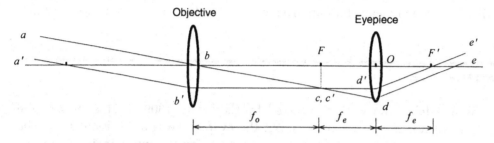

FIGURE 3.10 Schematic diagram of an astronomical telescope.

The *magnification*, or *magnifying power*, of a telescope is defined as the ratio between the angle subtended at the output end behind the eyepiece and the angle subtended at the input end in front of the objective. That is

$$M = \frac{\theta'}{\theta} \tag{3.10}$$

where M denotes the magnification, and θ and θ' represent the angles at the input and output ends, respectively.

To derive a representation of the magnification, we consider two specific light rays as shown in Figure 3.10. The ray $abcde$ passes through the center of the objective and therefore does not deflect. The second ray, $a'b'c'd'e'$, is suitably chosen such that it travels parallel to the optical axis after passing through the objective. These two rays meet at the focal plane and are transformed into a parallel orientation again by the eyepiece. Since the ray $a'b'c'd'e'$ is parallel to the optical axis between the two lenses, it passes the focal point of the eyepiece. Now, considering the right triangles bcF and $d'OF'$ in Figure 3.10, we have

$$\tan \theta = \frac{Fc}{f_o} \tag{3.11}$$

and

$$\tan \theta' = -\frac{Od'}{f_e} \tag{3.12}$$

where f_o and f_e are the focal lengths of the objective and eyepiece, respectively. The minus sign in Eq. 3.12 is due to the fact that the angles θ and θ' are subtended in different directions from the optical axis.

For small angles, $\tan \theta \approx \theta$ and $\tan \theta' \approx \theta'$. Substituting Eqs 3.11 and 3.12 into Eq. 3.10, applying the small angle assumption, and noting that $Fc = Od'$, we have

$$M = \frac{\theta'}{\theta} = -\frac{f_o}{f_e} \qquad (3.13)$$

The minus sign means that an inverted image is formed on the viewer's retina.

EXAMPLE 3.5

The first telescope was constructed by Galileo in 1609. In this system, the objective was a convergent lens, but the eyepiece was a divergent lens. The two lenses were placed in such a way that their focal points coincided after the eyepiece. Derive the magnification of this telescope.

A schematic diagram of the Galileo telescope is shown in Figure 3.11. The ray $abcd$ passes through the center of the objective without changing its direction. The other ray, $a'b'c'd'$, is deflected by the objective and travels parallel to the optical axis. Without the eyepiece, these two rays would meet at point e. After passing through the eyepiece, however, the ray $a'b'c'd'$ is deflected and appears to come from the focal point F. From the right triangles beF' and $Fc'O$, the input and output angles are given by

$$\theta \approx \tan \theta = \frac{eF'}{f_o}$$

and

$$\theta' \approx \tan \theta' = \frac{c'O}{f_e}$$

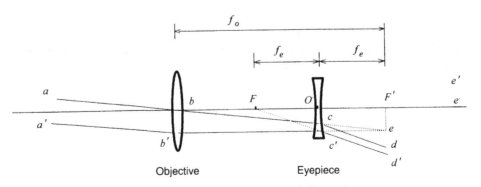

FIGURE 3.11 Schematic diagram of a Galileo telescope.

The magnification is

$$M = \frac{\theta'}{\theta} = \frac{f_o}{f_e}$$

The value of M is positive and, therefore, the image formed at the viewer's retina is an upright image.

There are many types of telescope. The astronomical and Galileo telescopes are just two examples. However, the fundamental principles of all these telescopes are same as those discussed above. The most popular telescopes are probably prism binoculars, which are in reality a pair of identical astronomical telescopes mounted side by side. To overcome the inconvenience of the inverted image, a pair of prisms placed between the objective and the eyepiece re-invert the image to the upright position. By reflecting the light back and forth between the prisms, the physical size (the length of the tube) is also reduced.

Although the telescope illustrated in Figure 3.9 is called the astronomical telescope, it is actually not used very often in astronomy nowadays. Later in this book we will learn that the resolution of an optical system is proportional to the size of its aperture. To distinguish between some double stars, the telescope has to have an objective of several meters in diameter. It is extremely difficult, if not impossible, to fabricate such a large lens. Instead, reflective mirrors with a suitably designed surface (e.g., parabolic) are used in an *astronomical reflective telescope*. The parallel light bundles coming from various stars are reflected by the mirror and focused at the focal plane as point images, which are recorded on the photographic film or detected by an array detector. The diameter of the largest objective mirror (it is generally called a primary mirror) at the present time is 6 m.

3.4 Microscope

The microscope helps a viewer to observe details of a close object. Like the telescope discussed in the preceding section, a microscope consists of two lenses, as shown in Figure 3.12. The first lens is of very short focus and is named the *objective*. The second one generally has a somewhat longer focus and is called the *eyepiece*. A tiny object is placed very close to the focal point of the objective, so that a magnified real image is formed. This real image becomes the object of the eyepiece. Since the distance between the real image and the eyepiece is less than (or, at most, equal to) the focal length of the

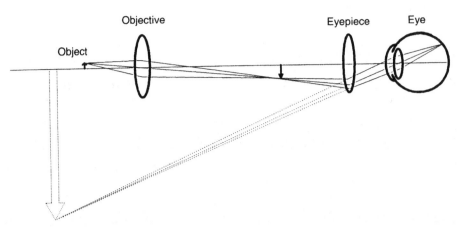

FIGURE 3.12 Schematic diagram of a microscope.

eyepiece, a virtual image is formed. This virtual image is then further imaged by the lens of the eye and a real image is formed on the viewer's retina.

It can be seen then that two magnification steps are involved in a microscope. The step for the objective is called the *linear magnification*, because it is measured in linear dimensions. The magnification of the eyepiece is called the *angular magnification*. The overall magnification of the microscope is the product of the linear and the angular magnifications.

The linear magnification of the objective is defined as the ratio between the dimensions of the image and the object. As shown in Figure 3.13 this can be written as

$$m_1 = \frac{y'}{y} = -\frac{x'+f}{x+f} \tag{3.14}$$

where the meaning of each parameter is as illustrated in Figure 3.13, and the minus sign signifies an inverted image. From the image equation, it is obvious that

$$\frac{1}{x'+f} + \frac{1}{x+f} = \frac{1}{f} \tag{3.15}$$

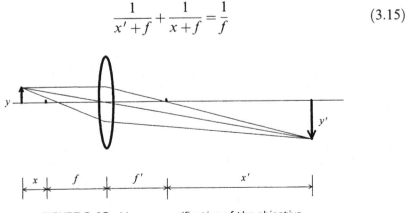

FIGURE 3.13 Linear magnification of the objective.

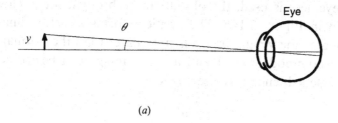

(a)

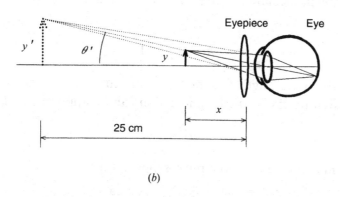

(b)

FIGURE 3.14 Angular magnification of the eyepiece.

Substituting Eq. 3.15 into Eq. 3.14, the linear magnification of the objective is expressed as

$$m_1 = -\frac{x'}{f} \tag{3.16}$$

The angular magnification of the eyepiece is defined as the ratio of the angle subtended at the eye by the image when an eyepiece is used, to the angle subtended by the object when no eyepiece is used. Since a microscope is always used to observe details of close objects, the distance from the object to the eye is chosen to be the normal reading distance (25 cm). It can be seen from Figure 3.14a that the angle subtended by an object at the normal distance to the naked eye is

$$\theta \approx \tan \theta = \frac{y}{25} \tag{3.17}$$

where the small angle approximation is applied and y is the height of the object.

When an eyepiece is used, the object can be brought much closer to the eye, as shown in Figure 3.14b. The eyepiece forms a virtual image of the object. As long as the virtual image is not nearer than the normal distance, the eye can accommodate to it and a sharp image can be formed on the retina. The angle subtended by the image is

$$\theta' \approx \tan \theta' = \frac{y'}{25} = \frac{y}{x} \tag{3.18}$$

where y' is the height of the image and x is the object distance, which can be obtained from the lens equation

$$\frac{1}{x} + \frac{1}{25} = \frac{1}{f} \tag{3.19}$$

With very little mathematical manipulation, the angular magnification can be obtained as

$$m_2 = \frac{25}{f} + 1 \tag{3.20}$$

In Eq. 3.20 the focal length f is measured in centimeters. Since f is usually small compared to $25\,\mathrm{cm}$, the magnification can be approximated as

$$m_2 = \frac{25}{f} \tag{3.21}$$

The overall magnification of the microscope is then

$$M = m_1 m_2 = -\frac{x'}{f_o} \frac{25}{f_e} \tag{3.22}$$

where f_o and f_e are the focal lengths of the objective and the eyepiece, respectively. Although the magnification of a microscope is measured by M, it is customary among manufacturers to label objectives and eyepieces according to their separate magnifications m_1 and m_2. The most common objectives have magnifications of 10, 20, or 40, denoted by 10^x, 20^x, and 40^x, respectively. The most commonly used eyepieces have magnifications of 5^x, 10^x, or 15^x.

EXAMPLE 3.6

Consider a microscope with a tube length (i.e., the distance between the objective and the eyepiece) of $18\,\mathrm{cm}$ and with focal lengths of the objective and the eyepiece used for an observation of 0.5 and $2.5\,\mathrm{cm}$, respectively. Determine the magnification of the microscope.

To determine the linear magnification, we need to know the distance between the focal point of the objective and the real image. Assume that the real image is formed at the front focal plane of the eyepiece. The distance x can be obtained as

$$x' = 18 - f_o - f_e = 18 - 0.5 - 2.5 = 15\,\text{cm}$$

and the linear magnification is

$$m_1 = -\frac{x'}{f_o} = -\frac{15}{0.5} = -30.$$

The angular magnification of the eyepiece is

$$m_2 = \frac{25}{f_e} = \frac{25}{2.5} = 10.$$

The overall magnification of the microscope is the product of m_1 and m_2

$$M = m_1 m_2 = 300.$$

Besides the magnification m_1, there is another important parameter for an objective. This is called the *numerical aperture* (NA) and is defined as

$$\text{NA} = \frac{D}{f_o} \tag{3.23}$$

where D is the diameter of the aperture and f_o is the focal length of the objective. You may have noticed that the numerical aperture is actually the reciprocal of the F-number, which is used to characterize the camera lenses. It is customary in the microscopy community to label an objective with its numerical aperture, instead of its F-number. As we learned in Section 2.5, the resolution of an objective, or any optical lens, is proportional to its numerical aperture. The numerical apertures of 10^x, 20^x, and 40^x objectives are usually 0.25, 0.50, and 0.65, respectively.

3.5 Projection Systems

A projection instrument illuminates a transparent object with bright light sources and forms a greatly enlarged image of the object on a screen. The most commonly used projection instruments in classrooms are slide projectors and overhead projectors. The working principles of these will be discussed in this section. The reader is encouraged to compare the real projector in the classroom with what we discuss in this section.

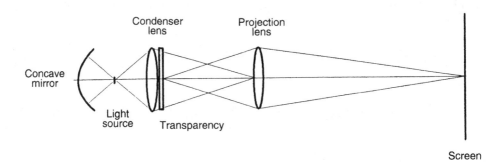

FIGURE 3.15 Optical system of a slide projector.

The optical system of a slide projector is illustrated schematically in Figure 3.15. The condensing lens collects the light from the light source and redirects it towards the projection lens. To collect as much light as possible, the condensing lens generally has a very short focal length but a large aperture. In other words, it generally has a large numerical aperture, as large as 1.0 in many cases. A concave mirror is often used to reflect the light propagating back to the condensing lens. In order to let all the light rays intercepted by the condenser lens pass through the projection lens and finally be projected onto the screen, the projection lens should be placed at the image plane of the condensing lens. Hence, the condensing lens forms an enlarged image of the light source (e.g., the lamp filament) at the surface of the projection lens, and the size of the image is roughly the same as the size of the aperture of the projection lens. Because of this need to fill the aperture of the projection lens, projection lamps always have several filaments, placed close together and parallel to one another. The lamp thus approximates a small, diffuse light source.

The transparency is placed directly behind the condensing lens. It is illuminated by the light from the projection lamp and is imaged onto the projection screen by the projection lens. Since the magnification is usually rather large, the distance from the transparency to the projection lens (i.e., the object distance) is very close to the focal length. A slight change in the object distance will lead to a large change in the image distance (i.e., the distance between the projector and the image screen). Therefore, a slight adjustment of the projection lens along the optical axis will focus the sharp image to a wide range of distances and hence meet the requirement for various magnifications.

EXAMPLE 3.7

Assume a slide projector is used in a classroom, in which the focal length of the projection lens is 50 mm. The magnification of the projector is usually 50^x. One day, an instructor brings transparencies consisting of very small letters. In order to let all the students see clearly the contents of these transparencies, a bigger magnification (say 100^x) is preferred. How should we adjust the projector in order to achieve a magnification of 100^x?

Let us denote the object distance and the image distance before moving the projector by l_o and l_i, respectively. From the Gaussian lens equation, we have

$$\frac{1}{l_o} + \frac{1}{l_i} = \frac{1}{f}$$

and

$$\frac{l_i}{l_o} = 50.$$

The solutions of these two equations are

$$l_o = 51\,\text{mm}$$

and

$$l_i = 2.55\,\text{m}.$$

Let the object distance and image distance after moving the projector be denoted by l_o' and l_i', respectively. Since the magnification is adjusted to 100^x, we have

$$\frac{1}{l_o'} + \frac{1}{l_i'} = \frac{1}{f}$$

and

$$\frac{l_i'}{l_o'} = 100.$$

The solutions are

$$l_o' = 50.5\,\text{mm}$$

and

$$l_i' = 5.05\,\text{m}.$$

Because the screen is not moved, we need to move the projector back about 2.5 m in order to adjust the image distance from 2.55 to 5.05 m. To obtain a sharp (focused) image on the screen, the projection lens needs to be moved about 0.5 mm along the optical axis, changing the object distance from 51 to 50.5 mm.

Although an overhead projector appears to be different from a slide projector, their operation is based on the same principles. In an overhead projector the light source and the condensing lens are packed in a box. The light

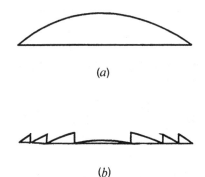

FIGURE 3.16   Conventional lens (a) and Fresnel lens (b).

illuminates the transparency from the bottom up. A mirror is incorporated with the projection lens to redirect the optical axis from vertical to horizontal and, consequently, an enlarged image is formed on the screen at the front of the classroom.

The condensing lens has no function other than to collect the light and fill up the aperture of the projection lens with the image of the light source. Therefore it does not need to be of high quality. In a overhead projector, however, a very large and yet flat condensing lens is needed. As we discussed previously, a relatively large numerical aperture is required for a condensing lens to collect efficiently the light from the light source. As the condensing lens becomes larger, so does the curvature of its surface. A lens of $25 \times 25\,cm^2$ with a focal length of 25 cm would be an impossibly thick and massive piece of glass. To overcome this drawback, the Fresnel lens can be used as the condensing lens in the overhead projector.

A conventional lens and a Fresnel lens are illustrated in Figure 3.16. We can imagine that a conventional lens is composed of many concentric rings. The cross-section of each ring consists of a rectangle and a curved triangle. It is the curvature in each of the triangles that refracts the light passing through it. If the rectangular portion is removed, the refraction property of the ring should change little, but the thickness and the mass of the lens can be reduced greatly. This idea was originally proposed by Fresnel in 1919. The first Fresnel lens was used in a beacon in 1922. Today, the Fesnel lenses used in overhead projectors are generally made of plastic. The advantages of using plastics include light weight, low cost, and simple manufacturing processes.

REFERENCES

1. W. G. Driscoll and W. Vaughan, *Handbook of Optics*, McGraw-Hill, New York, 1978.
2. E. Hect and A. Zajac, *Optics*, Addison-Wesley, Reading, MA, 1979.

3. F. A. Jenkins and H. E. White, *Fundamentals of Optics*, fourth edition, McGraw-Hill, New York, 1976.

4. M. Born and E. Wolf, *Principles of Optics*, sixth edition, Pergamon, Oxford, 1970.

5. F. T. S. Yu, *Optical Information Processing*, Wiley, New York, 1983.

PROBLEMS

3.1 If the diameter of the iris in an eye enlarges from 2 to 6 mm, what is the change in the intensity received by the retina?

3.2 A myopic eye needs a corrective lens with a focal length of 2 m. Find its far point.

3.3 If a boy cannot see clearly objects closer than 2 m, what kind of corrective lens does he need?

3.4 If the focal length of a camera is 50 mm and the size of the photographic film is $25 \times 36 \text{ mm}^2$, what is the field of view at a distance of 5 m?

3.5 A photographer changed the F-number of his camera lens from $f/16$ to $f/5.6$. In order to obtain an identical exposure on the film in both cases, how should he adjust the exposure speed?

3.6 Consider the H-and-D curve of a certain photographic film as shown in Figure 3.17.

   (a) Evaluate the γ value of the film.

   (b) Write a linear equation to represent the straight-line region of the H-and-D curve.

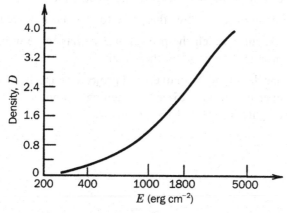

FIGURE 3.17

3.7 The objective lenses and eyepieces of a pair of binoculars have focal lengths of 26.25 cm and 25 mm, respectively. What is the angular magnification of the binoculars?

3.8 A student constructed a telescope in the laboratory. The lens he used as the objective was labeled as $D = 25$ mm, $f/16$. The lens he used as the eyepiece was labeled as $f = 20\,\text{mm}$. Determine the magnification of the telescope.

3.9 When the student looked through the telescope he had built, he found everything was upside down. How can he overcome this problem and obtain upright images?

3.10 If the student looks through the telescope backwards (i.e., looks through the objective end) what phenomena will he find? Can you explain the reasons for such phenomena?

3.11 The objective and the eyepiece of a microscope have focal lengths of 8.2 and 5.2 mm, respectively, and are located 180 mm apart. The real image formed by the objective is right at the focal plane of the eyepiece. Find:

(a) The distance from the object viewed to the objective.

(b) The linear magnification produced by the objective.

(c) The overall magnification of the microscope.

3.12 The objective of a microscope is labeled as $m_1 = 20^x$, NA $= 0.50$, and the focal length of the eyepiece is 25 mm. Find:

(a) The overall magnification of the microscope.

(b) The distance from the objective to the eyepiece.

3.13 The tube length of a microscope is 18 cm. The magnification is 300^x and the focal length of the objective is 5 mm. Find the focal length of the eyepiece.

3.14 A slide projector is placed 5 m from the screen. The focal length of the projection lens is 75 mm and the magnification is 60^x. When the projector is moved to a position 12 m from the screen, find:

(a) The magnification after the projector has been moved.

(b) The distance which the projection lens has to be moved in order to obtain a sharp image on the screen.

3.15 The Fresnel lens used in an overhead projector has a focal length of 30 cm and an aperture of $25 \times 25\,\text{cm}^2$. Assuming a point source is used, determine the light efficiency of the projector.

DETECTORS

In this chapter we shall discuss a few types of optoelectronic detector that are commonly used in optical radiation detection. The essence of these detectors converts a photon flux (or a spatially distributed array of photon fluxes) into electronic charges or a current, which is then amplified and processed with proper electronic circuits. In comparison with the human eye, which responds in a fraction of a second, some of these detectors have very short response times, which can be of the order of 10^{-9} to 10^{-12} s. In general, the sensitivity of available optoelectronic detectors is much higher than the human eye. Some detectors (e.g., photomultipliers) are even capable of detecting single photons. The spectral range of these detectors can in fact extend to the ultraviolet and infrared ranges.

4.1 Photoconductive Detectors

Photoconductive detectors are basically semiconductors (n-type or p-type) that are appropriately doped. An elementary schematic diagram of a photoconductive detector along with its external circuit is shown in Figure 4.1. The "dark" resistance of the detector (i.e., the resistance without any incident light) is assumed much higher than the load resistance R, by which the bias voltage is essentially applied across the detector chip.

By illuminating the detector chip, photons with sufficient ionizing energy excite electrons into the conduction band in n-type semiconductors, or generate holes in the valent band of p-type semiconductors, as illustrated in Figure 4.2. In either case, the incident light changes the number of carriers in the detector chip and reduces the resistance of the detector, which increases the voltage across the load resistor R.

Once an excess electron (or hole) has been created by the absorption of a photon, it will drift under the influence of the electric field toward the

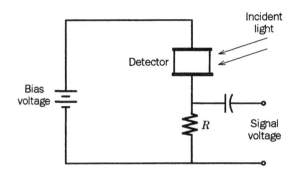

FIGURE 4.1 Photoconductive detector.

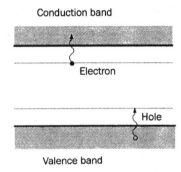

FIGURE 4.2 Two processes involved in a photodetective detector. (Top) Electron excitation (*n*-type); (bottom) hole excitation (*p* type).

appropriate contact and generate current in the external circuit. As soon as the first carrier is collected, the other contact emits another carrier. As long as photon absorption and photoexcited ionization continue, a current flows in the external circuit.

The energy required to excite electrons into the conduction band (or to create holes in the valence band) varies from material to material. The lowest required energy is about 0.04 eV, which corresponds to an optical wavelength of about 32 μm. The capability to detect long wavelength infrared radiation (up to 30 μm) is the principal advantage of photoconductive detectors over other photodetectors. However, this advantage is a double-edged sword. Due to the rather low energy requirement for creating photoelectrons, at room temperature ($T = 300\,\text{K}$) a large number of the acceptors in the detector chip would be ionized by thermal effects. To avoid such thermal noise, most photoconductive detectors working in the far infrared range have to be cooled to liquid nitrogen (77 K) or even liquid helium (4.2 K) temperature, and consequently are bulky and complicated structures.

When an optical beam of intensity I and frequency v is incident on a photoconductive detector, the rate of generation of photoexcited charge carriers can be expressed as:

$$G = \eta \frac{I}{h\nu}, \tag{4.1}$$

where h is Planck's constant, $h\nu$ is the energy of a single photon, and η is the quantum efficiency, which is defined as the probability of the excitation of a carrier by an incident photon. If the lifetime of the carriers is τ_o, and the time taken for a carrier to cross the detector chip is τ_d, the photocurrent generated in the external circuit is

$$i = \eta \frac{e}{h\nu} \left(\frac{\tau_o}{\tau_d} \right) I, \tag{4.2}$$

where e is the energy of the electron. Note that the factor τ_o/τ_d can be interpreted as a photoconductive gain. A larger carrier lifetime would lead to a higher photocurrent. However, a larger carrier lifetime also results in a slow response time. Therefore, trade-offs must be made in the detector design to tailor the specifications based on specific applications. The response time of photoconductive detectors is typically in the microsecond range.

In view of Eq. 4.2, we see that the photocurrent is proportional to the intensity of the incident light. In other words, it is proportional to the square of the electrical field of the light, $i \propto E_{light}^2$. Detectors with such characteristics are called *square-law detectors*. In fact, most photodetectors are square-law detectors, as will be seen later in this chapter.

EXAMPLE 4.1

A photoconductive detector and its external circuits are illustrated in Figure 4.1, in which the photocurrent generated by the incident light is assumed to be linearly proportional to the light intensity within a specific range. Determine the equivalent detector resistance R_t, which varies as a function of light intensity.

We assume that the photocurrent i is linearly proportional to the incident light intensity I, as given by

$$i = cI,$$

where c is the constant of proportionality. By referring to Ohm's law, the bias voltage V_b can be written as:

$$V_b = i(R_t + R) = cI(R_t + R).$$

Thus the equivalent detector resistance can be expressed as:

$$R_t = \frac{V_b}{cI} - R.$$

Notice that if there is no incident light (i.e., $I = 0$), the equivalent detector resistance would be very large $(R_t \rightarrow \infty)$. As the light intensity increases slightly, the detector resistance decreases. However, as the intensity of the incident light increases further (over a certain threshold), R_t becomes very small and most of the bias voltage is applied across the load resistor $R(R_t \rightarrow 0$ and $V_b \rightarrow cIR)$.

4.2   Semiconductor Photodiodes

There are limitations and difficulties associated with photoconductive detectors, such as severe thermal noise, bulky size, and relatively slow response time. These limitations may be overcome by semiconductor photodiodes based on semiconductor p-n junction technology. The commonly used semiconductor photodiodes include the simple p-n junction diode, the p-i-n diode, and the avalanche photodiode. These devices are rugged, sensitive, fast, easily produced, and easily biased. More importantly, they are compatible with the concept of integrated optics, in which an entire system (or subsystem), including the transmitter, receiver, multiplexer and processor, can be designed on a single chip of semiconductor substrate. In this section, we first discuss the principle of the p-n junction and then discuss the principle of simple junction photodiodes.

4.2.1   THE p-n JUNCTION

Before embarking on a description of the p-n diode detector, we need to understand the operation of the semiconductor p-n junction. A p-n junction is formed by bringing together a piece of donor-doped (n-type) semiconductor material, in which the charge carriers are dominantly electrons, and another piece of acceptor-doped (p-type) semiconductor material, in which the carriers are holes. The abrupt transition from the n-type region to the p-type region at the boundary of the two materials is known as the p-n junction.

As soon as the two pieces of material are put into contact, electrons flow from the n region to the p region, leaving behind the immobile donor ions. Similarly, holes flow from p to n region, leaving the negatively charged acceptors behind. These immobile donors and acceptors build up an internal electric field around the p-n junction, as shown in Figure 4.3a, which creates a potential barrier that prevents further gross migration of the carriers.

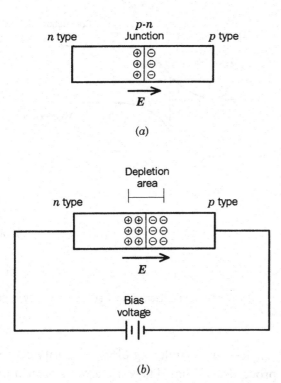

FIGURE 4.3 A *p-n* junction. (*a*) At zero applied bias, an internal space–charge field is built up to prevent further migration of carriers. (*b*) Under reversed bias, the mobile charge carriers around the *p-n* junction are swept by the bias electric field and a depletion area is formed.

Now let us consider the minority carriers; that is, holes in *n*-type materials and electrons in *p*-type materials. When a hole in the *n* region wanders into the vicinity of the *p-n* junction, it is immediately accelerated by the internal field and, therefore, *drifts* to the *p* side. Similarly, if an electron in the *p* region wanders into a position sufficiently close to the *p-n* junction, it is accelerated by the internal field and drifts into the *n* region. Both processes contribute to the drift current crossing the junction, which is balanced at zero bias by an opposite current due to the diffusion of majority carriers against the retarding field.

When a reversed bias is applied by an external circuit, however, almost all the majority mobile carriers in an area close to the *p-n* junction are swept by the bias electric field, with only the immobile donors and acceptors left behind, as shown in Figure 4.3*b*. Such an area is called the *depletion area*. Notice that the sweeping of majority carriers will stop when the internal electric field built up by the donors and acceptors reaches a sufficient level to cancel the bias field at the junction. This dynamic equilibrium is generally established in less than a nanosecond.

The increased internal electric field built up in the depletion area forms a potential barrier that greatly reduces the diffusion of majority carriers across

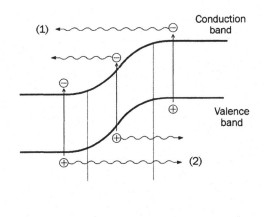

FIGURE 4.4 Various charge carrier excitation processes initiated by incident light in the vicinity of the *p-n* junction.

the junction. Nevertheless, the wandering of the minority carriers into both *n* and *p* regions still proceeds as usual. There are always some minority carriers, electrons in the *p* region and holes in the *n* region, that happen to move into the vicinity of the depletion area and to drift further into the other side of the *p-n* junction. The drift current under reversed bias is very small, generally being of the order of microamperes (μA).

4.2.2 JUNCTION PHOTODIODES

Upon illumination with optical radiation, the drift current crossing the *p-n* junction can be greatly increased. To allow access for the incident light, an optical window is provided in packaged junction photodiodes. The mechanisms that occur in a junction photodiode are illustrated in Figure 4.4. There are basically three types of charge carrier flow taking place within the diffusion length of the depletion layer, or in the depletion area. If the photon is absorbed in the *p* region, an electron–hole pair is created. The electron (as a minority carrier) will drift towards the depletion layer, cross the junction, and then contribute a charge $(-e)$ to the external current. Similarly, the hole created by photoabsorption in the *n* region will drift in the opposite direction and also contribute to the current. However, within the depletion area, the electron–hole pair will be separated and the electron and hole will drift away from each other in opposite directions, and thus both will contribute to the current flow.

It should be noted that only the electron–hole pairs generated in the depletion area or within a diffusion length of the depletion area contribute to the current flow. By "diffusion length" we mean the average distance a minority carrier traverses before recombining with a majority carrier. Those electron–hole pairs created deep in the neutral region, the region far away from the *p-n* junction, will recombine with carriers of opposite type before they ever diffuse to the junction, and thus do not contribute to the current flow in the external circuit.

If we consider only the photons absorbed within the diffusion length to the depletion area and neglect the recombination of the created electron–hole pairs, the current is given by:

$$i = e\eta\left(\frac{I}{h\nu}\right). \tag{4.3}$$

Comparing Eq. 4.3 with Eq. 4.2, we can see that there is no term corresponding to the conduction gain. This is due to the different physical mechanisms involved in photodiode and photoconductive detectors. In a *p-n* junction diode most of the applied bias voltage appears across the depletion area. Thus only those pairs formed within this area or capable of diffusing into this area can be influenced by the field and contribute to the external current.

The fact that a minority carrier may have to travel some distance (up to the diffusion length) before it is transported across the *p-n* junction results in a major disadvantage of simple junction diodes; namely, a relatively slow response time if most of the carriers are generated in the neutral regions. This is a particularly serious drawback in some semiconductor materials such as silicon, in which the depletion area is small compared to the diffusion length. As most of the electron–hole pair generation takes place in the neutral regions, the response time is slowed by the diffusion process. Nevertheless, this difficulty can be avoided by utilizing the *p-i-n* structure, as discussed in the next section.

The energy gap of most semiconductors is of the order of 1 eV. This corresponds to an optical wavelength of about 1 μm. In general, therefore, photodiodes can respond to light ranging from the ultraviolet to the near-infrared region.

EXAMPLE 4.2

Assume that the response time of a simple junction diode is 10 ns and that the diffusion velocity of the electrons (or holes) in a semiconductor diode is about one ten-thousandth of the speed of light. Find the diffusion length of the diode.

Since the response time of a junction diode is primarily limited by the diffusion time, it is reasonable to assume that the diffusion time τ_d is approximately equal to the response time. Thus, the diffusion length of the junction diode is

$$I_d = \tau_d v_d = \tau_d \left(\frac{c}{10000}\right) = 0.3\,\mu\text{m}.$$

In practice, however, the response time is also limited by other physical factors, and so the diffusion length may be less than $0.3\,\mu\text{m}$.

4.3 PIN and Avalanche Photodiodes

Figure 4.5a shows a diagram of the cross-section of a *p-n* junction photodiode, in which the top contact has windows through which the semiconductor materials are exposed to incident radiation. Although a photon may be

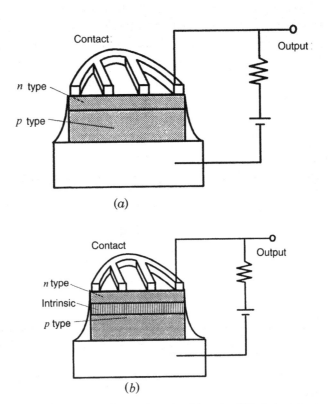

(a)

(b)

FIGURE 4.5 Structure of a *p-n* junction (a) and a PIN diode detector (b).

absorbed anywhere in the p-type or n-type material, as discussed in the preceding section only those photons that are absorbed in the depletion area or within a diffusion length of the depletion area can contribute to the external photocurrent. As the diffusion length is generally much larger than the width of the depletion region, most of the useful absorption and electron–hole pair production takes place outside the depletion area (but within the diffusion length). This substantially limits the response time of simple junction photodiodes.

4.3.1 PIN DIODES

In order to improve the response time of a photodiode, the carriers must be generated where the field is large, so that the charge transport is primarily due to fast drift rather than slow diffusion. This is accomplished by adding an intrinsic region that has a very high resistivity between the p and n layers, as shown in Figure 4.5b. Because of their sandwich structure, this type of diode is called a PIN (or p-i-n) photodiode.

In a PIN diode, the voltage drop occurs mostly across the intrinsic layer. If the intrinsic layer is made wide enough such that most of the incident light is absorbed within it, the long diffusion times associated with processes (1) and (2) as depicted in Figure 4.4 can be avoided. Therefore, PIN photodiodes respond much faster than do p-n junction diodes. Typically, the response time of a PIN diode is of the order of nanoseconds, whereas the response time of p-n junction photodiodes is tens of nanoseconds or even longer.

Besides the fast response, the PIN photodiode has other advantages over the p-n junction diode. For example, since the intrinsic layer separates the p and n layers, the capacitance of the device is greatly reduced. This further leads to a wider frequency response. Due to these merits, the PIN diode has become the most commonly used solid-state detector for laser radiation detection in the visible to near infrared region.

EXAMPLE 4.3

Assume that the thickness of the intrinsic layer of a PIN diode is 2.4 μm and that the average drift velocity of electrons is about one-tenth of the speed of light. Determine the response time of the diode.

The response time of a PIN diode is primarily limited by the carrier drift in the intrinsic layer. If we denote the average drift speed by V_d and the intrinsic thickness by I_d, the response time can be calculated as:

$$\tau = \frac{l_{in}}{v_d} = 0.08 \, \text{ns}.$$

In practice, the response time may also be limited by other factors, and thus it may be a little longer than 0.08 ns. Nevertheless, the response time of PIN diodes is generally in the nanosecond or subnanosecond range.

4.3.2 AVALANCHE PHOTODIODES

It can be seen from Eq. 4.3 that each photon absorbed around the depletion area creates only one electron–hole pair, and therefore there is no gain in the detection of the photodiode. However, by increasing the reverse bias voltage across the *p-n* junction, the field in the depletion area can be increased to a point at which carriers (electrons or holes) that are accelerated across the depletion area can gain enough kinetic energy to stimulate new electrons from the valence to the conduction band. These second-generation carriers move under the influence of the field and can create third-generation carriers. This process is referred to as *avalanche multiplication*. Diodes using this effect are called avalanche photodiodes (often denoted by APD).

If M new carriers are generated for each primary carrier created by the absorbed photon, Eq. 4.3 should be modified as:

$$i = e\eta M\left(\frac{I}{h\nu}\right). \tag{4.4}$$

The avalanche process greatly enhances the sensitivity of the device. Typical values of M range from 20 to 100. It must be noticed that arbitrarily large values of M should be avoided, since the avalanche multiplication also contributes additional noise. Under high reverse bias voltage, electron–hole pairs are not only created by the absorbed photons, but are also created by thermal processes. Thus thermal runaway is a distinct possibility at high values of M.

The structure of an avalanche photodiode is similar to that of an ordinary *p-n* junction diode, as shown in Figure 4.5a. However, because of the steep dependence of M on the applied field in the avalanche region, when fabricating avalanche photodiodes special care must be exercised to obtain highly uniform junctions. This can be readily achieved with the state-of-the-art solid state fabrication facilities.

4.4 Photomultipliers

The photomultiplier is one of the most sensitive photodetectors. It has been used to detect power levels as low as about 10^{-9} W. A schematic drawing of

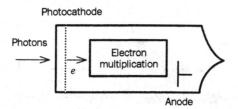

Photocathode

Photons

Electron
multiplication

e

Anode

FIGURE 4.6 Schematic diagram of a head-on-type photomultiplier tube.

the so-called 'head-on' type photomultiplier tube, which consists essentially of the cathode, the electron multiplier section, and the anode, is shown in Figure 4.6. The cathode consists of materials with a low *surface work function*, which is the energy needed to eject the electron from the material. Typically, these materials are made of compounds of silver–oxygen–cesium and antimony–cesium. The electron in the conduction band (where the electron is free) can be ejected from the surface into the vacuum, if it can overcome the work function W. Incident light at frequency ν, with quantized energy $h\nu$, will be able to eject the electron if $h\nu > W$. The work functions W for these cathode materials are typically of the order of an electron-volt (eV), corresponding to optical wavelengths in the ultraviolet (0.2 µm) to infrared (1.0 µm) region. The spectral responses of some commercial photomultipliers are shown in Figure 4.7.

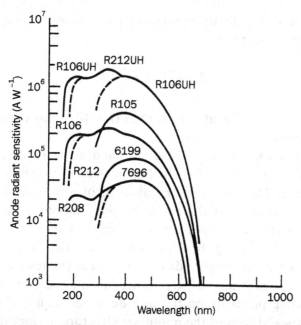

FIGURE 4.7 The typical spectral responses of photomultipliers using Sb-Cs as the antimony–cesium cathode material.

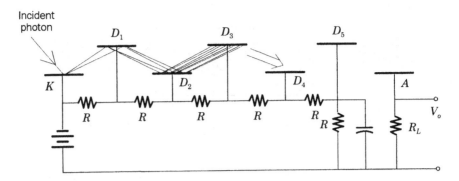

FIGURE 4.8 A five-stage photomultiplier.

The electron multiplication section consists of a series of electrodes called "dynodes," as shown in Figure 4.8. The dynodes are kept at progressively higher potentials with respect to the cathode, with a typical potential difference between adjacent dynodes of 100 V. The last dynode is the anode of the photomultiplier and is used to collect the electrons. When an electron is ejected from the photocathode, it is accelerated by the bias voltage and impinges on the first dynode. The kinetic energy of this electron enables it to cause secondary emission at the dynode. As the electron and the secondary electrons are accelerated down the chain, the process of electron multiplication is repeated, leading finally to substantial current being collected at the anode.

If one electron can generate m secondary electrons at each dynode, the current gain after N stages is

$$G = \frac{i_{out}}{i_{in}} = m^N. \tag{4.5}$$

As an example, if $N = 10$ and $m = 5$, then $G = 10^7$. This corresponds to a huge gain of 140 dB.

Because the electrons in the conduction band are only loosely bound to the cathode materials, some of them can be excited thermally and escape from the cathode. Thus, even in the absence of any optical radiation, a current may be generated at the anode by this random electron emission. Such a current is called the *dark current*, which is typically of the order of 10^{-11} to 10^{-12} A. By cooling the tube with dry ice, the dark current can be reduced by two orders of magnitude or even more. The response time of the photomultiplier is generally of the order of nanoseconds to subnanoseconds, depending primarily on the dynode materials used for each stage and the geometrical design of the multiplier structure, including the number of gain stages.

EXAMPLE 4.4

Calculate the maximum wavelength of light that has sufficient energy to eject electrons from a photocathode which has a work function of 2.5 eV.

The energy associated with light at frequency ν is $h\nu$, where h is Plank's constant. The wavelength λ is related to the frequency as $\lambda = c/\nu$. In order to eject electrons bound to the cathode with a work function of 2.5 eV, it is required that

$$h\nu = h\left(\frac{c}{\lambda}\right) > 2.5\,\text{eV}.$$

Thus we have

$$\lambda > \frac{hc}{2.5\,\text{eV}} = 497\,\text{nm}.$$

The photomultiplier responds to wavelengths longer than 497 nm, that is it responds to green, red, and infrared light.

4.5 Charge Coupled Devices

All the detectors we have discussed so far are single-element detectors suitable for detecting spatially uniform wavefronts. If the optical wave to be detected carries some spatially varying information, for instance, an image, then an array of small detectors must be used. Each small detector, the so-called *pixel* detects a corresponding part of the wavefront. The image is jointly represented by the outputs from all the pixels. In this section, we discuss a frequently used image detector, the charge coupled device (CCD).

The basic structure of a CCD is a shift register formed by an array of closely spaced potential-well capacitors. A potential-well capacitor is illustrated schematically in Figure 4.9. A thin layer of silicon dioxide (SiO_2) is grown on a silicon substrate. A transparent electrode is deposited over the silicon dioxide as a gate, to form a tiny capacitor. When a positive electrical potential is applied to the electrode, a depletion area or electrical potential well is created in the silicon substrate directly beneath the gate. Upon exposure to light electron–hole pairs are generated by the absorption of incident photons. The free electrons generated in the vicinity of the capacitor are stored and integrated in the potential well.

The quantity of electrons in the well is a measure of the incident light intensity and is detected at the edge of the CCD area by transferring the charge package. Figure 4.10 illustrates the principle of charger transfer

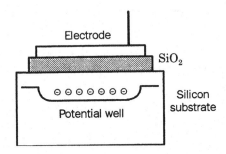

FIGURE 4.9　Structure of a CCD pixel array.

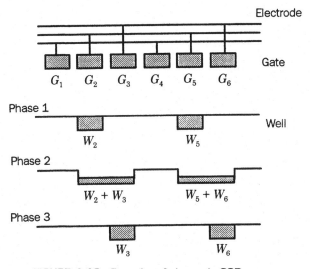

FIGURE 4.10　Transfer of charge in CCD arrays.

through the propagation of potential wells. In phase 1, gates G_2 and G_5 are turned on, while all other neighboring gates are turned off. Hence electrons are collected in wells W_2 and W_5. In Phase 2, G_2, G_3, G_5 and G_6 are on; therefore, wells W_2 and W_3 merge into a wider well. Similarly, W_5 and W_6 also form a single, wider well. In Phase 3, G_2 and G_5 are turned off, while G_3 and G_6 are left on. Thus, the electrons previously stored in W_2 are now shifted and stored in W_3, while the electrons stored in W_5 are transferred to W_6. By repeating this process, all charge packages will be transferred to the edge and be picked up there by external circuits.

CCD array detectors are available in 256×256 to 4096×4096 pixels. The center-to-center distance between adjacent pixels ranges from 10 to 40 μm. CCD cameras can retain a high modulation transfer function (MTF) up to Nyquist limits. At these limits, each pixel receives and detects data as a unique image element. MTFs of 60–75% are typical when the illumination is filtered to simulate human eye response. The linear dynamic range of CCD devices depends on the noise level. This is typically about 100 (40 dB), but a dynamic range of 40 000 (more than 90 dB) is possible. The sensitivity of video rate CCD cameras is of the order of 10^{-8} W/cm^2, which corresponds to about 0.05 lux. Moreover, higher sensitivity can be achieved by integrating and storing data over a relatively long period of time. When the device is cooled, the integration period can last as long as several minutes before background dark currents affect the quality of the image. New developments also make it possible for CCDs to control exposure shuttering, so that any level of charge integration can be obtained. When adequate light is available, exposure times of less than 1 ns can be set, thus allowing the capture of high-speed events without blur. State of the art CCDs can achieve nonuniformity of less than 3% and output signal-to-noise ratios of greater than 60 dB.

4.6 Noise and Sensitivity of Electro-optic Detectors

Photodetectors (including image detectors such as CCDs) are optoelectronic in nature and are always used along with their external electronic circuits. Their ultimate sensitivity depends on the noise level. Although some noise sources are unique to a particular device, many are common to all detectors. In this section we discuss the noise and sensitivity of photodetectors. Due to the introductory nature of this book, we shall confine the analysis primarily to the photomultiplier. However, many general descriptions obtained from the analysis of the photomultiplier can be extended to other detectors. The features of other detectors that are not similar to those of photomultipliers will be pointed out during the discussion. Interested readers can refer to references [4–6] for a more detailed analysis.

Photodetectors usually work in one of two detection modes. The first mode is called the *video detection mode* (or straight detection mode), in which the optical signal to be detected is modulated at a low frequency ω_m before being detected by the detector. The signal consists of an output current oscillating at ω_m, which has an amplitude proportional to the optical intensity. In the second detection mode, the signal to be detected is combined at the photodetector with a stronger optical wave. The output is then a current at the offset frequency between the signal wave and the local wave. This scheme is called the *heterodyne detection mode*. Noise and sensitivity in the video detection and heterodyne detection modes are discussed below.

4.6.1 VIDEO DETECTION MODE

In video detection the optical wave to be detected is modulated at a low frequency ω_m before it impinges on the photocathode. As a result of the square law detector characteristics of the photomultiplier, the photocurrent at the cathode is given by

$$i_p(t) \propto E_c^2 (1 + m \cos \omega_m t)^2 \cos^2 \omega_c t, \tag{4.6}$$

where m is the amplitude of modulation, E_c is the amplitude of the carrier wave, and ω_m and ω_c are the modulation frequency and carrier frequency, respectively. Since the optical frequency (of the order of 10^{14} Hz) of the carrier wave is far beyond the response range of any photodetector, the photocurrent is essentially averaged over many cycles and can be expressed as

$$i_p(t) = CE_c^2 \left[\left(1 + \frac{m^2}{2} \right) + 2m \cos \omega_m t + \frac{m^2}{2} \cos 2\omega_m t \right]. \tag{4.7}$$

To determine the proportionality constant C, consider the unmodulated situation (i.e., $m = 0$). If the power of the unmodulated carrier wave is P, the constant current associated with the carrier wave is

$$i_p = CE_c^2 = \frac{Pe\eta}{h\nu_c}, \tag{4.8}$$

where η is the efficiency of converting a photon to an electron. Substituting Eq. 4.8 into Eq. 4.7 we have

$$i_p(t) = \frac{Pe\eta}{h\nu_c} \left[\left(1 + \frac{m^2}{2} \right) + 2m \cos \omega_m t + \frac{m^2}{2} \cos 2\omega_m t \right]. \tag{4.9}$$

As a result of the multiplication of electrons in the multiplier tube by a gain factor G (see Eq. 4.4), the output signal from the photomultiplier is

$$i_s(t) = \frac{GPe\eta}{h\nu_c} (2m \cos \omega_m t). \tag{4.10}$$

Here we have assumed that the signal in which we are interested is at frequency ω_m, and the DC as well as the double-frequency terms are blocked by suitable filters.

There are two types of noise that contribute to the output photocurrent. One is associated with the random generation of charge carriers and is called *shot noise*. The other, so called *Johnson noise* or Nyquist noise, is associated with fluctuations in the voltage across the dissipative circuit elements, such as the output load R_L of the photomultiplier tube.

Shot noise has a mean square current amplitude given by

$$\overline{i_{sn}^2}(\nu) = 2e\overline{I} \, \Delta\nu, \tag{4.11}$$

where $\bar{I}$ is the average current associated with the electron and any charge carriers generating the shot noise, and $\Delta\nu$ is the bandwidth limit of the detector for the photomultiplier. The shot noise is caused by the dark current generated at the cathode. These thermally induced electrons also experience a gain G as they are accelerated through the amplifying stages. Therefore, for the photomultiplier tube,

$$\overline{i_{sn}^2} = 2G^2 e(\overline{i_p} + \overline{i_d})\Delta\nu, \tag{4.12}$$

where $\overline{i_p}$ and $\overline{i_d}$ are the average photoexcited current and average dark current at the cathode, respectively.

The derivation of the Johnson noise is quite complicated, yet the result is very simple. The mean square amplitude of the Johnson noise is given by

$$\overline{i_{Jn}^2} = \frac{4K_B T \Delta\nu}{R}, \tag{4.13}$$

where K_B is Boltzmann's constant $(K_B = 1.38 \times 10^{23}\,\text{J/K})$. Thus the electric power fluctuation across R (i.e., $R\overline{i_{Jn}^2}$) is of the order of the thermal energy $K_B T$ within the bandwidth interval $\Delta\nu$.

The signal-to-noise ratio is therefore given by

$$\frac{S}{N} = \frac{\overline{i_s^2}}{\overline{i_{sn}^2} + \overline{i_{Jn}^2}} = \frac{2G^2(Pe\eta/h\nu_c)^2}{2G^2 e(\overline{i_p} + \overline{i_d})\Delta\nu + (4K_B T\,\Delta\nu/R)}. \tag{4.14}$$

In the derivation of Eq. 4.14, we have assumed the unity modulation index $(m = 1)$. Due to the extremely large gain of the photomultipliers (more than 100 dB), the first term in the denominator of Eq. 4.14, which represents the amplified cathode shot noise, is much larger than the Johnson noise term. In many applications requiring a high level of sensitivity, the photomultiplier tubes are thermoelectrically cooled, and the dark current is reduced until it almost disappears. Neglecting the Johnson noise term and assuming $\overline{i_p} \gg \overline{i_d}$, the signal-to-noise ratio of a photomultiplier can be approximated as

$$\frac{S}{N} \approx \frac{2G^2(Pe\eta/h\nu_c)^2}{2G^2 e\overline{i_p}\,\Delta\nu} = \frac{P\eta}{h\nu_c\,\Delta\nu}. \tag{4.15}$$

In general, the signal can be distinguished from noise as long as the signal-to-noise ratio is greater than one. Based on this criterion, the sensitivity of photomultipliers can be characterized by the minimum detectable optical power:

$$P_{\min} = \frac{h\nu_c\,\Delta\nu}{\eta}. \tag{4.16}$$

This corresponds to the quantum limit of optical detection. In practice, to achieve such a limit in video detection is nearly impossible, since it requires near total suppression of the dark current and other extraneous noise sources, such as background radiation, from reaching the photomultiplier and causing shot noises. The quantum detection limit (Eq. 4.16) can, however, be achieved in the heterodyne mode of optical detection. This is discussed in the next section.

EXAMPLE 4.5

A photomultiplier is used to detect light from a HeNe laser. Assume the converting efficiency η is 10% and the bandwidth $\Delta\nu$ is 1 kHz. Determine the minimum detectable power level.

The wavelength of HeNe laser light is 633 nm; hence

$$\nu = \frac{c}{\lambda} = 4.74 \times 10^{14}\,\text{Hz}.$$

From Eq. 4.16, the minimum detectable power is

$$P_{\text{min}} = \frac{h\nu\,\Delta\nu}{\eta} = \frac{6.26 \times 10^{-34} \times 4.74 \times 10^{14} \times 1000}{0.1} \approx 3 \times 10^{-15}\,\text{W}.$$

EXAMPLE 4.6

Assume that the power of an Ar laser beam ($\lambda = 514.5$ nm) passing through the aperture window of a photomultiplier is 1 W. How many photons will hit the cathode of the photomultiplier in every second?

The energy of each photon is

$$E_{photon} = h\nu = h\frac{c}{\lambda} = \frac{6.26 \times 10^{-34} \times 3 \times 10^8}{514.5 \times 10^{-9}} = 3.65 \times 10^{-19}\,\text{J}.$$

The number of photons hitting the cathode in every second is

$$N = \frac{P}{E_{photon}} = \frac{1}{3.65 \times 10^{-19}} = 2.74 \times 10^{18}.$$

4.6.2 HETERODYNE DETECTION MODE

In heterodyne detection, the optical radiation is frequency modulated. The oscillating signal, $E_s = \cos\omega_s t$, is combined with the local field, $E_L = \cos\omega_L t$. The shift in frequency is much smaller than both the signal and local frequencies:

$$\Delta\omega = |\omega_s - \omega_L| \ll \omega_s, \omega_L.$$

The total field on the cathode of the photomultiplier is then

$$E_t = E_s \cos \omega_s t + E_L \cos \omega_L t. \tag{4.17}$$

The square-law detection characteristics of the photomultiplier again mean that the photoexcited current at the cathode is

$$i_p = C(P_L + P_s + 2\sqrt{P_L P_s} \cos \Delta \omega t), \tag{4.18}$$

where $P_L = E_L^2$ and $P_s = E_s^2$ represent the power of the local oscillator and the power of the optical signal to be detected, respectively. Again, the constant of proportionality C can be determined by setting the signal power to zero ($P_s = 0$), then

$$i_p = CP_L = \frac{P_L e\eta}{h\nu}. \tag{4.19}$$

Strictly speaking, in Eq. 4.19 ν is the local frequency. However, since the oscillating and local frequencies are very close, we can simply use ν for both frequencies, with good approximation.

In most applications requiring high sensitivity, the power of the local oscillator (derived from a laser) is much higher than that of the oscillating signal power. To a first-order of approximation, we have

$$i_p = \frac{P_L e\eta}{h\nu} \left(1 + 2\sqrt{\frac{P_s}{P_L}} \cos \Delta \omega t \right). \tag{4.20}$$

When we account for the gain factor G, the time-average signal current at the output anode is

$$\overline{i_s^2} = 2G^2 \left(\frac{P_L e\eta}{h\nu} \right)^2 \frac{P_s}{P_L}. \tag{4.21}$$

Similar to the video detection mode, the noise sources include the shot noise at the cathode and the Johnson noise occurring in dissipative circuit elements. The mean square representations for these noises are given by

$$\overline{i_{sn}^2} = 2G^2 e \left(\overline{i_d} + \frac{P_L e\eta}{h\nu} \right) \Delta \nu, \tag{4.22}$$

and

$$\overline{i_{Jn}^2} = \frac{4K_B T \Delta \nu}{R}. \tag{4.23}$$

The signal-to-noise ratio at the output is then given by

$$\frac{S}{N} = \frac{2G^2 (P_s P_L)(e\eta/h\nu)^2}{[2G^2 e(\overline{i_d} + P_L e\eta/h\nu) + 4K_B T/R]\Delta \nu}. \tag{4.24}$$

The advantage of heterodyne detection is that, instead of employing various means to suppress the dark current or other sources of noise, we can just increase the power P_L of the local oscillator until the denominator of Eq. 4.24 is dominated by the term $2G^2e(P_Le\eta/h\nu)$. Then, we have

$$\frac{S}{N} \approx \frac{P_s\eta}{h\nu\,\Delta\nu}. \qquad (4.25)$$

If we again set S/N to one, the lowest level of optical signal that can be detected is

$$P_{s,\text{min}} = \frac{h\nu_c\,\Delta\nu}{\eta}. \qquad (4.26)$$

It is identical to the lowest detectable level in the video detection mode given by Eq. 4.16. However, it is important to emphasize that Eq. 4.16 is obtained by requiring all other noise sources be absent or negligible, which is impossible in practice.

EXAMPLE 4.7

Typically, the dark current $\overline{i_d}$ in an uncooled photomultiplier tube is about 10^{-11} A. Assume the local optical beam comes from an Ar laser ($\lambda = 514.5$ nm) and the photonic-to-electronic converting efficiency is 10%. What is the minimum power of the Ar laser beam in order to suppress effectively the effect of the dark current?

The right-hand side of Eq. 4.22 indicates the contribution of dark current and local oscillator to the shot noise. In order to suppress the effect of the dark current, we need

$$\frac{P_Le\eta}{h\nu} \gg \overline{i_d}.$$

Practically, we can take the following criterion

$$\frac{P_Le\eta}{h\nu} > 10\overline{i_d}.$$

Therefore

$$P_L > \frac{10\overline{i_d}h\nu}{e\eta} = \frac{10 \times 10^{-11} \times 6.26 \times 10^{-34} \times 3.65 \times 10^{14}}{1.6 \times 10^{-19} \times 0.1} = 1.43 \times 10^{-9}\,\text{W}.$$

The power of the Ar laser beam should be at least 1.5 nW, which can be easily obtained in practice.

These two modes of optical signal detection, video and heterodyne, are also applied in photoconductive detectors and photodiodes. In photoconductive detectors, the main noise source is the shot noise associated with ionized impurities in the photoconductor. In photodiodes, the chief noise source is also the shot noise associated with the random generation of carriers. In general, the lowest level of optical power detected by photomultiplier and photoconductive detectors are of the order of 10^{-17} to 10^{-19} W, and for photodiodes 10^{-7} to 10^{-9} W. Nevertheless, photodiodes are relatively inexpensive and respond to a wide spectral range, from the ultraviolet to the infrared regions.

REFERENCES

1. W. G. DRISCOLL and W. VAUGHAN, *Handbook of Optics*, McGraw-Hill, New York, 1978.
2. F. T. S. YU, *Optical Information Processing*, Wiley, New York, 1983.
3. J. T. VERDEYEN, *Laser Electronics*, Prentice-Hall, Englewood Cliffs, NJ, 1989.
4. A. YARIV, *Optical Electronics*, Sanders, Philadelphia, 1991.
5. RCA, *Electro-Optics Handbook*, Technical Series EOH-11, RCA Corporation, New York, 1974.
6. R. W. BOYD, *Radiometry and the Detection of Optical Radiation*, Wiley, New York, 1983.

PROBLEMS

4.1 Conduct a literature search and write a brief report on the practical uses of one or two of the detectors discussed in this chapter.

4.2 Assume that the energy required to excite electron–hole pairs in a photoconductive detector is about 0.05 eV. Determine the cut-off wavelength at the long-wave end of the spectral response of the detector.

4.3 Assume that the quantum efficiency of a photoconductive detector is 10%, and that the lifetime of the carriers and the average time for a carrier to cross the detector chip are 10^{-5} and 10^{-7} s, respectively. Derive the equivalent resistance of the detector, R_t, as a function of the incident light intensity.

4.4 The diffusion length in silicon photodiodes is typically about 0.5 μm. If the diffusion velocity of electrons is 10^7 cm/s, estimate the response time of such detectors.

4.5 A photodiode is capable of detecting only radiation with photon energy $h\nu > E_g$, where E_g is the energy gap of the semiconductor. In order to detect light from a YAG laser at 1.06 μm wavelength, what is the maximum energy gap of the detector material?

4.6 What are the advantages and disadvantages of simple *p-n* junction photodiodes in comparison with photoconductive detectors and PIN photodiodes?

4.7 The thickness of the intrinsic layer in PIN photodiodes is typically about 2.5 μm. If the response time of a PIN photodiode is 0.05 ns, estimate the drift velocity of the electrons in the intrinsic layer.

4.8   Assume that the gain factor in Eq. 4.4 is given by

$$M = \frac{100}{1 + 10e^{-\frac{V}{h}}}$$

where V is the bias voltage and h is the thickness of the p-n junction. In order to achieve a photocurrent uniformity of 5% over the whole active area, what is the required uniformity in junction thickness h?

4.9   If the shortest wavelength of light that can be detected by a photomultiplier is 300 nm, what is the work function of the cathode material?

4.10   A photomultiplier has 10 multiplication stages. In the first four stages, each primary electron can stimulate four secondary electrons; whereas in the next six stages, each primary electron can stimulate five secondary electrons. What is the gain of the photomultiplier?

4.11   A student bought a CCD camcorder. In the manual, he found that the sensitivity is 0.1 lux. On a clear night, he measured the moonlight from the sky as being about 10^{-9} W/cm^2. Can he use the CCD camcorder to take videos of the moon?

4.12   A CCD camera with 512×512 pixels works at video rate (i.e., 30 frames/s). What is the data-transfer rate?

4.13   Show that the shot noise has the mean square amplitude given by Eq. 4.11, by considering the charges moving between two electrodes. (*Hint*: A model may be found in reference 4, Chapter 11.)

4.14   The light from a YAG laser ($\lambda = 1.06\,\mu$m) is detected with a photomultiplier. If the minimum detectable power is 10^{-18} W and the quantum efficiency is 8%, what is the bandwidth limit of the device $\Delta\nu$?

4.15   Calculating the energy associated with a quantum of light for the following typical laser wavelengths: (a) $\lambda = 488$ nm; (b) $\lambda = 514.5$ nm; (c) $\lambda = 930$ nm; (d) $\lambda = 1.06\,\mu$m; and (e)$\lambda = 10.6\,\mu$m.

4.16   Estimate the amount of current density emitted by a photocathode with a quantum efficiency of 10% when a weak laser light from a diode laser ($\lambda = 1.55\,\mu$m) is incident on it with an intensity of 10 mW/cm^2. Using a typical gain of 10^6, estimate the current at the anode.

4.17   Typically, the dark current in an uncooled photomultiplier tube is about 10–11 W. At what power level of the local oscillator (a laser, for example) is the cathode shot noise arising from the local oscillator equal to that arising from the dark current?

SPATIAL LIGHT MODULATORS

Spatial light modulators (SLMs) are electro-optic devices that can modulate certain properties of an optical wavefront, for example, the amplitude, intensity, phase, or polarization. These devices are critical components in many electro-optic systems, especially for applications to optical signal processing and optical computing, in which the devices are frequently used as input transducers, signal converters, spatial filters, page composers, etc. The information-bearing element for an SLM can be an electrical or optical signal, which leads to two major classes of SLM: namely, electrically and optically addressable SLMs. An electrically addressed SLM is usually constructed of a pixelated structure, by which the local optical wavefront is modulated at each individual pixel by means of electrical manipulation. The advantage of an electrically addressed SLM is its capability of interfacing with electronic units in an electro-optic system. The disadvantages include low efficiency due to a dead zone between the electrodes, and multiple diffraction patterns which are caused by the pixelated structure. On the other hand, an optically addressed SLM consists of a continuous structure for performing the modulation function. In general, the addressing optical image produces an electric charge distribution over the modulation material that generates secondary effects for electro-optic modulation. The major advantage of an optically addressed SLM is its ability to modulate a light beam by another light beam. One of the common disadvantages of optically addressed SLMs is their high fabrication cost. Although numerous applications of SLMs have been proposed in recent years, none of them actually fulfills all the requirements of speed, resolution, light efficiency, and spatial uniformity. There is no general agreement as to which device is best, and therefore, in practice, the selection of an SLM is highly dependent on the application involved.

5.1 Acousto-optic Modulators

The interaction of light waves with sound (acoustic) waves has been the basis of a large number of devices connected to various laser systems for display, information handling, optical signal processing, and numerous other applications requiring the spatial and temporal modulation of coherent light. The underlying mechanism of acousto-optic interaction is simply the induced change in the refractive index of an optical medium by the presence of an acoustic wave. When an acoustic wave exists in a medium, a snapshot of the medium will reveal a regular spatial alternation between regions where the density is higher than average and regions where it is lower than average. These density variations cause corresponding changes in the refractive index of the material. Because of the extremely high speed of light, the medium thus has the characteristics of a phase grating.

Acousto-optic modulators are usually used in laser systems for the electronic control of the intensity and position of the laser beam. When an acoustic wave is launched into the optical medium, it generates refractive index changes that give the effect of a moving sinusoidal grating. An incident laser beam passing through the device will be diffracted by the grating into several diffraction orders. The angular position of the selected diffraction order (e.g., first order) is linearly proportional to the acoustic frequency; thus, the higher the frequency, the larger the diffraction angle. The intensity of the diffracted light is proportional to the power of the acoustic wave, and thus the intensity can also be modulated.

The interaction between the light and the acoustic wave produces the so-called *Bragg diffraction effect*, as illustrated in Figure 5.1. The beam of light is incident on a plane acoustic wave in the acousto-optic cell. This acoustic wave is emitted by a transducer driven by an electric addressing signal. At a certain critical angle of incidence θ_i, the incident beam produces a coherent diffraction at angle θ_d, which is known as the *Bragg angle* and is given by

$$\theta_d = \sin^{-1}\left(\frac{\lambda}{2\Lambda}\right) \tag{5.1}$$

where λ and Λ are the wavelength of the light and the acoustic wave in the acousto-optic cell, respectively.

A variety of different acousto-optic materials are used, depending on the laser parameters such as wavelength, polarization, and intensity. For the visible and the near-infrared regions, the acousto-optic modulator is usually made from dense flint glass, tellurium oxide (TeO_2), or fused quartz. For the infrared region, germanium is often employed. Lithium niobate ($LiNbO_3$)

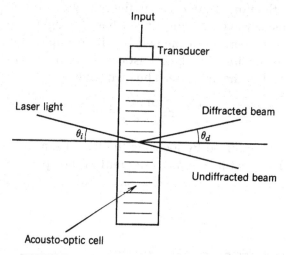

FIGURE 5.1 Operation of an acousto-optic cell.

and gallium phosphide (GaP) are used for high-frequency signal processing devices.

There are certain physical constraints on the types of signal waveform that can serve as the input and the way in which an acousto-optic modulator can be used. For example, acoustic waves emitted by the transducer into the medium usually cannot be seen if they are viewed by conventional optics. Moreover, the electric signals that drive the modulators must generally be band-pass in nature and have a frequency in the range 1 MHz to 1 GHz. Due to this restriction, input signals that are not originally band-pass in nature must be modulated by a carrier frequency before they are fed into the acousto-optic cell.

The rise and fall times of an acousto-optic modulator is limited by the transit time of the acoustic wave propagation across the optical beam. A typical rise time for a 1-mm diameter laser beam is around 150 ns. To achieve shorter rise times, it is necessary to focus the laser beam and decrease the acoustic transit time. In addition, the frequency of the output laser beam from an acousto-optic modulator is shifted by an amount equal to the acoustic frequency. This frequency shift can be used for heterodyne detection applications, where precise phase data are measured.

EXAMPLE 5.1

An acousto-optic cell is depicted in Figure 5.1. The width of the aperture is 3mm and the velocity of the acoustic wave in the crystal is 7500 m/s. What is the rise time of the device?

The rise time is the time delay between the time when the acousto-optic cell launches an acoustic wave and the time when the light is deflected by the acoustic wave. If the aperture of the cell is fully occupied by the light beam, the rise time equals the time that the acoustic wavefront takes to travel from one end to the other end of the aperture. Thus we have,

$$t = \frac{w}{V_A} = \frac{3\,\text{mm}}{7500\,\text{m/s}} = 0.4\,\mu\text{s}.$$

The rise time can be reduced by focusing the laser beam passing through the acousto-optic cell, so that it occupies only part of the aperture.

5.2 Magneto-optic Modulators

A magneto-optic spatial light modulator (MOSLM) is a two-dimensional electrically addressed SLM based on a magneto-optic effect known as the *Faraday effect*. The Faraday effect is a property of some transparent substances that causes the rotation of polarization of the light traversing through such substances when the material is subjected to a magnetic field. A MOSLM consists of a square grid of magnetically bistable mesas (pixels) that can be used to modulate incident polarized light by the Faraday effect. The state of each pixel can be electrically switched so that an object pattern can be written into the SLM using a computer. Thus, the device could function as a programmable SLM.

The basic structure of the MOSLM is a bismuth-doped, magnetic iron–garnic film that is epitaxially deposited on a transparent, nonmagnetic garnet crystal substrate. The film is then etched into a square grid of magnetically bistable mesas, and current drive lines are deposited between them. The resulting device is an $n \times n$ matrix of pixels, as illustrated in Figure 5.2.

When linearly polarized light is incident on the device, the axis of the polarization of the transmitted light will be rotated 45° clockwise for a magnetic state. The plane of polarization is rotated 45° counterclockwise for the opposite magnetic state, as shown in Figure 5.3. The magnetization state of a pixel can be changed by sending an electric current to its two adjoining drive lines. An analyzer can convert the polarization rotation into a useful output format. If the analyzer is set at a direction making a 45° angle with the original polarization axis, as shown in Figure 5.3, only the counterclockwise rotated light can pass through the analyzer. The clockwise-rotated light will be blocked by the analyzer. Thus intensity or brightness modulation of the

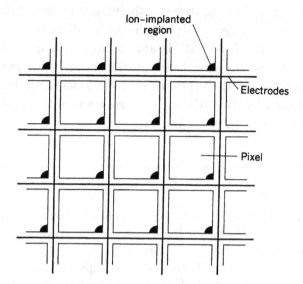

FIGURE 5.2 Structure of MOSLM pixels.

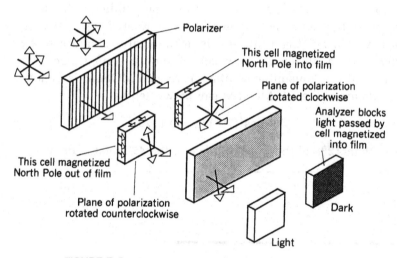

FIGURE 5.3 Operation of a MOSLM as a light valve.

incident light beam will be obtained. Alternatively, the axis of the analyzer can be set perpendicular to that of the polarizer. In this case, the light beams passing through pixels in different magnetized states will have equal output amplitudes but will polarize in opposite directions. In other words, the outputs from pixels with different magnetized states have a phase difference of 180°, which is desired for certain optical signal processing applications.

As the magnetization state of the substance is stable, the MOSLM has a storage capacity. The state is switched by means of an electric current. This current can generate heat, due to ohmic losses, which limits the performance of the MOSLM. The switching speed of the magnetic domain itself in these devices can be very fast, generally of the order of tens of nanoseconds. Currently, 256×256 pixels MOSLMs are commercially available. The center-to-center distance between pixels is typically about $70 \, \mu m$, the frame speed is $100–300 \, Hz$, and the contrast ratio is $300 : 1$ at a wavelength of $633 \, nm$. A major disadvantage of MOSLMs is their low transmittance, which is only about 5% for most laser wavelengths.

EXAMPLE 5.2

If the magneto-optic materials in an MOSLM can only rotate the light polarization by $10°$ and $-10°$, rather than by $45°$ and $-45°$ as shown in Figure 5.3, describe how the analyzer should be aligned.

The purpose of the analyzer is to block the light which has been rotated clockwise. Since the polarization of such light is rotated by only $10°$, the axis of the analyzer should be set at a direction making a $80°$ angle with the polarizer. The clockwise-rotated polarization is then perpendicular to the axis of the analyzer and is blocked. The counterclockwise-rotated light polarizes in the direction having a $70°$ angle to the axis of the analyzer, and a part of it can pass through the analyzer. A drawback of such a configuration is reduced transmittance. If the transmittance of a $45°$ rotated MOSLM is denoted by T_o, the transmittance of the $10°$ MOSLM can be written as

$$T = \cos^2(70°)T_o = 0.117 T_o.$$

This is less than 12% of the transmittance of the $45°$ rotated device.

5.3 Pockel's Readout Optical Modulators

A Pockel's readout optical modulator (PROM) is a two-dimensional optically addressed SLM based on an important electro-optic effect known as the *Pockel effect*. This effect is a linear electro-optic effect insofar as the induced birefringence is proportional to the applied electric field across an asymmetric crystal, by which the crystal can rotate the polarization of the light passing through it. The rotation of polarization is controlled by the applied electric field. There are two common configurations, referred to as *transverse* and *longitudinal*, depending on whether the applied electric field is perpendicular

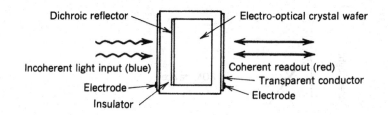

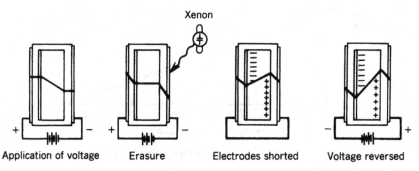

FIGURE 5.4 Composition of a PROM and its operation.

or parallel to the direction of light propagation, respectively. Most PROM devices work in the longitudinal mode.

PROMs are fabricated from various electro-optic crystals, such as ZnS, ZnSe, and $Si_{12}SiO_{20}$ (BSO). The basic configuration of a PROM device is shown in Figure 5.4. The electro-optic crystal wafer is sandwiched between two transparent electrodes and separated from them by an insulator. The crystal wafer is oriented in such a way that the field applied between the electrodes produces a longitudinal electro-optic effect.

The operation of the PROM is also illustrated in Figure 5.4. An applied dc voltage with an erase light pulse is used to create mobile carriers that cause the voltage V_o in the active crystal to decay to zero. The polarity of the applied voltage is brought to zero and then reversed. When a total voltage of $2V_o$ appears across the crystal, the device is exposed to the illumination pattern of blue light. The voltage in the area exposed to the bright part of the input pattern decays because of the optically created mobile carriers, but the voltage in the dark area remains unchanged, thus converting the optical intensity pattern into an electric voltage pattern. The relation between the input exposure and voltage across the crystal is given by

$$V_c = V_o e^{-KE}, \tag{5.2}$$

where V_o is the applied voltage and K is a positive constant. The readout is performed using red linearly polarized light (e.g., light from a HeNe laser). For BSO, the crystal is 200 times more sensitive in the blue region (around

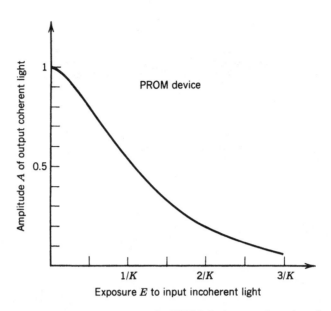

FIGURE 5.5   Amplitude of reflectance of a PROM device as a function of exposure.

400 nm) than in the red region (633 nm). Therefore, reading with an HeNe laser in real time does not produce significant voltage decay over a relatively short period of time.

Readout is performed by reflection of the reading beam. In his readout mode, the area of the crystal across which the voltage has not been affected by the input light intensity acts like a half-wave retardation plate. The angle of polarization of the linearly polarized laser light reflected by such an area is therefore rotated by 90°. Thus the light reflected by the area corresponding to the bright region has a polarization perpendicular to that of the dark region. The reflected reading beam then passes through a polarizer, and the amplitude of the transmitted light is attenuated according to the polarization plane of the light reflected from the crystal.

The theoretical curve for the amplitude A of the output coherent light versus the exposure E of the device is plotted in Figure 5.5. It can be seen that the transfer characteristic is similar to that of a photographic film, with a linear region between $E = 2/K$ and $E = 0$, and a bias point at $E = 1/K$.

The PROM device has some drawbacks, however. For example, the fabrication of the device still needs refinement to improve the uniformity and consistency of the crystal. A narrower spectral sensitivity than what these devices can offer at present is often desired. In addition, any extended exposure to high intensity readout light causes substantial decay of the crystal voltage, thus reducing the output amplitude of the device. Due to such a destructive readout process, the PROM cannot be read over an extended period of time under strong illumination.

PROMs can be used for incoherent to coherent conversion, amplitude and phase modulation, and optical parallel logic operation. A typical PROM device requires $5–600\,\mu J/cm^2$ optical writing energy. The typical resolution is about 100 lines/mm. The contrast ratio can be as high as $10\,000:1$. The write–read–erase cycle can be repeated indefinitely, with operation at video frame rates (30 frames/s). The write and erase times are generally less than 1 ms.

5.4 Liquid Crystal Light Valves

Another example of a two-dimensional optically addressed SLM is the liquid crystal light valve (LCLV). Most of us are familiar with liquid crystal materials in their use in the display panels of digital wristwatches and calculators. In fact, liquid crystal displays have found their way into notebook personnel computers, portable laboratory equipment, kitchen appliances, automobile dashboards, and many other applications requiring good visibility, low power consumption, and compact geometry. Most of these applications utilize the liquid crystal in twisted nematic cells. Each cell is either opaque or transparent, depending on the magnitude of the applied electric voltage. However, the LCLV combines the property of the twisted nematic cell in the off state with the property of electrically tunable birefringence to control transmittance over a wide continuous range in the on state.

A simplified sketch of a transmission-type twisted nematic cell is shown in Figure 5.6a. A thin layer (1–20 μm) of nematic liquid crystal is sandwiched between two transparent electrode-coated glass plates. The electrode surfaces are treated to obtain the preferred direction of alignment of the liquid crystal molecules. The plates are so arranged that, with no electric voltage applied across the layer of liquid crystal molecules, the layer is twisted continuously by 90°. A polarizer and an analyzer are placed in front of and behind the sandwiched cell. The axis of the polarizer is parallel to the direction of molecule alignment at the front electrode surface. As the light passes through the twisted liquid crystal layer, its polarization is also twisted by 90°. If the axis of the analyzer is parallel to that of the polarizer (i.e., perpendicular to the direction of molecule alignment at the back electrode surface), the polarized light will be blocked by the analyzer and cannot exit the cell. Thus the device works in its dark state.

An interesting phenomenon occurs when a voltage is applied across the twisted nematic liquid crystal layer. The molecules tend to align themselves in

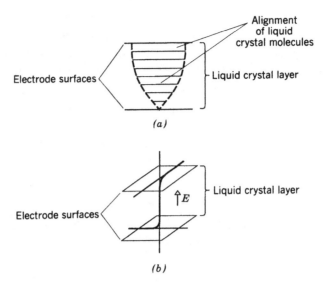

FIGURE 5.6 (a) Twisted nematic liquid crystal layer with no applied electric field. (b) Molecular alignment with the applied field E.

the direction of the applied electric field; that is, perpendicular to the electrode surfaces. The resulting splay and bending of the molecules is shown in Figure 5.6b. Consequently, if the axes of the polarizer and analyzer are parallel, the polarized light will pass unaffected through the liquid crystal layer as well as the analyzer, which is the on state for a liquid crystal alphanumeric display. However, this on state behavior is not exhibited in the LCLV.

The LCLV takes advantage of the pure birefringence of the liquid crystal materials in order to modulate a laser beam while in the on state. The operation of the LCLV is illustrated in the side view shown in Figure 5.7. The input pattern is projected by the writing beam onto the photoconductor (CdS–CdTe) layer to gate the applied alternate voltage to the liquid crystal layer in response to the input at every point in the input space. The writing beam can be either coherent or incoherent light. The molecule alignment has a 45° twist between the two surfaces of the liquid crystal layer. To obtain the output, a coherent light beam from a laser illuminates the rear side of the LCTV. While this reading beam is reflected by the device, it is also spatially modulated by the birefringence of the liquid crystal according to the voltage distribution across the liquid crystal layer. The dielectric mirror plays an important role by providing optical isolation between the input writing beam and the output reading beam. As with the MOSLM, phase modulation can be generated by properly orienting the analyzer.

The LCLV was first developed in the 1960s as an incoherent to coherent converter to convert the incoherent output of some display devices (e.g.,

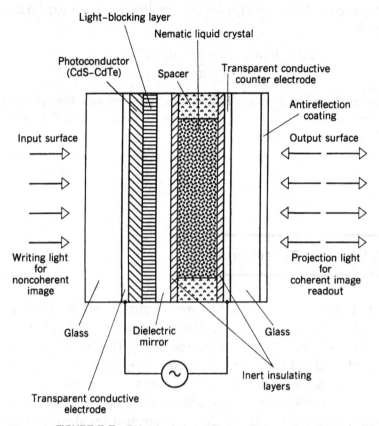

FIGURE 5.7 Side view of an LCLV and its operation.

cathode ray tube) into a coherent format. If the writing beam is coherent, the LCLV can also be used as the real-time square low device in many optical signal processing systems, such as the joint transform correlator discussed in Chapter 10. In addition, the LCLV is also used as a wavelength converter. An infrared scene projected onto an LCLV can modulate the visible readout beam. This application is required for the test of military and remote sensing equipment, which generally operate in the medium infrared (3–5 μm) and long infrared (8–12 μm) ranges. It is worth noting that the resolution of commercially available LCLVs is about 60 lines/mm. The readout time is about 10 ms, which is primarily limited by the slow response of the liquid crystal and has little potential to be improved further. The required writing energy is typically $10 \, \mu K/cm^2$.

EXAMPLE 5.3

What is the required laser power to write a $5 \times 5 \, mm^2$ grating pattern onto an LCLV with $6 \, \mu J/cm^2$ sensitivity and a 10 ms response time?

In order to accumulate enough energy in the illumination area, the minimum laser power is

$$P = \frac{EA}{T} = \frac{6(\mu J/cm^2) \times 5 \times 5 (mm^2)}{10 \, (ms)} = 0.15 \, mW.$$

In practice, the laser beam generally does not match exactly the aperture of the input pattern. Therefore, some of the energy is wasted, and the laser power required is higher than 0.15 mW.

5.5 Liquid Crystal Television

The liquid crystal television (LCTV) was originally produced as a pocket television set. With proper modification, a commercially available LCTV can be used as an electronically addressed SLM. The basic structure of the LCTV is illustrated in Figure 5.8a. Two polarizing sheets are attached to the substrates with adhesive, one sheet acting as the polarizer and the other as the analyzer. A plastic diffuser provides diffused illumination on one side, and a clear plastic window protects the device from dust on the other side. The pocket LCTV can be converted to an SLM as follows:

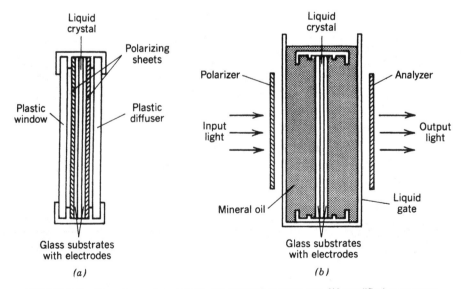

FIGURE 5.8 Structure of an LCTV: (a) original structure; (b) modified structure.

(1) Open the housing and break the hinge so that the LCTV screen can be opened fully.
(2) Disassemble the aluminum frame of the LCTV screen and remove the plastic diffuser and window from the frame.
(3) Peel off the two polarizing sheets from the glass substrates.
(4) Clean the substrates carefully with acetone.
(5) Submerge the modified screen in an index-matching liquid gate, which will be used as an SLM, as illustrated in Figure 5.8b.

The last step is necessary to remove the phase noise distortion caused by the thin glass substrate. As consumer electronic technology rapidly progresses, new types of LCTV screen, which can be modified into SLMs, will appear. Two examples are LCTV display panels for projection TV, and LCTV screens for camcorder finders. These newly introduced products generally show better contrast, higher resolution, and a larger space–bandwidth product.

The operation of an LCTV is illustrated in Figure 5.9. Each liquid crystal cell (i.e., each pixel) of the LCTV screen is a nematic liquid crystal twisted by 90°. The cell is individually controlled by electronic signals. When no electric field is applied, the plane of polarization for linearly polarized light is rotated by 90° by the twisted liquid crystal molecules. Thus no light can be transmitted through the analyzer, as illustrated in the upper part of the figure. Under an applied electric field, however, the twist and tilt of the molecules are altered, and the liquid crystal molecules attempt to align parallel with the electric field, which results in partial transmission of light through the

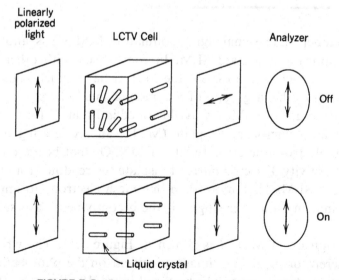

FIGURE 5.9 Basic operation of an LCTV as an SLM.

analyzer. As the electric field increases further, all the liquid crystal molecules align in the direction of the applied field. The tilted molecules do not affect the plane of polarization, so all the light passes through the analyzer, as shown in the lower part of the figure. In general, an analyzer parallel with (0°) and orthogonal to (90°) the polarizer will produce a positive and negative image, respectively. If the analyzer is set at 45°, phase modulation can be obtained.

Varying the applied voltage at each liquid crystal cell allows light transmission to be varied. If the applied voltages of the LCTV screen are generated by a computer, a camcorder, or a VCR, gray scale images can be produced.

The major advantages of the LCTV are its low cost (due to mass production) and its programmability, which makes the device a very practical SLM for many optical signal processing and optical computing applications. However, since LCTVs are made for consumer video display, the frame rate is limited to 60 Hz, although the contrast ratio, resolution, and number of pixels are being significantly improved. At present, LCTVs with 680×440 pixels are commercially available. The center-to-center distance between pixels is typically about 50 μm. The contrast ratio is $100:1$, which can be improved further by applying high-quality polarizers.

5.6 Microchannel Plate Spatial Light Modulators

The microchannel plate signal light modulator (MSLM) is another two-dimensional optically addressed SLM. In comparison with other optically addressed SLMs, such as the LCLV, the MSLM has a much higher optical sensitivity owing to the large gain of the microchannel plate.

The basic structure of an MSLM is illustrated in Figure 5.10. It consists of a photocathode, a microchannel plate (MCP), an accelerating mesh electrode, and an electro-optic crystal plate of $LiNbO_3$ that bears, on its inner side, a high resistivity dielectric mirror to isolate the readout from the write-in-side. Some MSLMs have more than one mesh electrode to improve the electrostatic imaging and focusing. All these components are sealed in a vacuum tube.

The operation of an MSLM is shown in Figure 5.11. The writing light (bearing coherent or incoherent images) incident on the photocathode generates a photoelectron image, which is multiplied 10^5 times by the MCP,

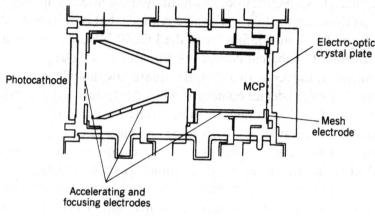

FIGURE 5.10 Structure of an MSLM.

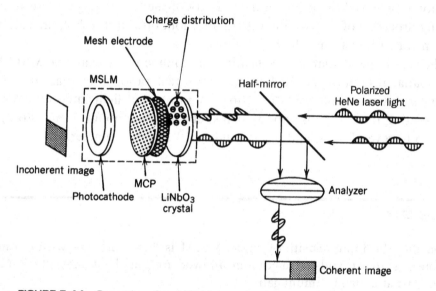

FIGURE 5.11 Operation of an MSLM as an incoherent to coherent converter.

accelerated by the mesh electrode, and then deposited on the dielectric mirror of the LiNbO$_3$ crystal plate. The process of electron multiplication is similar to that in an image intensifier tube. The resulting charge distribution, in combination with the bias voltage, creates a spatially varying electric field within the crystal plate in the direction of the optical axis. This field, in turn, modulates the refractive indices of the crystal plate. Because of the birefringence property of the LiNbO$_3$ crystal, its refractive indices in the horizontal and vertical planes are modulated differently. Thus, after twice passing through the crystal plate, the readout light, which was originally polarized in a plane bisecting the x and y axes of the crystal, will have a relative phase retardation between its x and y components. The higher the charge density, the greater the phase retardation. If an analyzer is inserted in the readout light path, as shown in the figure, a coherent image that is proportional to the input image can be obtained. Similar to the LCLV, MSLMs can be used not only as incoherent to coherent converters, but also as wavelength converters, input or output transducers, or real-time square-law modulators.

The versatility of the device comes from its architecture. As just described, photoelectron generation, intensification, transfer, deposit, readout light modulation, and detection are performed by each functional component, which allows a wide variety of modes of manipulation. An important feature of an MSLM is that the device can hold an image memory (charge distribution) for as long as several days. Since the output image is produced by the charge distribution, the MSLM can perform addition and subtraction by superimposing the charge distribution generated by two input images. Subtraction is achieved by reversing the polarity of the bias voltage between the writing processes of the two images. In addition, optical thresholding can be demonstrated by varying the bias voltage.

The typical performance specifications of commercially available MSLMs are: spatial resolution 20–50 lines/mm, contrast ratio more than 1000:1, input sensitivity around $30\,\text{nJ/cm}^2$, maximum readout light intensity $0.1\,\text{W/cm}^2$, writing time response about 10 ms, erasing time about 20 ms, storage time up to several days, and input window 10–20 mm diameter.

EXAMPLE 5.4

Given that the input sensitivity of an MSLM is $30\,\text{nJ/cm}^2$, the writing time response is 10 ms, and its maximum allowed readout light intensity is $0.1\,\text{W/cm}^2$, what is its maximum gain?

To accumulate the required writing energy during the response time period, the minimum writing light intensity is

$$I_{in} = \frac{E}{T} = \frac{30\,(\text{nJ/cm}^2)}{10\,(\text{ms})} = 3\,\mu\text{J/cm}^2.$$

When the readout beam has the maximum allowed intensity, the gain is

$$G = \frac{I_{out}}{I_{in}} = \frac{0.1\,(\text{W/cm}^2)}{3\,(\mu\text{W/cm}^2)} = 3.3 \times 10^4.$$

The large gain of the MSLM makes it a very attractive device for many applications. The gain of currently available MSLMs ranges from 10^3 to 10^5.

5.7 Photoplastic Devices

Unlike the SLMs described in the previous sections, the photoplastic device does not modulate the amplitude or intensity, but the phase distribution of an optical beam. The basic components of a photoplastic device are shown in Figure 5.12. The device is composed of a glass substrate that is coated with a transparent conductive layer (tin oxide or indium oxide), on the top of which is a layer of photoconductive material, followed by a layer of thermoplastic. For the photoconductor, poly-n-vinyl carbazole (PVK) sensitized with trinitrofluorenone (TNF) can be used with an ester resin thermoplastic (Hurculus Floral 105). The five steps in a typical operation cycle are shown in Figure 5.13. Before exposure, the device is charged either by a corona discharge or with a charging plate made of another transparent conductive material. The charging plate is separated from the device by a strip of 100 µm thick Mylar tape. After the charging process, the device can be exposed to the signal light, which is usually an interference pattern of holographic fringes and causes a variation in the charge pattern proportional to the intensity of the input light. The illuminated region displaces the charge from the transparent conductive layer to the photoplastic interface, which in turn reduces the surface potential of the outer surface of the thermoplastic, as shown in the figure. The device is then recharged to the original surface potential and developed by raising the temperature of the thermoplastic to the softening point and then rapidly reducing it to room temperature. The temperature reduction can be accomplished by passing an ac voltage pulse through the conductive layer. Surface deformation caused by the electrostatic force then produces a phase recording of the intensity of the input light. This recording can be erased by raising

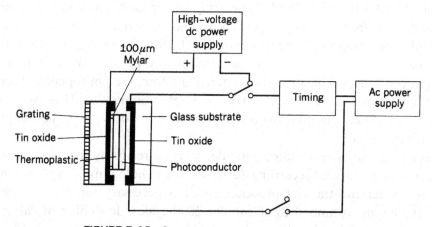

FIGURE 5.12 Composition of a photoplastic device.

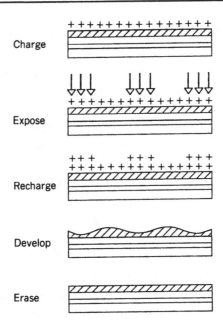

FIGURE 5.13 Operation of a photoplastic device in recording light intensity patterns.

the temperature of the thermoplastic above the melting point, thus causing the surface tension of the thermoplastic to flatten the surface deformation and erase the recording.

The spatial frequency response of the photoplastic is poor at low frequencies. To use the device for recording signals with low spatial frequencies, we must modulate the signal with a sinusoidal signal having a spatial frequency matching the peak of the frequency response of the device, which is typically 50 cycles/mm. This can be done by putting a sinusoidal grating in front of the thermoplastic device, as shown in Figure 5.12.

Unlike other optically addressed SLMs discussed in the preceding sections, such as the LCLV and MSLM, the writing and reading processes in photoplastic devices are performed sequentially. The input pattern cannot be read out while it is being written into the device. Instead, the input pattern must firstly be fixed by thermodevelopment, and then to perform the modulation the light beam passes through the device. Therefore, the photoplastic device is not a real-time SLM, but is similar to photographic film. However, the input pattern (i.e., the modulation characteristics) can be refreshed in the photoplastic device.

The advantage of the photoplastic device is its relatively low cost, which makes it a very practical recording device for use in holographic applications. It does not require the wet processing that is necessary for photographic films. Its lifetime is limited to about 300–2000 cycles, depending on laboratory conditions. The signal-to-noise ratio is relatively low because of the random thickness variation of the thermoplastic plate. A diffraction efficiency of 10% with exposures of $60\,\mu J/cm^2$ have been reported, which is comparable to the sensitivity of a high resolution photographic emulsion. The resolution can be better than 2000 lines/mm.

5.8 Deformable Mirror Array Devices

The deformable mirror array device (DMD) is another example of a two-dimensional phase modulator. Unlike the thermoplastic device discussed in the preceding section, a DMD can be addressed by electric signals, and is thus a programmable device. The frame speed is higher than 100 frames/s. DMDs were originally developed for use as scanners in laser printers. Recently, they have found application in projection display and optical signal processing.

The operation of a DMD involves the mechanical deformation of its modulators in response to an electrostatic force. The modulating elements of the DMD are tiny metal mirrors. These mirrors are fabricated to form a cantilevered beam structure over an underlying silicon address structure. A cross-section of an individual mirror element is shown in Figure 5.14. The mirror and underlying addressing structure form an air-gap capacitor, which is typically 2–3 μm wide. The addressing structure, which is a transistor array, allows a prescribed amount of charge to be deposited below each tiny mirror. The amount of deflection by the mirror is determined by electrostatic attraction. In other words, a two-dimensional input signal is first converted into the electric charge distribution across the device, and is then exhibited by the tilt of each tiny mirror at the corresponding position. When a plane wavefront is

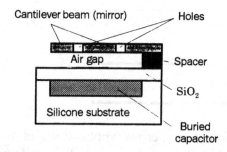

FIGURE 5.14 Side view of a DMD pixel element.

projected onto the DMD, mirrors that are not tilted by the addressing charges will reflect the light in one direction and mirrors that are addressed by certain amounts of input charge will reflect the light in other directions due to their tilts. In this way, the uniform phase of the original plane wave is modulated according to addressing electric input signal.

Prototype DMD devices with 128×128 pixels have been manufactured. Larger arrays up to 1000×1000 pixels are feasible. Each pixel in such a prototype DMD device consists of four (2×2) cantilevered-beam mirrors, each of which is approximately $12.7 \times 12.7\,\mu m^2$. The resolution is about 10 lines/mm. It can be addressed by a single analog input with a data rate as high as 20 MHz, which corresponds to a frame rate of 150 frames/s. Although it is the phase or optical path length of the light beam that is directly modulated by the DMD device, intensity modulation can be achieved by using a schilieren optical system or by spatial filtering. At approximately 2 : 1 the contrast ratio of the schilieren output is generally very low. However, the contrast ratio of a spatially filtered output could be as high as 100 : 1.

EXAMPLE 5.5

Intensity modulation of a DMD can be achieved by spatial filtering; that is, by placing a pinhole at the back focal plane of an optical lens. If the deflection of the tiny mirrors addressed by signal 1 is 1° and the focal length of the lens is 100 mm, what is the diameter of the pinhole?

Assume that the optical axis of the readout beam is perpendicular to the surface of the DMD. The bright pixels (i.e., the pixels addresssed by signal 0) will reflect the incident light back along its original axis. However, the dark pixels addressed by signal 1 will reflect the incident light in a direction at 2° to the optical axis. At the back focal plane, the light rays from the bright pixels and dark pixels are focused on two spots. The distance between these two spots is

$$D = L \tan \theta = 100 \tan 2° = 3.5\,mm.$$

In order to block the light from all the dark pixels, the diameter of the pinhole in the focal plane of the lens should be less than 3.5 mm.

5.9 Optical Discs

Optical discs have found numerous applications in the past decade. In particular, audio optical discs or compact discs (CDs) had been rapidly

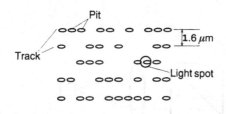

FIGURE 5.15 Structure of a compact disc.

penetrating the consumer market. An optical disc can be utilized as an SLM to modulate the phase of the light beams in the same way that a DMD does (see Section 5.8). In the field of optical data storage, the optical disc system has been used for a computer peripheral device called the CD-ROM (read-only memory). In principle, information is stored using a string of pits on the surface of the CD.

Figure 5.15 illustrates schematically a small surface area of an optical disc. The recorded signal is encoded in the length of the pit and the spacing of the pits along the track. The distance between two adjacent tracks (track pitch) is 1.6 μm. The width of a pit is equal to a recording spot size of 0.5–0.7 μm. The light source in the optical disc system is usually GaAlAs semiconductor laser diodes with a wavelength of 0.78–0.83 μm. The spot size of the readout beam is determined by the numerical aperture (NA) of the objective lens. Typically, $\lambda/NA = 1.55$ is chosen, so that the effective diameter of the readout spot is approximately 1 μm. The spot size is larger than the width of a pit, but a single readout spot does not cover two tracks (see Figure 5.15).

The optical pickup is shown schematically in Figure 5.16. When there is no pit, as shown in Figure 5.16a, the light is fully reflected to the detector. However, when there is a pit the focused beam covers both the pit and the surrounding area, as depicted in Figure 5.16b. Both the pit and land are coated with high reflectivity material (e.g., aluminum), and so the light is reflected from both of them. The depth of the pit is made such that the phase difference between the reflected light from a pit and a land is π. Consequently, a destructive interference occurs at the detector and low light intensity is detected.

We are interested, however, in the SLM aspect of an optical disc rather than its sequential data storage aspect. To function as an SLM, the optical disc must be read out in parallel. As a consequence, the disc could be considered as a pixelated SLM with a 0.7 μm pixel diameter and a cross-track pixel spacing of 1.6 μm. While cross-track pixel spacing is set by a manufacturing standard (1.6 μm), along-track pixel spacing can be varied; therefore, a similar 1.6 μm spacing can be selected. In contrast to other SLMs, the pixel format is not a Cartesian coordinate. The pixels (pits) are distributed on the tracks, which are concentric circles. However, if the readout collimated beam

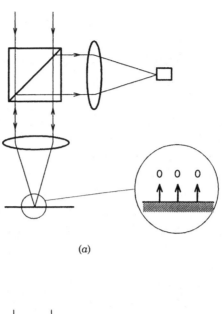

(a)

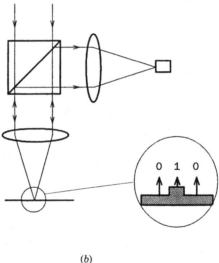

(b)

FIGURE 5.16   Pickup operation from a compact disc: (a) bright reflection; (b) dark reflection.

illuminates only a small sector of track, the polar format would be very close to the Cartesian coordinate. This small sector of an optical disc could consist of a large number of pixels compared with that in an ordinary SLM. Since the pixel modulates the reflected light by a phase shift of 0 or π, an optical disc can be used as a binary phase-modulating SLM. It can be employed to record and reconstruct a computer-generated hologram. Further potential applications to optical neural network processors have been proposed. Nevertheless, the CD is inconvenient for use by optical engineers or researchers, because the disc must be put through a mastering process.

While CDs are primarily produced for the consumer market for which copyright of the content (music, movie, or graphic and text) needs protection, an SLM optical disc must be easily writeable by the user. An optical disc called a WORM (write once, read many) has become commercially available. A WORM disc consists of either a polycarbonate or hardened-glass substrate and a recording layer made of a highly reflective substance (dye polymer or tellurium alloy). As with the CD, the recording layer is covered by clear plastic to protect the recording medium. WORM disc systems use a laser beam to record data sequentially. A write beam burns a hole (pit) in the recording medium to produce a change in the reflectivity. In contrast to CDs, however, the WORM optical disc modulates the reflected intensity of the readout beam. A WORM optical disc system can be used to perform optical correlation. The primary drawback of the system is that the disc is neither erasable nor rewritable.

An erasable and rewritable optical disc system is best represented by a magneto-optic disc akin to the MOSLM mentioned previously (see Section 5.2). The magneto-optic medium makes use of a recording material which, at room temperature, is resistant to changes in magnetization. The reverse magnetic field required to reduce the magnetization of the recording material to zero is called the *coercivity*. In other words, the coercivity of the magneto-optic recording material at room temperature is quite high. The coercivity can be altered only at high temperature, which is defined as the *Curie point*. The typical Curie point of the magneto-optic material used is 150°C. The heat of the write laser beam brings the recording material to the Curit point. A bias magnet then reverses the magnetization of the heated area that represents a bit of information. As with MOSLM, a low-power linearly polarized laser beam can be used to read the data stored in the disc. According to the Kerr magneto-optic effect, the polarization of the readout beam is rotated to the left or right, depending on whether the magnetization of the recording material is upwards or downwards. At present, the typical rotation is about 1°. In principle, the intensity modulation can be obtained by applying an analyzer.

5.10 Photorefractive Crystals

To understand the photorefractive effect, we begin with a discussion of the electro-optic effect that involves changes in the refractive index in a medium caused by an applied electric field. For the first-order electro-optic effect, known as the *Pockels effect*, the change in the refractive index is proportional to the applied electric field,

$$\Delta \eta = \frac{1}{2} \eta_o^3 \gamma_{eff} E, \tag{5.3}$$

where η_o is the normal refractive index of the material, γ_{eff} is the effective Pockels coefficient, and E is the applied electric field. However, for the second-order electro-optic effect, known as the *Kerr effect*, the change in the refractive index is proportional to the square of the applied electric field and is expressed as

$$\Delta \eta = \frac{1}{2} \eta_2 E^2, \tag{5.4}$$

where η_2 is the second-order nonlinear index coefficient.

We note that the Pockels effect occurs in solids without inversion of symmetric centers, whereas the quadratic Kerr effect occurs in materials with any symmetry and is commonly exhibited in liquids composed of anisotropic molecules. In general, the effective Pockels coefficient γ_{eff} is much larger than the second-order nonlinear index coefficient η_2.

The electro-optic effect is commonly used in light modulators, in which the change in the refractive index is utilized for phase (or frequency) modulation. Owing to the extremely high modulation frequency (up to 100 GHz), these electro-optic modulators have become key components in broadband optical fiber communication networks.

The photorefractive effect in solid materials has received a remarkable amount of attention because of its application in low power visible lasers. Although most nonlinear optical effects take megawatts of power to induce, the photorefractive effect requires only a few microwatts. The nonlinear behavior is spatially and temporally nonlocal, and the effects are independent of intensity.

To understand the basic nonlinear optical interaction in a photorefractive material, consider the photorefractive process depicted in Figure 5.17, in which we assume that two coherent plane waves interfere within the photorefractive crystal. The first graph represents the intensity of the interference fringes. It is well known that the presence of light intensity releases charge due to the photoionization effect, and these charges can migrate due to diffusion (or draft) until they are trapped. The $\rho(x)$ curve shows the position dependence of the charge distribution induced by the light interference. Note that the majority carriers accumulate in the valleys of the interference pattern, and that their vacancies are most prominent at the peak of the fringe intensity. As a result of the charge separation and migration, an internal electric field $E(x)$ is created, as shown in the third graph. Note that the peak of the electric field is half way between the positive and negative charges

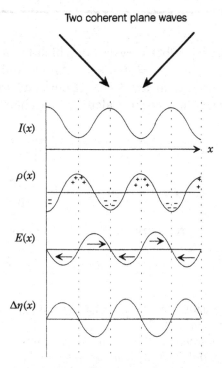

Two coherent plane waves

FIGURE 5.17 Principle of the photorefractive effect.

and that $E(x)$ is (by symmetry) zero at the maxima and minima of the charge distribution. For this reason, the electric field grating is 90° out of phase with respect to the interference fringes. This shift leads to the very important aspect of two-beam coupling. As the final step of the photorefractive process, the periodically varying electric field modulates (through the Pockels effect) the refractive index of the material, as can be seen in the last curve in Figure 5.17. This completes the photorefractive effect from the interference fringes to form a refractive index grating.

In general, the build-up of the refractive index grating exhibits an exponential behavior, as given by

$$\Delta\eta = \Delta\eta_{max}(1 - e^{-t/\tau}), \tag{5.5}$$

where $\Delta\eta_{max}$ is the maximum refractive index change (often referred as the dynamic range) of the photorefractive crystal, τ is the photorefractive response time constant of the material, and t is the exposure time. Finally, we note that the photorefractive effect has offered a wide range of applications in photonic devices; for instance as applied to massive holographic storage, novel spatial filtering, and optical conjugation.

EXAMPLE 5.6

Consider a lithium niobate (LiNbO$_3$) crystal, in which the largest Pockels coefficient is along the 33 direction, $\gamma_{33} = 30.8 \times 10^{-12}$ m/V, and $\eta_o = 2.29$. Assume that the dimension of the crystal are $1 \times 1 \times 1$ cm^3 and that a dc voltage of 10 000 V is applied across the crystal. Calculate the change in the refractive index within the crystal.

By referring to Eq. 5.3, the refractive index change induced by the electric field can be shown as

$$\frac{\Delta\eta}{\eta_o} = \frac{1}{2}\eta_o^2\gamma_{33}\frac{V}{L} = \frac{1}{2} \times 2.29^2 \times 30.8 \times 10^{-12} \times \frac{10^4}{10^{-2}} = 0.018.$$

Generally, the change in the refractive index is of the order of 0.1–1%.

REFERENCES

1. F. T. S. YU and S. JUTAMULIA, *Optical Signal Processing, Computing and Neural Networks*, Wiley, New York, 1983.
2. F. T. S. YU, *Optical Information Processing*, Wiley, New York, 1983.
3. A. D. FISHER (ed.), *Spatial Light Modulators and Applications Technical Digest*, Optical Society of America, Washington DC, 1993.
4. A. D. FISHER, Spatial light modulators: functional capacities, applications, and devices, *International Journal of Optoelectronics*, **5** (1990) 125.
5. J. A. NEFF, R. A. ATHALE and S. H. LEE, Two-dimensional spatial light modulators: a tutorial, *Proceedings of the IEEE*, **78** (1990) 826.
6. P. GUNTER and J. P. HUIGUARD, *Photorefractive Materials and their Applications*, Springer-Verlag, Berlin, 1988.

PROBLEMS

5.1   The rise time of an acousto-optic cell is 450 ns and the width of its aperture is 3.6 mm. Determine the speed of the acoustic wave in the acousto-optic crystal.

5.2   The acousto-optic cell in Problem 5.1 is used to modulate a digital optical signal.

         (a) What is the bandwidth of the signal if the light beam fully illuminates the device aperture?

         (b) Comment on how to reduce the rise time and thus to increase the signal bandwidth. What are the trade-offs? Show how a 1 GHz bandwidth can be obtained using the same device.

5.3   Describe how an acousto-optic cell can be used as a scanner.

5.4   Assume a MOSLM rotates light by 45° and −45° for a binary input of 0 or 1, respectively. How can binary phase modulation be realized using the device?

5.5　If the MOSLM in Problem 5.4 rotates light by only 20° and −20°, how can the phase modulation be implemented? What is the light transmittance of such a phase modulator?

5.6　Do we have to align the polarization of the readout light to a specific axis of the MOSLM? If so, explain why. What will happen if we rotate the analyzer by 90°?

5.7　The Kerr law of a PROM device is given by

$$\eta_1 - \eta_2 = k\lambda E^2,$$

where η_1 and η_2 are the refractive indices of light with planes of polarization parallel and perpendicular to the applied electric field E, respectively, k is the Kerr constant, and λ is the wavelength of light. Show that the applied voltage V_h required to produce a half-wave phase retardation, that is to rotate the polarization plane by 90°, is

$$V_h = \sqrt{(m + 1/2)d/2k},$$

where m is an integer and d is the thickness of the crystal. (*Hint*: To produce half-wave retardation, the phase difference of the ordinary wave (η_1) and the extraordinary wave (η_2) is $\Delta\phi = \pi,\ 3\pi,\ 5\pi,\ldots$, at the exit of the retardation plate.)

5.8　Refer to the PROM described in Problem 5.7, if one changes the wavelength of the readout light, does the applied voltage have to be tuned? Explain.

5.9　Referring to the Kerr law as given in Problem 5.7:

(**a**) What would happen if the readout light is polarized parallel with respect to the polarization of the ordinary and extraordinary light, respectively;

(**b**) What conditions are needed for the polarization of the readout beam to operate a PROM device properly?

5.10　An LCLV can generate a large screen monochrome display image from a TV monitor. Draw a schematic diagram for this application.

5.11　Describe how the systems shown in Figure 5.18 can be used as an optical memory.

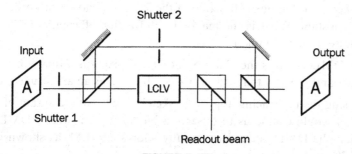

FIGURE 5.18

5.12 Can one use the system shown in Figure 5.19 as an optical memory? Explain.

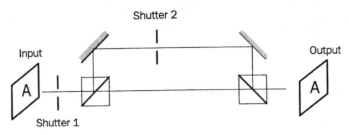

FIGURE 5.19

5.13 What is the power required for a laser to write a 10×10 mm transparency onto an LCTV with $10 \, \mu J/cm^2$ sensitivity and a 10 ms response time?

5.14 Assume the scanning process of the LCTV is exactly the same as that of a commercial TV monitor, which works by the point-by-point process. Can we obtain the Fourier transform of the pattern shown on the LCTV panel? Describe under what conditions the answer is yes or no.

5.15 Without an analyzer, can we observe:

(a) The pattern written on the LCTV?

(b) The Fourier spectra of the pattern displayed on the LCTV? Explain.

5.16 Assume that the input sensitivity of a MSLM is $25 \, nJ/cm^2$ and the writing time response is 10 ms. To have a gain of $10\,000$, what would be the readout light intensity?

5.17 A monochrome slide is projected on the input side of an MSLM. If a white-light beam is used for the readout, a colour output is obtained. Explain why this happens.

5.18 Why is a photoplastic device good for making holograms but not suitable for generating photographic images?

5.19 Assume that a thin phase grating is made by a photoplastic device, as given by

$$t(x, y) = \exp[i\alpha \cos(2\pi by)] = \sum i^n J_n(\alpha) \exp(in2\pi by)$$

where J_n is the Bessel function of the first kind and the nth order, and α is a constant. What is the maximum diffraction efficiency? (*Hint*: Use the Bessel function table.)

5.20 A DMD changes the phase of individual pixels as a function of position in x and y. Show that this device can be used as an amplitude modulator.

5.21 Using the ray tracing approach, sketch a schematic diagram of a system that would enable us to observe a binary pattern with a DMD. Notice that the DMD is an electronically addressed SLM, as shown in Figure 5.20.

DMD

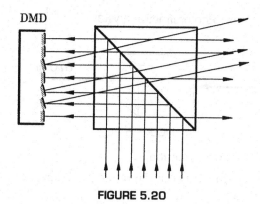

FIGURE 5.20

5.22 Why does the size of the readout beam have to be larger than the width of the pit in a nonerasable CD? On the other hand, why does the beam size have to be smaller than the size of the pit in a WORM and a magneto-optic disc?

5.23 When a 120 mm CD is employed as an SLM, what is the number of pixels in the $1\,mm^2$ area close to the edge of the CD?

5.24 Figure 5.21 shows the surface structure of a CD. Determine the pit depth d in terms of wavelength λ and the refractive index of the clear coating η. If $\lambda = 0.78\,\mu m$ and $\eta = 1.5$, what is d?

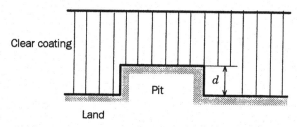

FIGURE 5.21

5.25 Consider a specially doped lithium niobate (Ce:Fe:LiNbO$_3$) crystal. The dynamic range of the material is given as 4×10^{-3} and the photorefractive response time constant τ is given as $0.5\,s$.

   (a) Determine the exposure time required to obtain a refractive index change of 10^{-6}.

   (b) Evaluate the induced electric field.

   (c) Calculate the saturated electric field within the crystal.

5.26 One promising application of photorefractive materials is volume holographic memory, in which the information is stored in the volume of the photorefractive medium in the form of phase gratings. Figure 5.22 shows a reflection type volume hologram which is constructed by two counter-propagating waves (reference and signal). Assume that the writing wavelength is 670 nm, the normal refractive index of the photorefractive crystal is 2.29, the dynamic range is about 4×10^{-3}, and the effective coefficient is $30.8 \times 10^{-12}\,m/V$. Calculate:

Photorefractive crystal

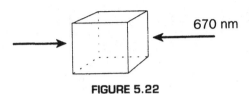

FIGURE 5.22

(a) The phase grating period.

(b) The space shift between the refractive index grating and the interference fringes.

(c) The average charge transport distance.

(d) The saturation internal electric field.

5.27 A photorefractive crystal can be considered as a thick phase grating such as an acousto-optic cell. What is the maximum diffraction efficiency?

LASERS $\boxed{6}$

The word "laser" is an acronym for *light amplification by the stimulated emission of radiation*. It is widely accepted that laser is one of the greatest inventions in the last half of the twentieth century. Together with the satellite, the computer, and the integrated circuit, laser is a symbol of high technology. Since its invention in 1960, the laser has had a profound impact on almost all fields of optical science and engineering and has opened up new realms beyond the traditional optics that had been primarily dealing with optical instruments such as microscopes and telescopes. With laser light, one can now record three-dimensional holograms, project ever-changing light patterns, spot flaws in a centuries-old painting, or play crystal-clear digital music recorded on a compact disc. Lasers can also send signals through miles of optical fiber cable, print computer output, read printed codes in the supermarket, diagnose and cure disease, cut and weld materials, and make ultraprecise measurements. In view of the rapid advances in laser technology, it is impossible to cover the vast domain of this field. The objective of this introductory chapter is to present the fundamental principles of lasers. More advanced discussions can be found in the references listed at the end of this chapter.

6.1   Quantum Behavior of Light

Early physicists had long been debating the nature of light. Newton held that light was basically composed of tiny particles, but Huygens believed that light was made up of waves, vibrating up and down in a direction perpendicular to the direction of travel of the light. Since Huygens' wave theory provided an excellent explanation of experiments on phenomena such as refraction, diffraction, and interference, by the end of the eighteenth century the theory was eventually accepted by most physicists. Our discussions in this book, especially those on the diffraction and interference phenomena, are based on the wave theory of light. The propagation of a light wave is governed by the Maxwell equations, by which the speed of light is given by

$$c = \lambda\nu, \tag{6.1}$$

where λ denotes the wavelength and ν stands for frequency of the light wave.

At the end of the last century, however, the wave theory of light had been challenged by the observation of the photoelectric effect. When a light beam impinges on a piece of metal, electrons may be emitted from the metal surface. It was observed that there is a *cut-off* frequency ν_o, below which no photoelectric effect occurs, no matter how intense the illuminating beam is. To explain this phenomenon, Einstein assumed that light is a form of energy, and used the *photon* to represent the minimum energy unit. A photon can in fact be considered as a chunk or quantum of energy, which is localized in a small volume of space and remains localized as it moves from the light source with velocity c. Each photon has an amount of energy proportional to the frequency of the light wave, as given by

$$E = h\nu, \tag{6.2}$$

where h is *Planck's constant*. The higher the frequency of a photon, the more energy it possesses.

Einstein also assumed that in the photoelectric process one photon is completely absorbed by one electron in the metal. The kinetic energy of the electron emitted from the metal surface is given by

$$k = h\nu - w, \tag{6.3}$$

where w is the work function required to remove an electron from the metal. In order for an electron to be released from the metal surface, the kinetic energy in Eq. 6.3 should be greater than zero, that is

$$h\nu \geq w, \tag{6.4}$$

which explains the existence of the cut-off frequency. If the frequency is reduced below a certain value, the individual photons, no matter how many of them there are (i.e., no matter how intense the illumination), will not have enough energy to eject photoelectrons.

Today, we know that light is neither purely a wave phenomenon nor merely a stream of particles (photons). Its behaviour is wave-like under some circumstances, but particle-like in other situations. This is known as the *wave–particle duality* of light. The propagation, diffraction, and interference of light are better explained by the wave nature; whereas the interaction between light and matter is better interpreted using the concept of the photon. When light interacts with matter, the energy exchange can take place only at certain discrete values, for which the photon is the minimum energy unit that the light can give or accept. In this chapter, the quantum nature of light is used to explain the absorption and emission of light, as well as the principle of lasers.

EXAMPLE 6.1

Photon energy is often measured in terms of electron-volts (eV). If the wavelength of a light wave is $1\,\mu m$, calculate the photon energy in terms of the electron-volt.

From Eq. 6.2, we have

$$E = h\nu = \frac{hc}{\lambda} = \frac{6.63 \times 10^{-34}(\text{J} \cdot \text{s}) \times 3 \times 10^{8}(\text{m/s})}{1 \times 10^{-6}\,(\text{m})} = 2 \times 10^{-19}\,\text{J}.$$

one electron-volt is the energy required to raise the potential of an electron by one volt. Since the charge of an electron is 1.6×10^{-19} coulomb (C), $1\,\text{eV}$ corresponds to

$$1.6 \times 10^{-19}(\text{C}) \times 1\,(\text{V}) = 1.6 \times 10^{-19}\,\text{J},$$

and the photon energy can be expressed as

$$E = \frac{2 \times 10^{-19}(\text{J})}{1.6 \times 10^{-19}(\text{J})} = 1.24\,\text{eV}.$$

To estimate the photon energy for a given wavelength λ, the following equation can be used:

$$E = \frac{1.24}{\lambda},$$

where the energy is measured in electron-volts and the unit of wavelength is micrometers.

6.2 Spontaneous and Stimulated Emission

In the preceding section, we have learned that light energy cannot take arbitrary values, but must be multiples of the photon energy $h\nu$. Similarly, the electrons of an atom cannot have arbitrary amounts of energy associated with them; only discrete (or *quantized*) energy levels are possible. The energy levels are organized in a system of shells and subshells, each of which can contain a maximum number of electrons. Usually the electrons of an atom

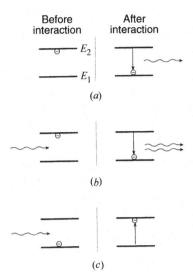

FIGURE 6.1 Interaction between photons and electrons: (a) spontaneous emission; (b) stimulated emission; (c) absorption.

occupy the position with the lowest possible energy. Let us take the silicon atom as an example. It has 14 electrons and three shells with capacities of 2, 8, and 8 electrons, respectively. We would find that two electrons occupy the lowest energy level (corresponding to the first shell), eight occupy the next level (second shell), and four have the energy of the third shell. When the electrons are in this configuration, the atom is said to be in the *ground state*. If, by any means, one or more electrons are moved to a higher energy level, the atom is said to be in its *excited state*.

Photons are capable of inducing transitions of electrons between different energy levels. If the photon has enough energy, an electron may absorb the energy and move to a higher level. On the other hand, an electron in an excited state may move down to a lower energy level and releases energy in the form of a photon. Consider a particular transition from a higher level E_2 to a lower level E_1, as shown in Figure 6.1. The energy of the emitted photon is exactly equal to the difference between the energy levels, such that the emission frequency can be written as

$$\nu = \frac{E_2 - E_1}{h}. \tag{6.5}$$

There are two types of emission, namely spontaneous and stimulated emission, which are illustrated schematically in Figures 6.1a and 6.1b, respectively. Spontaneous emission occurs all by itself, and is generally the source of light we see in our daily life, including light from the sun, stars, incandescent bulbs, fluorescent lamps, and television sets. If an electron is located at an energy level above the ground state, it can spontaneously drop to a lower level without any external intervention, as illustrated in Figure 6.1a. The excess energy is released as a photon of light. In fact, all the excited electrons in atoms or molecules have a spontaneous emission lifetime, which is defined as the average time for which they would remain in the upper energy level before spontaneous emission occurs. If the population (i.e., number of atoms) of energy level E_2 is denoted by N_2, the decay of this population can be written as

$$\left.\frac{dN_2}{dt}\right|_{sp} = -\frac{N_2}{t_{sp}},\tag{6.6}$$

where t_{sp} is the lifetime of the spontaneous emission. Since spontaneous emission takes place in a random manner, the emitted photons have different propagation directions, polarizations, and initial phases. Consequently, the light generated by spontaneous emission is incoherent.

The form of emission associated with lasers, stimulated emission, is fundamentally different from spontaneous emission. When a photon with suitable energy ($h\nu = E_2 - E_1$) hits an excited atom, it can stimulate an electron at the higher energy level to transit to the lower level and emit another photon, as shown in Figure 6.1b. The emitted photon has exactly the same frequency, phase, polarization, and propagation direction as the incident photon, for which the stimulated emission is always coherent. The emission rate of photons can be calculated using

$$\frac{dN_2}{dt} = -\beta_{12}\rho(\nu)N_2,\tag{6.7}$$

where $\rho(\nu)$ is the spectral energy density of the electromagnetic field and β_{12} is known as the Einstein β coefficient for stimulated emission.

It is certainly also possible that the incident photon hits an atom in its ground state, instead of its excited state. In this case, an electron in the atom may absorb the photon energy and transit to the excited state, as depicted in Figure 6.1c. The absorption process increases the population at the excited energy level E_2. The increasing rate is proportional to the population of the lower energy level E_1 and can be expressed as

$$\frac{dN_2}{dt} = -\frac{dN_1}{dt} = \beta_{12}\rho(\nu)N_1,\tag{6.8}$$

where N_1 and N_2 are the populations of the ground and excited energy levels, respectively.

In practice, the absorption and spontaneous emissions always occur together with stimulated emission. As long as some atoms are in the excited state, spontaneous emission occurs. The emitted photons may either hit other excited atoms, leading to stimulated emission, or be absorbed by atoms at their ground state. In normal conditions, atoms and molecules tend to be at their lowest possible energy level. At thermodynamic equilibrium, the population N_1 is much larger than N_2. Therefore, a photon has a much higher probability of being absorbed than of stimulating an excited atom. After an atom absorbs a photon, it will stay at the higher energy level for an average lifetime t_{sp}. If the excited atom is not stimulated by another photon during the short lifetime, which is very likely, it will revert to the lower energy level and spontaneously emit a photon.

There is, however, a way to make stimulated emission become the dominant process. If more electrons are in the excited state than in the ground state (i.e., $N_2 > N_1$), photons are more likely to stimulate emission than to be absorbed. Such a condition is called *population inversion*, because it is the inverse of the normal situation. Under the population inversion condition, stimulated emission can produce a cascade of light. The first spontaneously emitted photon stimulates the emission of more photons, and those stimulated-emission photons can likewise stimulate the emission of still more photons, like a chain reaction. As long as there are more atoms with the higher energy than with the lower energy, stimulated emission is more likely than absorption, and the cascade of photons grows; or, in other words, the light is amplified. When the ground state population becomes larger than the excited state population, population inversion ends, and spontaneous emission again dominates. It is therefore apparent that, in order to produce and maintain population inversion, atoms must be constantly pumped up from their ground state to the excited state, as will be discussed in the next section.

EXAMPLE 6.2

Show that at thermodynamic equilibrium the ground state population N_1 is always higher than the excited state population N_2.

By combining Eqs 6.6–6.8, the rate of change of N_2 is given by

$$\frac{dN_2}{dt} = -\frac{N_2}{t_{sp}} - \beta_{12}\rho(\nu)N_2 + \beta_{12}\rho(\nu)N_1.$$

At thermodynamic equilibrium, the population N_2 should be approximately constant. By equating the preceding equation to zero, we have

$$\frac{N_2}{N_1} = \frac{\beta_{12}\rho(\nu)}{\beta_{12}\rho(\nu) + 1/t_{sp}} < 1$$

Thus, the excited population N_2 is always less than the ground level population N_1. When the energy difference between E_2 and E_1 is about $1\,eV$, the N_2/N_1

ratio is typically of the order of 10^{-17}. This means that at thermodynamic equilibrium virtually all the atoms are in the ground state.

6.3 Population Inversion

In the preceding section, we saw that a population inversion is required in order to make the stimulated emission dominant; in other words, to make laser operation possible. It should be noted that an atom or molecule can occupy many energy levels and a transition can occur between any pair of levels. The minimum requirement for lasing is that at least one high energy level has a higher population than a lower level.

The most straightforward way to generate a population inversion is to put an excess of atoms in the higher energy state. However, it is also possible to depopulate the lower level, or pick a system where a lower level is unstable, so that it contains few or no atoms. If the laser is to operate continuously, both populating the upper level and depopulating the lower level are important, because accumulation of too many atoms in the lower level can end the population inversion and consequently stop the laser action.

6.3.1 THREE- AND FOUR-LEVEL LASERS

Lasers are commonly classified into two categories: three-level systems and four-level systems. A model of a three-level laser is illustrated schematically in Figure 6.2. For simplicity, we assume that all the atoms start out in the ground state. Then most of them are atoms *pumped* up to a higher energy

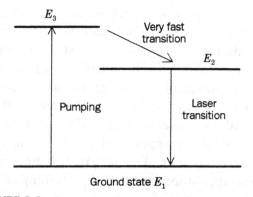

FIGURE 6.2   Energy level diagram of a three-level laser.

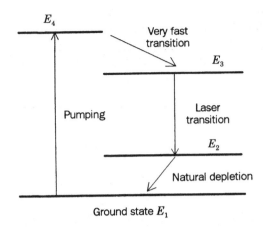

Ground state E_1

FIGURE 6.3   Energy level diagram of a four-level laser.

level E_3. They quickly fall from the short-lived level E_3 to the metastable upper level E_2, which has a much longer lifetime (typically thousands of times longer). The result is the population inversion between energy level E_2 and the depopulated level E_1, which are the laser transition levels. Owing to this population inversion, more electrons are available in the upper level for stimulated emission than in the lower level for absorption; therefore, stimulated emission can dominate on the laser transition.

The four-level laser scheme is depicted in Figure 6.3. As in the three-level laser, the excitation energy pumps electrons from the ground state to a short-lived highly excited level E_4. The atoms then drop quickly to a metastable upper laser level E_3. The laser transition then takes these atoms to a lower level E_2, rathre than all the way down to the ground state as in three-level lasers. After they have dropped to the lower level E_2, the atoms eventually lose the rest of their excess energy by spontaneous emission or other processes, and finally fall to the ground state.

In contrast to three-level lasers, the lower laser transition level in four-level lasers is not the ground state. We know that normally most atoms are in the ground state. In order to produce a population inversion in a three-level laser, most of the atoms in the ground state must be pumped up to the upper laser level. This requires an intense burst of pumping energy. In a four-level laser, however, the lower laser level is not the ground state. As soon as some atoms are pumped to the upper laser level, population inversion is achieved. This obviously requires less pumping energy than does a three-level laser. In almost all four-level lasers, the lifetime of the lower level E_2 is much shorter than that of the upper level E_3. Thus the atoms in level E_2 quickly drop to the ground state. This steady depletion of the lower laser level helps sustain the population inversion, by avoiding an accumulation of atoms in the lower level.

It should be mentioned that the first laser (ruby laser), invented by Maiman in 1960, was a three-level laser. Since the population inversion in a three-level system is very difficult to sustain, it worked in pulsed mode. Today, many four-level lasers have been developed. They require far less power to create and to sustain the population inversion, and can work in the continuous wave mode.

6.3.2 OPTICAL AND ELECTRIC PUMPING

The excitation of atoms to produce a population inversion is called pumping. There are generally two types of pumping process: optical pumping and electrical pumping.

Optical pumping uses photons to excite atoms. Both three-level and four-level lasers can be pumped optically by illuminating the laser material with light of suitable wavelengths to raise the atoms (or molecules) from the ground state to the top level. These atoms will then drop to the metastable upper laser level to realize the population inversion.

It appears from Figures 6.2 and 6.3 that the pumping light must be of a specific wavelength, because the transitions between the isolated levels are narrow. In fact this is usually not the case. The ground and, particularly, the excited levels, generally span a range of energies. Typically, there are multiple excited levels which all decay to the metastable upper laser level (i.e., E_2 in three-level lasers and E_3 in four-level lasers). Since atoms in the laser material can absorb light at wavelengths corresponding to transitions between the ground state and any level of the excited states, a laser can be optically pumped by a light source emitting a broad spectral range, such as a flash-lamp. However, because the atoms must be pumped from the bottom level to the topmost level, the pumping photon must have higher energy (i.e., higher frequency or, equivalently, shorter wavelength) than the emitted laser light. This is in fact one of the factors that limits the efficiency of a laser.

Optical pumping can be used for any laser medium that is transparent to the pump light. In practice, it is mostly used for solid state crystalline lasers and liquid tunable dye lasers. The optical pumping energy may be delivered steadily or in pulses, although many optically pumped lasers cannot produce continuous wave laser light.

To understanding the electrical pumping process, we consider an electrically excited gas laser. An electric current flows through the gas in a fluorescent tube, transferring energy to the atoms and molecules in the gas. Those energized atoms and molecules are pumped to the excited state in the three-level or four-level systems. A high voltage pulse initially ionizes the laser gas so that it conducts electricity. A much lower voltage is then used to sustain the current and maintain the pumping process.

Electrical pumping can only be used with laser materials that can conduct electricity without destroying the lasing activity. In practice, this is limited to gases and semiconductors. Some electrically pumped gas lasers produce continuous wave laser light as a constant current passes through the gas, while others produce pulses of light after intense electrical pulses pass through the gas. Some high-power lasers are excited by electron beams, the electrons firing directly into the laser gas.

EXAMPLE 6.3

The ruby laser is an optically pumped three-level laser, in which the energy between the ground level and the excited level is 2.254 eV, and the energy between the ground level and the metastable level is 1.784 eV. Determine the wavelengths of the pumping light and the laser light.

The pumping frequency should match the energy difference between the excited and the ground states. The wavelength can be calculated using the equation given at the end of the Example 6.1:

$$\lambda_{pump} = \frac{1.24}{E_3 - E_1} = \frac{1.24}{2.254} = 0.55 \, \mu m.$$

Similarly, the laser wavelength can be found as

$$\lambda_{laser} = \frac{1.24}{E_2 - E_1} = \frac{1.24}{1.786} = 0.694 \, \mu m,$$

which is at the red end of the visible range. The excited state actually spans a spectral range. The central pumping wavelength is around 0.55 μm and matches the output of a xenon flashlamp very well. Therefore, a xenon flashlamp was used in the first laser invented in 1960.

6.4   Optical Resonant Cavity

Although population inversion makes the laser radiation possible, it does not guarantee laser action. To extract energy efficiently from a medium with population inversion and radiate a laser beam, an optically resonant cavity is required to build up (or amplify) stimulated emission by feedback. To understand the role of the resonant cavity in a laser, let us start by looking at the process of amplification.

We have seen that a photon with energy corresponding to a laser transition can stimulate the emission of a cascade of other photons of the same wavelength, polarization, phase, and propagation direction. If the initial photon is taken as a signal, it is amplified by the process of stimulated emission. Within a laser medium, the first photon may come from spontaneous emission between the two laser transition levels. When this first photon encounters an atom in the upper laser level, it stimulates another photon. If it encounters an atom in the lower level, it could certainly be absorbed. However, owing to the population inversion, more atoms are in the upper level than in the lower level. Consequently, the light is more likely to stimulate emission than to be absorbed. After the first stimulated emission, there are two photons having the same characteristics. Each of them is more likely to encounter an atom in the upper level than in the lower level. Thus, they too are likely to produce stimulated emission. Such a chain reaction takes place until these photons travel out of the laser medium.

The degree of amplification is measured as *gain*, which is the increase in intensity when a light beam passes a medium, and is expressed as

$$G = \frac{1}{I}\frac{dI}{dx}.\qquad(6.9)$$

The intensity after the light beam travels a distance x can then be written as

$$I(x) = I_o\, e^{Gx},\qquad(6.10)$$

where I_o is the initial intensity at $x = 0$. The intensity increases exponentially as the light beam propagates through the amplification medium.

Most laser materials have a very low gain, typically between 0.01 and 0.0001/cm. In order to produce a large amplification, the light has to pass a long length of the laser medium. For example, if the gain is 0.001/cm, a light beam needs to travel over 69 m in order to be amplified 1000 times. Such a long distance is obviously not practical.

The most practical way to get the light to pass through a long length of the laser medium is by putting mirrors on both sides of the laser tube or laser rod. These two mirrors, together with the laser materials in between, form an optically resonant cavity. The light is bounced back and forth between the mirrors and makes many passes through the laser medium. The amount of stimulated emission grows on each pass until it reaches an equilibrium level. Assume that the length of the laser medium within an optical cavity is L, the reflectances of the two mirrors are R_1 and R_2, respectively, and the initial light intensity is I_0, then, the intensity of the light beam after a round trip in the cavity is given by

$$I(2L) = R_1 R_2 I_o\, e^{G2L}.\qquad(6.11)$$

To maintain the amplification of the stimulated emission, it is required that

$$I(2L) \geq I_o. \tag{6.12}$$

This leads to

$$G \geq -\frac{1}{2L} \ln(R_1 R_2). \tag{6.13}$$

This inequality (Eq. 6.13) is known as the condition for lasing. In practice, however, one of the cavity mirrors reflects essentially all the light that reaches it. The other one reflects most, but not all, of the light back into the laser cavity; the remaining light is transmitted through the mirror as the laser output beam. Due to the low gain of most laser materials, only a very small fraction of the light is allowed to transmit through the mirror.

EXAMPLE 6.4

Assume that the length of a laser tube is 150 mm and the gain factor of the laser material is 0.0005/cm. If one of the cavity mirrors reflects all the light that impinges on it, determine the required reflectance of the other cavity mirror.

By referring to the lasing condition (Eq. 6.13), we have

$$R_2 \geq \frac{1}{R_1\, e^{G2L}} = \frac{1}{100\% \times e^{0.0005 \times 30}} = 0.985.$$

The reflectance of the output mirror should be higher than 98.5%.

It is interesting to notice that laser power levels are higher inside the cavity than outside. As shown by the preceding example, the output mirror reflects at least 98.5% of the light back into the cavity and transmits, at most, 1.5%. Hence the output power is no more than 1.5% of the power within the cavity. If the output power is assumed to be 1 mW, the intracavity power would be as high as 67 mW.

It is not necessary for the mirrors in an optical cavity to be flat. In fact, most practical laser cavities contain at least one curved mirror. Several laser cavity configurations are illustrated schematically in Figure 6.4. The cavity consisting of two flat mirrors shown in Figure 6.4a, although conceptually simple, is unstable. The problem comes in aligning the two mirrors precisely parallel, so that they reflect light rays directly back at each other. Otherwise, the light rays passing back and forth through the laser medium will miss the mirror after several round trips, before passing through a sufficient length and building up enough stimulated emission. The misalignment needs not to be large to cause problems. For example, if one mirror is only 0.25° out of alignment, a light ray striking it from the center of the other mirror in a 15-cm

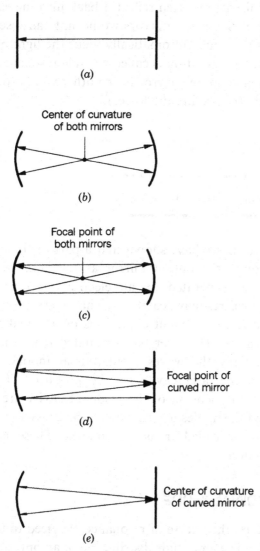

FIGURE 6.4 Major laser cavity configurations: (a) plane parallel cavity; (b) concentric cavity; (c) confocal cavity; (d) hemifocal cavity; (e) hemispherical cavity.

long cavity would be reflected to a point 1.3 mm from the center of the other mirror. It may then miss the edge of a cavity mirror 2 mm in diameter. Successive reflections would magnify the effects of misalignment, and increase the amount of light lost.

These light-leakage losses can be avoided if one or both of the cavity mirrors are curved, as in most of the designs shown in Figure 6.4. Following the light rays as depicted in Figure 6.4b–e, one can see how the focusing nature of the curved mirror keeps light within the laser medium. If the light is not emitted precisely along the axis of the tube, but does hit the curved mirror, the mirror can often reflect it back into the cavity. Likewise, slight misalignment of the cavity mirrors would not cause severe problems, because their curvature would automatically focus the light back toward the other mirror. Such a configuration is called a stable resonant cavity, because the light rays reflected from one mirror to the other will keep bouncing back and forth indefinitely (neglecting the losses).

6.5 Modes of Laser Beam

In the preceding section, we have shown that a part of the light in the laser cavity emerges through the output mirror as the laser beam. Similar to electromagnetic waves generated by an electrical resonator, the optical waves within an optical resonant cavity are characterized by their resonant modes, which are discrete resonant conditions determined by the physical dimensions of the cavity. The laser beam radiated from the laser cavity is thus not arbitrary. Only the waves oscillating at modes that match the oscillation modes of the laser cavity can be produced. The laser modes governed by the axial dimension of the resonant cavity are called the longitudinal modes, and the modes determined by the cross-sectional dimension of the laser cavity are called transverse modes. These laser modes are discussed in this section.

6.5.1 LONGITUDINAL MODE

In order to understand the nature of resonance, we need to turn back to the wave picture of light. The amplitude distribution of an optical wave along the axis of the cavity is illustrated schematically in Figure 6.5. In particular, the

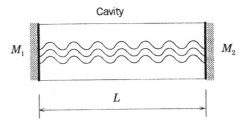

FIGURE 6.5 Light wave resonance in a cavity. Notice that the length of the laser cavity is equal to an integral number of wavelengths.

figure shows what happens when twice the cavity length is an integral multiple of the wavelength. Because all the light waves in the cavity are coherent (due to stimulated emission), they are all in phase. Thus, each of these waves is in the same phase when it is reflected from one of the cavity mirrors. For example, if light starts out at a wave peak when it is reflected from the output mirror, it will travel an integral number of wavelengths before it reaches the output mirror once more, where it will again be at the wave peak, as will the light waves stimulated by that wave. Hence all of the waves will add in amplitude, by constructive interference.

Suppose, however, that twice the cavity length is not an integral multiple of wavelengths. Then each wave will reach the cavity mirror in a different phase. The waves will add in amplitude. However, since they are not in phase, destructive interference will cancel them out.

The result is the condition of resonance: light waves are amplified strongly if, and only if, they satisfy the equation

$$2\eta L = N\lambda, \tag{6.14}$$

where L is the length of the cavity, η is the refractive index of the laser medium within the cavity, ηL is the so called *optical path*, N is an integer, and λ denotes the wavelength.

The integer N cannot be an arbitrary number. The gain of the laser materials, discussed in the last section, is also a function of wavelength, $G(\lambda)$. Laser oscillation can only take place when the gain is large enough to maintain the resonance. Consequently, as shown in Figure 6.6, the actual profile of wavelengths emitted by a laser is the product of the envelope of longitudinal oscillation modes and the gain profile.

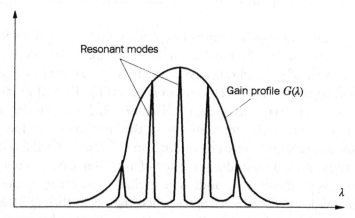

FIGURE 6.6 Several resonant modes can fit within the gain profile.

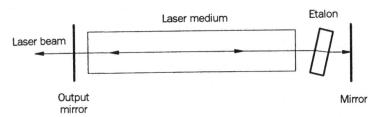

FIGURE 6.7 Single longitudinal mode selection using a Fabry–Perot etalon.

EXAMPLE 6.5

The nominal wavelength of a HeNe laser is 632.8 nm and the half-width of the gain profile of He and Ne gases is about 2×10^{-3} nm. If the cavity length of the laser is 30 cm, how many longitudinal modes can be excited?

By referring to Eq. 6.14, separation of the spectral lines between adjacent longitudinal modes can be calculated as

$$\Delta\lambda_{mode} = \frac{\lambda^2}{2\eta L} = \frac{632.8 \times 632.8}{2 \times 1.0 \times 30 \times 10^7} = 0.66 \times 10^{-3} \, \text{nm},$$

where we have assumed that the refractive index of the gas within the cavity is 1.0. Since the separation between adjacent spectral lines is approximately one-third of the half-width of the gain profile, three longitudinal modes can be excited.

As indicated by this example, several wavelengths satisfying the oscillation condition may fall within the gain profile of the active medium. Each of these wavelengths represents a distinct longitudinal mode. The coexistence of multiple modes reduces the monochromaticity of the laser light and should be avoided in certain applications. This leads to the need for single mode selection.

There are many ways of limiting the laser oscillation to the width of a single longitudinal mode. The most common technique is to insert a Fabry–Perot etalon, a pair of parallel reflective surfaces, between the laser mirrors to form a second resonant cavity. The etalon is generally tilted at an angle θ to the axis of the laser cavity, as shown in Figure 6.7. Since the length of the etalon resonator is much shorter than that of the laser cavity, its longitudinal modes are so far apart that only one oscillation will fall under the laser's gain curve. Of the multiple longitudinal modes of the laser cavity, only the one that coincides with the etalon mode can oscillate; all other modes will be canceled out by destructive interference when they pass through the etalon. The oscillation mode of the Fabry–Perot etalon can be tuned to coincide with a laser cavity mode by tilting the etalon to a suitable angle with respect to the axis of the laser cavity.

6.5.2 TRANSVERSE MODE

The configuration of the optical cavity also determines the transverse modes of the laser output, which characterizes the intensity distribution across the cross-section of the laser beam. This is simply a natural consequence of electromagnetic waves being confined within a cavity and restricted by the boundary conditions. In general, the allowed modes in an optical cavity are designated as TEM_{mn}, where T, E, and M stand for transverse, electric, and magnetic modes, respectively, and m and n are integers. The simplest, or the lowest-order transverse mode is TEM_{00}, which has a smooth cross-section profile with a peak in the middle, as shown in Figure 6.8. The complex amplitude of the TEM_{00} mode is expressed as

$$E(r, z) = E_0 \frac{w_0}{w(z)} e^{-\frac{r^2}{w^2(z)}} e^{-i[kz - \tan^{-1}(z/z_0)]} e^{-i\frac{kr^2}{2R(z)}}, \qquad (6.15)$$

where E_0 is the initial complex amplitude at $z = 0$, w_0 is the beam waist at the point where the cross-section of the laser beam has the smallest diameter, $R(z)$ is the radius of curvature of the beam, and $w(z)$ is the beam waist in the z plane, which can be written as

$$w^2(z) = w_0^2 \left(1 + \frac{z^2}{z_0^2} \right). \qquad (6.16)$$

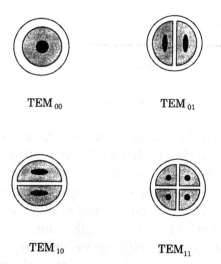

TEM$_{00}$

TEM$_{01}$

TEM$_{10}$

TEM$_{11}$

FIGURE 6.8 Beam profiles of transverse modes.

The cross-section intensity profile is a Gaussian distribution as given by

$$I(r,z) = I_0 \frac{w_0^2}{w^2(z)} e^{-\frac{2r^2}{w^2(z)}}, \tag{6.17}$$

for which the TEM_{00} mode is often called the Gaussian beam.

The integers m and n associated with the TEM_{mn} mode represent the numbers of zeros or minima between the edges of the beam in two orthogonal directions. As shown in Figure 6.8, a TEM_{01} beam has a single minimum dividing the beam into two bright spots. A TEM_{11} beam has two perpendicular minima (one in each direction), dividing the beam into four quadrants. The larger the values of m and n, the more bright spots are contained in the laser beam. The TEM_{00} mode is desirable because the spread it experiences due to diffraction approaches a theoretical minimum value. In Chapter 8, we will show that the far field diffraction pattern is equal to the Fourier transform of the amplitude distribution at the input plane. Since the Fourier transform of a Gaussian function is again a Gaussian function, the profile of the Gaussian beam does not alter as the beam propagates. To achieve a single transverse mode operation, one can place a small, circular aperture that is slightly larger than the spot size of the TEM_{00} mode in the middle of the laser cavity. The aperture will have no effect on the resonance of the TEM_{00} mode, but will filter out or attenuate all other modes. Hence only the TEM_{00} mode is able to oscillate. The same results can be obtained by making the bore diameter of the laser tube just slightly larger than the spot size of the TEM_{00} mode. In practice, most stable laser resonators with low gain are designed to produce only the TEM_{00} mode. However, some stable laser resonators do operate in one or more higher order modes, especially when they are designed to maximize output power.

6.6 Spectral Bandwidth and Coherence Length

Laser light is usually considered monochromatic, meaning single-colored or oscillation at a single wavelength. In fact, this is not true. Light from all kinds of lasers has a finite, although normally very narrow, spectral bandwidth. As we have discussed in the preceding section, light waves are strongly amplified when twice the cavity length is an integral multiple of wavelengths. What would happen, then, if twice the cavity length is not exactly, but very close to, an integral multiple of wavelengths? Since the phase of the wave after making a round trip in the cavity is very close to that of the initial wave, the two waves will still add in amplitude by constructive interference. This results in a finite oscillation bandwidth for each longitudinal mode, as shown in Figure 6.6. The oscillation bandwidth of a mirrored cavity can be obtained using interference theory, as given by

$$\Delta \lambda = \frac{\lambda^2}{2\pi L} \left(\frac{1 - R}{\sqrt{R}} \right), \tag{6.18}$$

where L is the length of the cavity, and R is the reflectance of the output mirror. Alternatively, the spectral bandwidth is often expressed in terms of frequency as

$$\Delta \nu = \frac{c}{2\pi L} \left(\frac{1 - R}{\sqrt{R}} \right), \tag{6.19}$$

where c stands for the speed of light. It should be noted that the actual spectral bandwidth may not be as simple as described by Eqs 6.18 and 6.19. For example, if an etalon is inserted in a laser cavity, the spectral bandwidth will be dominated by the performance of the etalon. Since the thickness of an etalon is generally much smaller than the length of the cavity, the use of an etalon can significantly reduce the bandwidth of the laser output.

In addition to the spectral bandwidth, the degree of monochromaticity of a laser beam is often described by another term, the *coherence length*. The coherence length measures how long light waves remain in phase as they travel. Light waves become incoherent as differences in their optical path make them drift out of phase. The coherence length depends on the nominal wavelength and the spectral bandwidth of the laser beam, as given by

$$l_c = \frac{\lambda^2}{2 \Delta \lambda}, \tag{6.20}$$

or, alternatively,

$$l_c = \frac{c}{2 \Delta \nu}. \tag{6.21}$$

The coherence length is a very useful measure of temporal coherence because it tells us how far apart two points along the light beam can be and remain coherent with each other. In interferometry and holography, the coherence length is the maximum optical path difference that can be tolerated between the two interfering beams.

Now let us compare the coherence lengths of some commonly encountered light sources. Light bulbs emit light from the visible to well into the infrared. If we assume the range of wavelengths is 400–1000 nm and take an average wavelength of 700 mm, the light bulb has a coherence length as short as 400 nm. An inexpensive semiconductor laser may have a wavelength of 850 nm and wavelength range of 1 nm, giving a coherence length of

0.36 mm. An ordinary HeNe laser has a narrow bandwidth of 0.002 nm at 632.8 nm wavelength, corresponding to a coherence length of 100 nm. Stabilizing a HeNe laser so that it emits in a single longitudinal mode limits the line width to about 2×10^{-6} nm, and thus extends the coherence length to about 100 m.

EXAMPLE 6.6

Assume that the length of a HeNe laser is 300 mm and the reflectance of the output mirror is 99%. Calculate the spectral bandwidth of each longitudinal mode and the coherence length when the laser works in the single-mode operation.

Referring to Eq. 6.18, the line width of a longitudinal mode is

$$\Delta\lambda = \frac{\lambda^2}{2\pi L}\left(\frac{1-R}{\sqrt{R}}\right) = \frac{632.8\,(\text{nm}) \times 632.8\,(\text{nm})}{2 \times 3.14 \times 300\,(\text{mm})}\left(\frac{1-0.99}{\sqrt{0.99}}\right) = 2.12 \times 10^{-6}\text{nm}.$$

The corresponding single-mode coherence length is

$$l_c = \frac{\lambda^2}{2\,\Delta\lambda} = \frac{632.8\,(\text{nm}) \times 632.8\,(\text{nm})}{2 \times 2.12 \times 10^{-6}\,(\text{nm})} = 94.4\,\text{m}.$$

In practice, the spectral line of a laser cavity can be broadened by various factors, causing a reduction in coherence length. Most small HeNe lasers, with cavity lengths ranging from 200 to 500 mm, work in multiple longitudinal modes. Their coherence lengths are usually of the order of 100 mm.

6.7 Types of Lasers

There are many ways to define the type of a laser. A laser can be classified as an optically pumped laser or an electrically pumped laser, based on its pumping scheme. On the basis of the operation mode, lasers fall into classes of either continuous wave lasers or pulsed lasers. According to the materials used to produce laser light, lasers can be divided into three categories: gas lasers, solid state lasers, and semiconductor lasers. The last classification is used in our discussion here.

TABLE 6.1 Wavelength and power of major gas lasers

Type	λ (nm)	Power range (W)	Operation
Argon–fluoride excimer	193	0.5–50 (avg)	Pulsed
Krypton–fluoride excimer	249	1–100 (avg)	Pulsed
Xenon–chloride excimer	308	1–100 (avg)	Pulsed
Helium–cadmium (UV lines)	325	0.002–0.05	Continuous
Nitrogen	337	0.001–0.01 (avg)	Pulsed
Argon ion (UV lines)	350	0.001–2	Continuous
krypton ion (UV lines)	350	0.001–1	Continuous
Xenon-fluoride excimer	351	0.5–30 (avg)	Pulsed
Helium–cadmium	442	0.001–0.05	Continuous
Argon ion	488–514.5	0.001–20	Continuous
Copper vapor	510, 578	1–50 (avg)	Pulsed
Xenon ion	540	–	Pulsed
Helium–neon	543	0.0001–0.001	Continuous
Gold vapor	628	1–10 (avg)	Pulsed
Helium–neon	632.8	0.0001–0.05	Continuous
Krypton ion	647	0.001–6	Continuous
Iodine	1300	–	Pulsed
Hydrogen fluoride	2600–3000	0.01–150	Pulsed or CW
Deuterium fluoride	3600–4000	0.01–100	Pulsed or CW
Carbon monoxide	5000–7000	–	Pulsed or CW
Carbon dioxide	9000–11000	0.1–15000	Pulsed or CW

CW, continuous wave; UV, ultraviolet.

6.7.1 GAS LASER

Gas lasers generally have a wide variety of characteristics. For example, some gas lasers emit feeble power below 1 mW, but other commercial gas lasers emit power of the order of kilowatts. Some lasers can emit continuous beam for years; others emit pulses lasting a few nanoseconds. Their outputs range from deep in the vacuum ultraviolet through the visible and infrared to millimeter waves. Table 6.1 lists the wavelength and output power ranges for major commercial gas lasers.

Of the gas lasers, the most commonly used is the HeNe laser with its familiar red beam. The laser medium is a mixture of helium and neon gases. An electrical discharge, in the form of direct current or radiofrequency current, is used to excite the medium to a higher energy level. The pumping action takes place in a complex and indirect manner. First, the helium atoms are excited by the discharge to the 2^1s and 2^3s, two of the excited energy levels. As shown in Figure 6.9, these two levels happen to be very close to the $3s$ and $2s$ levels of the neon atoms. When the excited helium atoms collide with the neon atoms, energy is exchanged, pumping the neon atoms to the

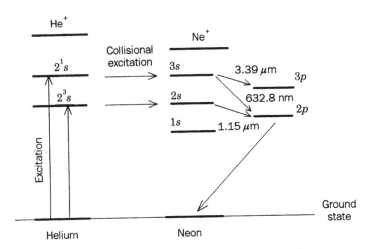

FIGURE 6.9   Helium and neon laser energy levels.

respective levels. The atoms at the neon $3s$ level eventually drops down to the $2p$ level, as a result of stimulated emission, and light of wavelength 632.8 nm is emitted. The atoms at the $2s$ level, on the other hand, drop to the $2p$ level by emitting light at 1.15 μm. However, the atoms at the $3s$ level may instead drop down to the $3p$ level, by emitting light at 3.39 μm. In fact, stimulated emission tends to occur between the $3s$ and $3p$ levels. Thus laser action would normally take place at 3.39 μm, instead of at the desired 632.8 nm in the visible range. This problem can be overcome by attenuating the 3.39 mm radiation in a way that would not affect the 632.8 nm emission. This is commonly achieved by utilizing glass elements that strongly absorb the 3.39 μm infrared radiation.

The argon ion (Ar^+) laser is another popular gas laser. It can provide a coherent beam at a power level many times higher than that of the HeNe laser. Although only one of the laser lines emitted by the HeNe laser is in the visible region, as many as 10 visible lines ranging from deep blue to light green can be obtained from the Ar^+ laser. The Ar^+ laser can be operated using either pure argon gas or a mixture of helium and argon gases. The gas medium is excited by highly energized electrons through an electrical discharge, either dc or radiofrequency current. The argon atoms become ionized, and the ions are excited until they emit several lines of intense radiation. The transitions of the Ar^+ laser are depicted in Figure 6.10. A single oscillation line can be selected by inserting a prism or grating in the laser cavity.

6.7.2 SOLID STATE LASER

It should be noted that the term *solid state* has different meanings in laser science and electronic technology. A solid state laser is one in which the

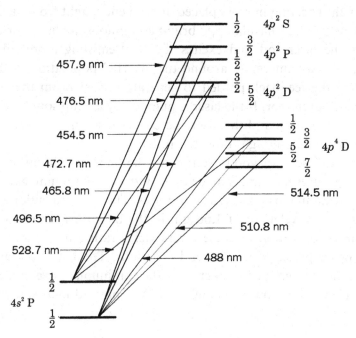

FIGURE 6.10 Argon ion laser energy levels.

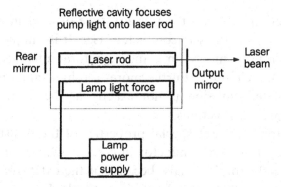

FIGURE 6.11 A generic solid state laser.

atoms that emit light are fixed within a crystal or a glassy material. It does not include the semiconductor laser, even though semiconductors are crystalline materials. In the community of laser scientists, semiconductor lasers belong to a separate category, as will be described in the next section.

The first laser invented by Maiman in 1960, the ruby laser, was a solid state laser. Although the operation of solid state lasers has been refined greatly since then, the same basic principles underlie the operation of the entire family of solid state lasers. A generic solid state laser is illustrated in Figure 6.11. The atoms that emit light in solid state lasers are dispersed in a crystal or a piece of glass that contains many other elements. The crystal is shaped into a rod, with reflecting mirrors placed at each end. Light from an external source (such as a pulsed flashlamp, a bright continuous arc lamp, or another laser) enters the laser rod and excites the light-emitting atoms. The two mirrors form a resonant cavity around the inverted population in the laser rod, providing the feedback needed to generate a laser beam that emerges through the output mirror. If the laser is pumped by a lamp source, as shown in Figure 6.11, the lamp and the laser rod are enclosed in a reflective cavity that focuses the pumping light onto the rod.

The Nd:YAG laser is a good example of the most commonly used solid state lasers. The laser medium is made up of yttrium–aluminum–garnet, with trivalent neodymium ions present as impurities. The laser transition involved corresponds to a wavelength of $1.06 \, \mu m$, in the near infrared region. Typically, the pulsed operation of these layers yields rather high power (megawatts and above) pulses of nanosecond durations. The Nd:YAG lasers can also be continuously pumped to operate in the continuous wave mode, which provides excellent high power (as high as $10 \, W$) infrared light sources.

6.7.3 SEMICONDUCTOR LASER

A unique, and perhaps the most important, type of laser in terms of optoelectronic applications is the semiconductor laser. It is unique because of its small dimensions (mm $\times$ mm $\times$ μm), and its natural integration capabilities with microelectronic circuitry. Furthermore, the light amplification by the process of stimulated emission is not exactly in the form that we have described in the previous sections.

A semiconductor laser uses special properties of the transition region at the junction of a p-type semiconductor in contact with an n-type semiconductor. In semiconductor materials, because of the extensive interaction of energy between atoms, the energy levels form bands. Energy band diagrams for an n-type and a p-type semiconductor are depicted in Figure 6.12. The energy gap between the valence band and the conduction band is designated by E_g and is measured in electronvolts, eV_g. The Fermi level E_f is the level that divides the occupied from the unoccupied levels.

In a p-n junction, as shown in Figure 6.13, the energy levels readjust in accordance with thermodynamics so that the E_f band is the same throughout the junction. The valence band E_v and the condition band E_c of the p-type semiconductor are higher than the corresponding bands of the n-type semiconductor. If a positive voltage is applied on the p side (the so-called positive bias), the electrons on the n side will be attracted by the applied voltage and will cross into the junction region. There they recombine with the holes that have been pushed into the junction region by the positive bias. This process

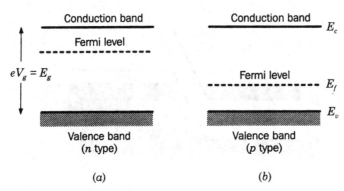

FIGURE 6.12   Energy band diagrams: (a) an *n*-type semiconductor; (a) a *p*-type semi-conductor.

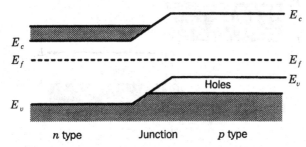

FIGURE 6.13   A *p-n* junction energy band diagram.

will continue as long as the external circuit is on, because the electrons and holes that have recombined are continuously replenished.

When the electrons and holes recombine, they emit energy in the form of photons. The junction transition region in which this takes place is therefore the source of radiation, and may be viewed as equivalent to the E_2 and E_1 transition levels which we discussed earlier. The frequency of the light emitted from the recombination is given by

$$\nu = \frac{E_2 - E_1}{h} = \frac{eV_g}{h}. \qquad (6.22)$$

To obtain stimulated emission and amplification from this region, the equivalent of a population inversion needs to be created, for which a high density of electrons and a high density of holes must exist *simultaneously* in the junction region. To achieve this, heavily doped *p-n* junctions are used in semiconductor lasers. Figure 6.14*a* shows the resultant energy levels. When a positive bias is applied on the *p* side, there is a transition region with a high concentration of electrons and holes, as shown in Figure 6.14*b*. This region serves as a population-inverted medium, which amplifies the radiation emitted within it through electron–hole recombination.

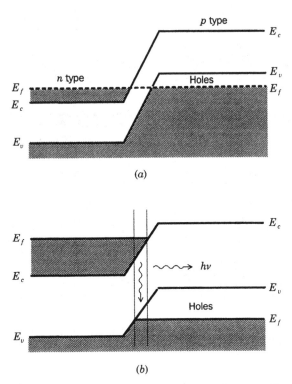

FIGURE 6.14   Energy band diagrams: (a) a heavily doped *p-n* junction; (b) with an applied bias voltage.

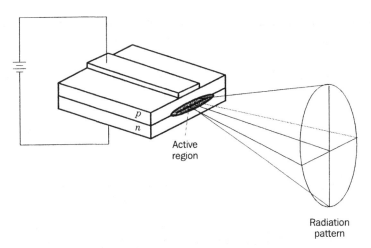

FIGURE 6.15   A generic *p–n* junction semiconductor laser.

The tiny, planar dimensions of the semiconductor laser cause a unique, but undesirable output pattern. A *p-n* junction semiconductor laser is illustrated schematically in Figure 6.15. The shaded area is the transition region where the laser action takes place. This region is about 1–2 μm thick, and tens of

TABLE 6.2 Major semiconductor laser materials

Material	Wavelength range	Comments
GaInPAs/GaAs	600–800 nm	Very hard to grow
AlGaInP/GaAs	580–690 nm	Very hard to grow
GaInP/GaAs	670–680 nm	Easy to grow
GaAlAs/GaAs	720–900 nm	Short lived at short wavelengths
GaAs/GaAs (pure)	904 nm	
InGaAs	1060 nm	
PbCdS	2.7–4.2 μm	Require cooling
PbSSe	4.2–8 μm	Require cooling
PbSnTe	6.5–30 μm	Require cooling
PbSnSe	8–30 μm	Require cooling

micrometers long. As a result, the emission is squeezed into a thin plane, leading to an elliptical cross-section of the beam.

Many semiconductor materials can be used to build semiconductor lasers. Presently, developers have concentrated primarily on four families of semi-conductor laser material: GaAlAs, InGaAsP, AlGaInP, and lead salt materials. The wavelength ranges of commonly used semiconductor lasers are listed in Table 6.2.

REFERENCES

1. J. T. VERDEYEN, *Laser Electronics*, Prentice-Hall, Englewood Cliffs, NJ, 1989.
2. A. YARIV, *Optical Electronics*, W. B. Saunders, Philadelphia, PA, 1991.
3. E. HECHT and A. ZAJAC, *Optics*, Addison-Wesley, Reading, MA, 1979.
4. A. E. SIEMAN, *Lasers*, University Science Books, Mill Valley, CA, 1986.
5. K. A. JONES, *Introduction to Optical Electronics*, Harper & Row, New York, 1987.

PROBLEMS

6.1 Assume that the energy required to remove an electron from the surface of a metal (the so-called work function) is 2.3 eV. Determine the cut-off frequency for the photoelectric effect.

6.2 Can a human eye detect photons with an energy of 2 eV? If so, why?

6.3 Calculate the photon energy for the following laser wavelengths: 488, 514, 632.8, and 1.06 μm. What are the colors corresponding to these wavelengths?

6.4 The coefficients for spontaneous emission and stimulated emission are related as follows:

$$t_{sp}\beta_{12}\rho(\nu) = \frac{1}{\exp(h\nu/kT) - 1},$$

where h is Planck's constant, ν is the frequency, k is the Boltzmann constant, and T is the temperature in kelvin (the absolute scale). If the difference between the two energy levels E_2 and E_1 is 1 eV, determine the population ratio at thermodynamic equilibrium at room temperature.

6.5 Under the conditions given in Problem 6.4, what is the ratio of the spontaneous emission intensity and the stimulated emission intensity I_{sp}/I_{st}?

6.6 Assume that the wavelength of the emission is 0.6 μm and the lifetime t_{sp} is 10^{-6} s. Calculate the coefficient for the stimulated emission $\beta_{12}\rho(\nu)$.

6.7 There are actually two excited bands in a ruby laser. One is discussed in Example 6.4; the other is centered on the energy level at about 3.1 eV above the ground state. What pumping wavelength would be needed for the second excited band? Can a xenon lamp be used to pump atoms into this excited band?

6.8 Assume that a narrow-band light source with a central wavelength of 0.4 μm is used to pump a ruby laser, and all the pumping energy is converted into the laser emission. What is the energy conversion efficiency?

6.9 What are the pumping modes of the most commonly used lasers, such as the HeNe laser, Ar^+ laser, and Nd:YAG laser?

6.10 What are the advantages and disadvantages of the three-level and four-level laser systems?

6.11 Assume that the gain of the laser material within a 15 cm long cavity is 0.005/cm. If the reflectance of both cavity mirrors is 100%, how many photons can be excited by a single initial photon in 1 μs?

6.12 A student is constructing a cavity using two mirrors having reflectances of 100% and 99%. If the gain of the laser material is 0.005/cm, how far apart should he place the two mirrors?

6.13 Assume that the length of the plane parallel cavity shown in Figure 6.4a is 300 mm and the diameter of two flat mirrors is 2 mm. In order to keep a light ray bouncing back and forth within the cavity for at least 100 round trips before it misses the cavity mirror, determine the maximum tolerable misalignment of the two cavity mirrors.

6.14 The half-width of the gain profile of a HeNe laser material is about 2×10^{-3} nm. In order to have a single longitudinal mode oscillation, what would be the maximum length of the HeNe laser cavity?

6.15 The propagation of a Gaussian beam is given by Eq. 6.16. Derive the expression for the beam divergence.

6.16 How can you experimentally determine the longitudinal and transverse modes of a laser beam?

6.17 If you want to design a HeNe laser with a coherence length of over 200 m, how many available approaches can you pursue to achieve your objective? What are they?

6.18 The strongest spectral line of a mercury arc lamp is about 546.1 nm. A narrow band interference filter limits the bandwidth of such a spectral line to about 1 nm. Calculate the coherence length of the light passed through the filter.

6.19 Can you experimentally measure the coherence length of a laser beam? If so, how?

6.20 Assume that the width of the transition region of a 850 nm semiconductor laser is about 2 μm. Estimate the divergence of the beam by employing diffraction theory.

6.21 List the types, operation modes, wavelengths, and power (or energy) levels of all the lasers you have ever seen or used.

7 LINEAR SYSTEM TRANSFORMS

Generally speaking, an optical system can be represented by an input–output block diagram. In most instances the concept of the system theory can be easily applied to optical imaging and processing systems. There are, however, no general techniques to solve nonlinear optical systems, except for a few special cases. Nonlinear problems are usually solved by means of graphical and approximation procedures. Although no optical system is strictly linear, a great number of optical systems can be treated as linear, and these can be approached by linear system analysis.

The analysis of optical systems usually involves linear spatially invariant concepts. It is the aim of this chapter to introduce some of the fundamentals. Since most of the optical systems can be described by two-dimensional spatial variables, we shall, in various parts of this book, use two-dimensional notation.

7.1 Linear Spatially Invariant System

It is well known that the behavior of a physical system may be described by the relationship of the output response to the input excitation, as shown in Figure 7.1. We assume that the output response and the input excitation are measurable quantities. Let us say that the output response $g_1(x,y)$ is caused by an input excitation $f_1(x,y)$, and a second response $g_2(x,y)$ is produced by $f_2(x,y)$. The input–output relationships are written in the following forms,

$$f_1(x,y) \rightarrow g_1(x,y) \tag{7.1}$$

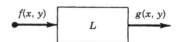

FIGURE 7.1 Block diagram representation of a physical system.

and

$$f_2(x,y) \to g_2(x,y), \tag{7.2}$$

where (x,y) represent a two-dimensional spatial coordinate system.

If the sum of these two input excitations are applied to the input of the physical system, the corresponding output response would be a linear combination of $g_1(x,y)$ and $g_2(x,y)$, that is,

$$f_1(x,y) + f_2(x,y) \to g_1(x,y) + g_2(x,y). \tag{7.3}$$

Equation 7.3, in conjunction with Eqs 7.1 and 7.2, represents the well-known *additivity* property of a linear system, which is also known as the *superposition* property. In other words, a necessary condition for a physical system to be linear is that the *principle of superposition* be held. The principle of superposition implies that the presence of one excitation in the system does not affect the response caused by other excitations.

If we multiply the input excitation by an arbitrary constant K, that is $Kf_1(x,y)$, and the output response is equal to $Kg_1(x,y)$,

$$Kf_1(x,y) \to Kg_1(x,y), \tag{7.4}$$

then the system is said to possess the *homogeneity* property. In other words, a linear system is capable of preserving the scale factor of an input excitation. Thus, if a physical system is to be a linear system, the system must possess the additivity and homogeneity properties given in Eqs 7.3 and 7.4.

There is, however, another important physical aspect that characterizes a linear system with constant parameters. This physical property is *spatial invariance*, which is analogous to the *time invariance* of an electrical network. If the output response of a physical system remains unaltered (except with appropriate translation) with respect to the input excitation, the physical system is said to possess the spatially invariant property. The qualification of a spatially invariant system is that if

$$f(x,y) \to g(x,y),$$

then

$$f(x - x_0, y - y_0) \to g(x - x_0, y - y_0), \tag{7.5}$$

where $f(x,y)$ and $g(x,y)$ are the corresponding input excitation and output response, respectively, and x_0 and y_0 are some arbitrary constants.

Thus, if a linear system possesses the spatially invariant property of Eq. 7.5, the system is known as a *linear spatially invariant system*.

We note that the concept of linear spatial invariance is rather important in the analysis of the electro-optic systems. This property would simplify otherwise rather complicated formulations.

EXAMPLE 7.1

Given an ideal linear-phase low-pass filter, the amplitude and phase distributions are

$$A(\omega) = \begin{cases} A, & |\omega| \leq |\omega_c| \\ 0, & |\omega| > |\omega_c| \end{cases}$$

and

$$\phi(\omega) = -t_0\omega,$$

where ω is in radians per second and t_0 is an arbitrary positive constant. To show the linearity of this filter, we let the input signals $f_1(t)$ and $f_2(t)$ be

$$f_1(t) = A\sin\omega_1 t, \qquad |\omega_1| \leq |\omega_c|$$

and

$$f_2(t) = B\cos\omega_2 t, \qquad |\omega_2| \leq |\omega_c|.$$

The corresponding output responses can be shown as

$$g_1(t) = A^2 \sin\omega_1(t - t_0)$$

and

$$g_2(t) = AB\cos\omega_2(t - t_0).$$

To prove that the filter is a linear filter, we would let the input signal be the sum of $K_1 f_1(t)$ and $K_2 f_2(t)$,

$$\begin{aligned} f(t) &= K_1 f_1(t) + K_2 f_2(t) \\ &= K_1 A\sin\omega_1 t + K_2 B\cos\omega_2 t, \end{aligned}$$

where K_1 and K_2 are arbitrary constants. The corresponding output response is therefore

$$\begin{aligned} g(t) &= K_1 A^2 \sin\omega_1(t - t_0) + K_2 AB\cos\omega_2(t - t_0) \\ &= K_1 g_1(t) + K_2 g_2(t). \end{aligned}$$

From this equation we see that the filter possesses the properties of homogeneity and additivity. It is therefore a linear filter.

EXAMPLE 7.2

If the input excitation in Example 7.1 is delayed by a time factor of τ,

$$f_1'(t) = A\sin\omega_1(t - \tau) = f_1(t - \tau),$$

the output response can be shown as

$$g_1'(t) = A^2 \sin\omega_1(t - \tau - t_0) = g_1(t - \tau),$$

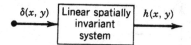

FIGURE 7.2 Description of a spatial impulse response.

which is delayed by the same τ. Thus, the ideal low-pass filter is a linear time-invariant filter.

Notice that the same linearity concept can be applied in the spatial domain as in the time domain.

One of the apparent distinctions of a linear spatially invariant system is that the system's *transfer function* can be uniquely described by the *spatial impulse response*. In contrast with the temporal (i.e., time) impulse response $h(t)$, the spatial impulse response $h(x, y)$ describes a function of two spatial variables x and y. In other words, if we assume that an impulse excitation $\delta(x, y)$ is applied to a linear spatially invariant system, as shown in Figure 7.2, then the output response must be the spatial impulse response of the system. The important aspect of a spatial impulse response of a linear spatially invariant system is that the system response can be uniquely described by an impulse response. The Fourier transform of a spatial impulse response describes the *system transfer function* $H(p, q)$:

$$H(p, q) = \int\int_{-\infty}^{\infty} h(x, y) \exp[-i(px + qy)] \, dx \, dy, \tag{7.6}$$

where p and q are the angular spatial-frequency variables, in radians per unit distance, as opposed to the temporal frequency ω, in radians per unit time. The concept of Fourier transformation will be discussed in the following sections.

EXAMPLE 7.3

If the input excitation of the idea low-pass filter in Example 7.1 is $\delta(t)$, with

$$\delta(t) \begin{cases} \infty, & t = 0 \\ 0, & \text{otherwise,} \end{cases}$$

then the impulse response of the filter can be shown as

$$h(t) = \frac{A\omega_c}{\pi} \frac{\sin \omega_c(t - t_0)}{\omega_c(t - t_0)},$$

as sketched in Figure 7.3.

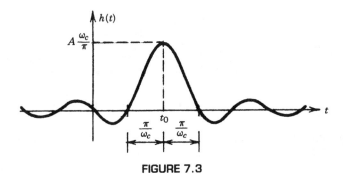

FIGURE 7.3

EXAMPLE 7.4

Let the spatial impulse response of a linear optical system be

$$h(x,y) = \begin{cases} 1, & |x| \leq |a|, \quad |y| \leq |a| \\ 0, & |x| > |a|, \quad |y| > |a|, \end{cases}$$

where a is an arbitrary constant. Determine the system transfer function of the optical system.

To evaluate the system transfer function, we would substitute the spatial impulse response of $h(x,y)$ into Eq. 7.6,

$$H(p,q) = \int_{-a}^{a} \int_{-a}^{a} h(x,y) \exp[-i(px+qy)] \, dx \, dy,$$

which can be shown as

$$H(p,q) = 4a^2 \frac{\sin ap}{ap} \frac{\sin aq}{aq}.$$

7.2 Fourier Transformation and Fourier Spectrum

In the analysis of a linear spatially invariant optical system, the use of the spatial frequency domain and Fourier transformation are critically important. In other words, the analysis can conveniently be carried out in the frequency domain with Fourier transformation.

Let us consider a complex function $f(x,y)$ that satisfied the following sufficient conditions.

(1) $f(x,y)$ must be sectionally continuous, but with a finite number of discontinuities in every finite region of the (x,y) domain.

(2) $f(x,y)$ must be absolutely integrable over the spatial domain (x,y), that is,

$$\int\int_{-\infty}^{\infty} |f(x,y)| \, dx \, dy < \infty. \tag{7.7}$$

Then $f(x,y)$ can be written as

$$f(x,y) = \frac{1}{4\pi^2} \int\int_{-\infty}^{\infty} F(p,q) \exp[+i(px+qy)] \, dr \, dq, \tag{7.8}$$

where

$$F(p,q) = \int\int_{-\infty}^{\infty} f(x,y) \exp[-i(px+qy)] \, dx \, dy, \tag{7.9}$$

and (p,q) is the angular spatial-frequency coordinate system, in radians per unit distance.

Equations 7.8 and 7.9 are the well-known *Fourier transform pair*: Eq. 7.9 is called the *direct Fourier transform* and Eq. 7.8 the *inverse Fourier transform*. Equations 7.7 and 7.8 can be written in briefer forms:

$$f(x,y) = \mathscr{F}^{-1}[F(p,q)] \tag{7.10}$$
$$F(p,q) = \mathscr{F}[f(x,y)], \tag{7.11}$$

where $\mathscr{F}^{-1}$ and $\mathscr{F}$ denote the inverse and direct Fourier transformations, respectively.

We note that $F(p,q)$ is, in general, a complex quantity. It can be represented by an amplitude and a phase factor, that is

$$F(p,q) = |F(p,q)| \exp[-i\phi(p,q)], \tag{7.12}$$

where $|F(p,q)|$ is generally referred to as the *amplitude spectrum*, $\phi(p,q)$ is known as the *phase spectrum*, and $F(p,q)$ is called the *Fourier spectrum* or *spatial-frequency spectrum* of $f(x,y)$.

It is, however, interesting to show that a Fourier transform is a *linear transformation*. The linearity of a mathematical transformation is essentially that of a linear system; namely it has the same additivity and homogeneity properties. For example, given that C_1 and C_2 are two arbitrary complex constants and that $f_1(x,y)$ and $f_2(x,y)$ are Fourier-transformable functions, that is,

$$\mathscr{F}[f_1(x,y)] = F_1(p,q) \tag{7.13}$$

and

$$\mathscr{F}[f_2(x,y)] = F_2(p,q) \tag{7.14}$$

we can show that

$$\mathscr{F}[C_1 f_1(x,y) + C_2 f_2(x,y)] = \int\limits_{-\infty}^{\infty}\!\!\int [C_1 f_1(x,y) + C_2 f_2(x,y)] \exp[-i(px+qy)] \, dx \, dy$$

$$= C_1 \int\!\!\int f_1(x,y) \exp[-i(px+qy)] \, dx \, dy$$

$$+ C_2 \int\!\!\int f_2(x,y) \exp[-i(px+qy)] \, dx \, dy$$

$$= C_1 F_1(p,q) + C_2 F_2(p,q), \tag{7.15}$$

which has the properties of additivity and homogeneity. Thus, the Fourier transformation is, in fact, a linear transformation.

Engineering students are made familiar with the concept of time frequency, but spatial frequency is rarely mentioned in most engineering courses. It is our aim to elaborate the space and spatial-frequency concepts.

In optical signal processing in particular, image functions are generally represented by a two-dimensional spatial variable. For example, the complex quantity of a wave field can be described by an (x,y) orthogonal coordinate system, that is, $f(x,y)$. The corresponding Fourier transform $F(p,q)$ represents another complex wave field that describes a spatial frequency domain (p,q), rather than a temporal frequency domain, in cycles per second. Spatial frequency is, in fact, a concept taken directly from temporal frequency.

EXAMPLE 7.5

Consider a two-dimensional spatial Fourier transformer, as shown in Figure 7.4. If the input spatial function is given as

$$f(x,y) = \sin(20\pi x)\cos(30\pi y),$$

determine the corresponding output Fourier transform.

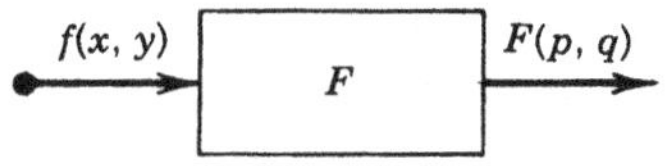

FIGURE 7.4

By substituting $f(x,y)$ in Eq. 7.9, we have

$$F(p,q) = \int\limits_{-\infty}^{\infty}\!\!\int \sin(20\pi x)\cos(30\pi y)\exp[-i(px+qy)] \, dx \, dy.$$

The preceding equation can be written as

$$F(p,q) = \int_{-\infty}^{\infty} \sin(20\pi x)e^{-ipx}\,dx \int_{-\infty}^{\infty} \cos(30\pi y)e^{-iqy}\,dy.$$

Since

$$\sin\theta \triangleq \frac{e^{i\theta} - e^{-i\theta}}{2i} \quad \text{and} \quad \cos\theta \triangleq \frac{e^{i\theta} + e^{i\theta}}{2}$$

the equation just given becomes

$$F(p,q) = \left\{ \frac{1}{2i}\int_{-\infty}^{\infty} \exp[-i(p-20\pi)x]\,dx - \frac{1}{2i}\int_{-\infty}^{\infty} \exp[-i(p+20\pi)x]\,dx \right\}$$
$$\times \left\{ \frac{1}{2}\int_{-\infty}^{\infty} \exp[-i(q-30\pi)y]\,dy + \frac{1}{2}\int_{-\infty}^{\infty} \exp[-i(q+30\pi)y]\,dy \right\}.$$

This equation can be further reduced to

$$F(p,q) = \frac{1}{4i}[\delta(p-20\pi) - \delta(p+20\pi)][\delta(q-30\pi) + \delta(q+30\pi)]$$
$$= \frac{1}{4i}[\delta(p-20\pi, q-30\pi) - \delta(q+20\pi, q-30\pi)$$
$$+ \delta(p-20\pi, q+30\pi) - \delta(q+20\pi, q+30\pi)],$$

where δ denotes the Dirac delta function, as will be defined in Section 7.3.

EXAMPLE 7.6

Show that the inverse Fourier transformation of Eq. 7.8 is also a linear transformation. For simplicity of notation, let

$$f(x,y) = \mathscr{F}^{-1}[F(p,q)].$$

We assume that

$$f_1(x,y) = \mathscr{F}^{-1}[F_1(p,q)]$$

and

$$f_2(x,y) = \mathscr{F}^{-1}[F_2(p,q)].$$

To show that $\mathscr{F}^{-1}$ is a linear transformation, we have

$$\mathscr{F}^{-1}[C_1F_1(p,q) + C_2F_2(p,q)] = C_1\mathscr{F}^{-1}[F_1(p,q)] + C_2\mathscr{F}^{-1}[F_2(p,q)]$$
$$= C_1f_1(x,y) + C_2f_2(x,y).$$

Since the inverse transformation possesses the homogeneity and the additivity properties, the inverse Fourier transformation is a linear transformation.

EXAMPLE 7.7

Calculate the spatial-frequency content of $f(x, y)$, which is referred to in Example 7.5.

By inspecting the corresponding Fourier spectrum of $F(p, q)$, we can identify the spatial-frequency content of $f(x, y)$ as

$$f_x = 10, \qquad f_y = 15;$$
$$f_x = -10, \qquad f_y = 15;$$
$$f_x = 10, \qquad f_y = -15;$$
$$f_x = -10, \qquad f_y = -15,$$

which represent four discrete spectra points; variables f_x and f_y are the spatial frequencies in cycles per unit distance.

7.3 Dirac Delta Function

We shall now define a two-dimensional Dirac delta function $\delta(x, y)$, as described in the (x, y) spatial domain. Similar to the temporal Dirac delta function $\delta(t)$, $\delta(x, y)$ is defined as

$$\delta(x, y) \triangleq \begin{cases} \infty, & \text{for } x = y = 0 \\ 0, & \text{otherwise} \end{cases} \tag{7.16}$$

and

$$\int\int_{-\infty}^{\infty} \delta(x, y)\, dx\, dy = 1. \tag{7.17}$$

Thus, for $\delta(x - x_0, y - y_0)$, it is defined as

$$\delta(x - x_0, y - y_0) \triangleq \begin{cases} \infty, & \text{for } x = x_0, \quad y = y_0 \\ 0, & \text{otherwise} \end{cases} \tag{7.18}$$

and

$$\int\int_{-\infty}^{\infty} \delta(x - x_0, y - y_0)\, dx\, dy = 1, \tag{7.19}$$

where x_0 and y_0 are arbitrary constants. In other words, $\delta(x - x_0, y - y_0)$ exists only at (x_0, y_0), and has zero value over the x, y plane elsewhere, as sketched in Figure 7.5.

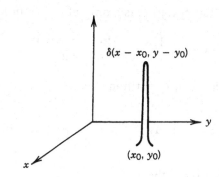

FIGURE 7.5 Sketch of $\delta(x - x_0, y - y_0)$.

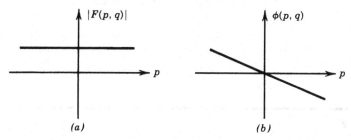

FIGURE 7.6 One-dimensional representation of a Fourier spectrum.

With reference to Eq. 7.9, the Fourier transformation of the delta function can be obtained as follows:

$$F(p, q) = \mathcal{F}[\delta(x - x_0, y - y_0)] = \exp[-i(px_0 + qy_0)]. \qquad (7.20)$$

The amplitude and phase spectra of a delta function can be identified as

$$|F(p, q)| = 1 \qquad (7.21)$$

and

$$\phi(p, q) = -(px_0 + qy_0). \qquad (7.22)$$

Equation 7.21 shows the amplitude spectral distribution, which is extended uniformly with a unit height over the spatial-frequency plane, as can be seen in Figure 7.6a. Equation 7.22 shows a linear-phase spectral distribution over the spatial frequency domain, as shown in Figure 7.6b.

EXAMPLE 7.8

Prove that the following relationship is true:

$$\int\!\!\!\int_{-\infty}^{\infty} \delta(x, y)f(x, y)\, dx\, dy = f(0, 0).$$

Equation 7.16 indicates that $\delta(x, y)$ exists only at $x = 0$ and $y = 0$, that is,

$$\delta(x, y) = \begin{cases} \infty, & \text{for } x = y = 0 \\ 0, & \text{otherwise.} \end{cases}$$

Thus, the integral equation can be written as

$$\int\int_{-\infty}^{\infty} \delta(x, y) f(x, y) \, dx \, dy = f(x, y) \Big|_{\substack{x=0 \\ y=0}} \int\int_{-\epsilon}^{\epsilon} \delta(x, y) \, dx \, dy.$$

Using Eq. 7.17, we have proved that

$$\int\int_{-\infty}^{\infty} \delta(x, y) f(x, y) \, dx \, dy = f(0, 0)$$

is true.

EXAMPLE 7.9

Show that the relationship

$$\int\int_{-\infty}^{\infty} \delta(x - x_0, y - y_0) f(x, y) \, dx \, dy = f(x_0, y_0)$$

holds, where $f(x, y)$ is assumed continuous at (x_0, y_0). By replacing the following independent variables, $x' = x - x_0$ and $y' = y - y_0$, we can write the integration equation as

$$\int\int_{-\infty}^{\infty} \delta(x - x_0, y - y_0) f(x, y) \, dx \, dy = \int\int_{-\infty}^{\infty} \delta(x', y') f(x' + x_0, y' + y_0) \, dx' \, dy'.$$

Since

$$\int\int_{-\infty}^{\infty} \delta(x, y) f(x, y) \, dx \, dy = f(0, 0),$$

the preceding equation is reduced to

$$\int\int_{-\infty}^{\infty} \delta(x - x_0, y - y_0) f(x, y) \, dx \, dy = f(x + x_0, y + y_0) \Big|_{\substack{x=0 \\ y=0}}$$

$$= f(x_0, y_0).$$

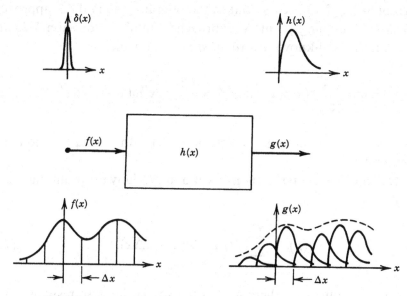

FIGURE 7.7 Response from a linear spatially invariant system.

7.4 Convolution and Correlation

7.4.1 CONVOLUTION

The concept of convolution can be easily illustrated by a block diagram of an input–output linear system, like that shown in Figure 7.7. Let us assume that a band-limited input excitation $f(x)$ is applied at the input end of a linear system. Since $f(x)$ is band-limited, $f(x)$ can be approximated by a sampling functions, such as

$$f(x) = \sum_{n=-N/2}^{N/2} f(x)\delta(x - n\,\Delta x),\qquad (7.23)$$

where N is the total sampling points, $\delta(x)$ denotes the Dirac delta function, and Δx is the sampling distance known as the *Nyquist sampling interval*. The Nyquist sampling interval can be obtained from the sampling frequency, that is,

$$\Delta x = \frac{1}{f_s},\qquad (7.24)$$

where f_s is *Shannon's sampling frequency*. The sampling frequency satisfies the following inequality,

$$f_s \geq 2f_m,\qquad (7.25)$$

where f_m is the highest frequency limit of input excitation $f(x)$.

Referring to Eq. 7.23, we see that as the sampling interval Δx approaches zero, the total sampling point N approaches infinity. Equation 7.23 then converges on the well-known *convolution integral*, which is,

$$f(x) = \lim_{\substack{\Delta x \to 0 \\ N \to \infty}} \sum_{-N/2}^{N/2} f(x)\delta(x - n\,\Delta x) = \int_{-\infty}^{\infty} f(x')\delta(x - x')\,dx'. \qquad (7.26)$$

Thus, we see that convolution of a function $f(x)$ with a delta function yields the function itself.

Since the block diagram system is linear and spatially invariant, the output excitation is

$$g(x) = \sum_{-N/2}^{N/2} f(n\,\Delta x)h(x - n\,\Delta x), \qquad (7.27)$$

as sketched in Figure 7.7, where $h(x)$ is the spatial impulse response of the system. Again, we show that, as $\Delta x \to 0$, $N \to \infty$, Eq. 7.27 will converge on the following convolution integral:

$$g(x) = \lim_{\substack{\Delta x \to 0 \\ N \to \infty}} \sum_{-N/2}^{N/2} f(n\,\Delta x)h(x - n\,\Delta x) = \int_{-\infty}^{\infty} f(x')h(x - x')\,dx'. \qquad (7.28)$$

To simplify this equation, we write

$$g(x) = f(x) * h(x),$$

where the asterisk denotes the convolution operation. Thus, for a linear, spatially invariant system, the output response is the convolution of the input excitation with respect to the spatial impulse response of the system.

A simple example of the convolution integral is shown in the sketches in Figure 7.8. Figures 7.8a and 7.8b show the functions $f(x')$ and $h(x')$ as defined in the x' spatial domain. The function $h(-x')$, defined in Figure 7.8c, is the image of $h(x')$ with respect to the vertical axis at $x' = 0$. The function $h(x - x')$ represents the translation of $h(-x')$ on the x' axis for a given x, as shown in Figure 7.8d. We further note that, since x is a spatial variable, then $h(x - x')$ would translate over the x' domain. The product of $f(x')h(x - x')$, for a given x, is plotted in Figure 7.8e, and the shaded area, bounded by $f(x')h(x - x')$, represents the convolution integral,

$$\int_{-\infty}^{\infty} f(x')h(x - x')\,dx',$$

for a given x.

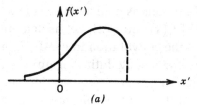

(a)

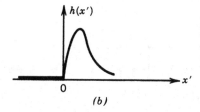

(b)

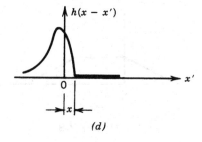

(c)

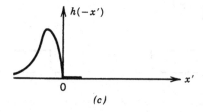

(d)

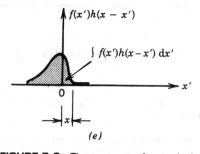

(e)

FIGURE 7.8 The concept of convolution.

EXAMPLE 7.10

Given two arbitrary functions as depicted in Figures 7.9a and 7.9b, evaluate the convolution integral of $f_1(x)$ and $f_2(x)$. Let us demonstrate the usefulness of the graphical approach to the convolution integral. Figures 7.9c and 7.9d show the appropriate function $f_2(x)$ convoluting against $f_1(x)$, and the shaded areas represent the convolution of these two functions for a given x. Function $g(x)$ represents the result of the convolution integral as depicted in Figure 7.9e.

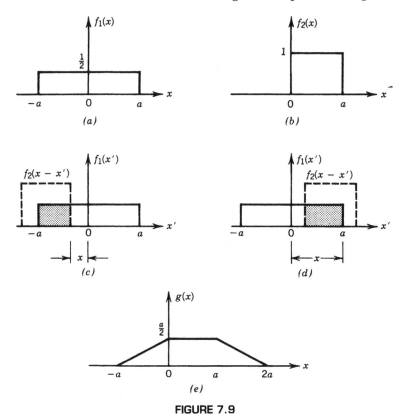

FIGURE 7.9

EXAMPLE 7.11

Show that the convolution of a function $f(x)$ against $\delta(x - x_0)$ is equal to $f(x - x_0)$, that is,

$$\int_{-\infty}^{\infty} f(x')\delta(x' - x + x_0)\, dx' = f(x - x_0).$$

If we replace $x - x_0$ by τ, the convolution integral becomes

$$\int_{-\infty}^{\infty} f(x')\delta(x' - \tau)\, dx'.$$

Since the convolution of any function with a delta function yields the function itself, we conclude that

$$\int_{-\infty}^{\infty} f(x')\delta(x' - \tau)\,dx' = f(\tau) = f(x - x_0).$$

7.4.2 CORRELATION

The concept of correlation can also be interpreted by an input–output linear system, as illustrated in Figure 7.10, where $h'(x) = h(-x)$ represents the spatial impulse response of the system.

Since the output response can be written as

$$g(x) = \sum_{-N/2}^{N/2} f(x)h'(x + n\,\Delta x) = \sum_{-N/2}^{N/2} f(x)h(-x - n\,\Delta x) \qquad (7.29)$$

and then, because $\Delta x \to 0$ and $N \to \infty$, we show that

$$g(x) = \lim_{\substack{\Delta x \to 0 \\ N \to \infty}} \sum_{-N/2}^{N/2} f(x)h(-x - n\,\Delta x) = \int_{-\infty}^{\infty} f(x')h(x + x')\,dx'. \qquad (7.30)$$

The preceding integral equation is known as the *correlation integral* of functions $f(x)$ and $h(x)$. The operation given in Eq. 7.30 can be represented by the sketches in Figure 7.11. Figures 7.11a and 7.11b show the functional distribution of $f(x')$ and $h(x')$, and Figure 7.11c illustrates the translation of $h(x')$ over the x' axis. Figure 7.11d shows the functional distribution of the product of $f(x')$ and $h(x + x')$, and the shaded area in this figure represents the correlation integral for a given x. It is beneficial for the reader to comprehend the concept of autocorrelation, because its operation plays a key role in the extraction of the signal from random noise (e.g., radar detection). To illustrate the autocorrelation operation, we assume that the impulse response in Figure 7.11 is

$$h'(x) = f(-x).$$

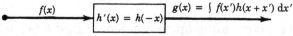

FIGURE 7.10 Convolution operation $h'(x)$ represents the spatial impulse response.

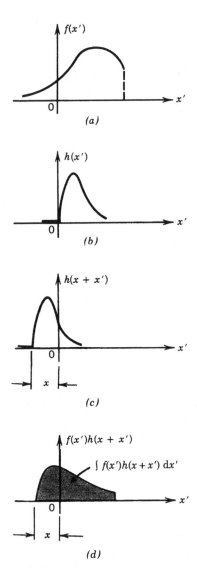

FIGURE 7.11 The concept of correlation.

Then the convolution integral of Eq. 7.30 becomes

$$R_{11}(x) = \int_{-\infty}^{\infty} f(x')f(x'+x)\,dx', \qquad (7.31)$$

where $R_{11}(x)$ denotes the *autocorrelation function*, and the integral is known as the *autocorrelation integral*. To simplify Eq. 7.31, we write it as

$$R_{11}(x) = f(x) \circledast f(x),$$

with $\circledast$ denoting the correlation operation. Similarly, if $h'(x) = u(-x)$, where $u(x)$ is any arbitrary function other than $f(x)$, Eq. 7.30 becomes

$$R_{12}(x) = \int_{-\infty}^{\infty} f(x')u(x'+x)\,dx', \qquad (7.32)$$

which is the *cross-correlation function* of $f(x)$ and $u(x)$; the subscripts 1 and 2 denote the cross-correlation operation. Equation 7.32 is known as the *cross-correlation integral* and can be simplified to read

$$R_{12}(x) = f(x) \circledast u(x). \qquad (7.33)$$

Properties of an Autocorrelation Function
It is worthwhile to state some of the basic properties of an autocorrelation function:

$$R_{11}(x) = R_{11}(-x) \qquad (7.34)$$
$$R_{11}(0) > 0 \qquad (7.35)$$
$$R_{11}(0) \geq R_{11}(x). \qquad (7.36)$$

To show that the autocorrelation function is an even function, we utilize the definition given by Eq. 7.31,

$$R_{11}(-x) = \int_{-\infty}^{\infty} f(x')f(x'-x)\,dx'.$$

By letting $x' - x = \alpha$, we have

$$R_{11}(-x) = \int_{-\infty}^{\infty} f(\alpha+x)f(\alpha)\,d\alpha = R_{11}(x).$$

To show that $R_{11}(0)$ is a positive quantity, we use the definition of $R_{11}(x)$,

$$R_{11}(x) = \int_{-\infty}^{\infty} f(x')f(x'+x)\,dx'.$$

Then, with $x = 0$, we have

$$R_{11}(0) = \int_{-\infty}^{\infty} f^2(x')\,dx'.$$

Since $f^2(x)$ is a non-negative quantity, we conclude that

$$R_{11}(0) \geq 0.$$

To show that $R_{11}(0) \geq R_{11}(x)$ is true, we note that

$$\int_{-\infty}^{\infty} [f(x') - f(x'+x)]^2\,dx' \geq 0.$$

Expanding this expression, we have

$$\int_{-\infty}^{\infty} f^2(x')\,dx' + \int_{-\infty}^{\infty} f^2(x'+x)\,dx' - 2\int_{-\infty}^{\infty} f(x')f(x'+x)\,dx' \geq 0.$$

Since the first two terms have the same value, each being equal to $R_{11}(0)$, it is therefore trivial that

$$R_{11}(0) \geq R_{11}(x).$$

EXAMPLE 7.12

Given two rectangular functions as defined in the Figures 7.12a and 7.12b, evaluate the correlation function between them. We again use the graphical approach to obtain the correlation function.

Figure 7.12c shows that the appropriate function $h(x)$ correlates against $f(x)$, with the shaded area representing the correlation of these two functions for a given x. The term $R(x)$ is the correlation function obtained with this graphical approach, as shown in Figure 7.12d.

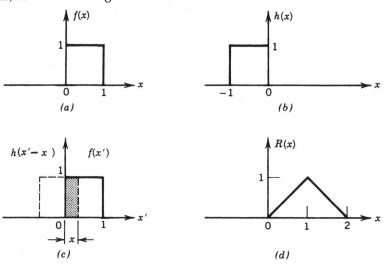

FIGURE 7.12

EXAMPLE 7.13

Given a periodic function, such as

$$f(x) = \cos(px + \phi),$$

where ϕ is an arbitrary phase shift, calculate the corresponding autocorrelation function.

By substituting $f(x)$ into Eq. 7.31, we can obtain the autocorrelation function:

$$R_{11}(x) = \int_{-\infty}^{\infty} \cos(px' + \phi) \cos(px' + \phi + px)\, dx'.$$

But

$$\cos A \cos B = \tfrac{1}{2}[\cos(A + B) + \cos(A - B)].$$

The autocorrelation can therefore be shown as

$$R_{11}(x) = \frac{1}{2}\int_{-\infty}^{\infty} [\cos(2px' + 2\phi + px) + \cos(px)]\, dx'$$
$$= \tfrac{1}{2}\cos px,$$

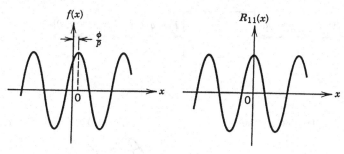

FIGURE 7.13

which is again a periodic function of the same period, as sketched in Figure 7.13. Notice that $R_{11}(x)$ is independent of the phase factor ϕ of $f(x)$.

7.5 Properties of Fourier Transformation

Fourier transformation offers myriad ways for analyzing linear, spatially invariant systems. In this section we describe the basic properties of the Fourier transform that are frequently used in optical engineering.

7.5.1 FOURIER TRANSLATION PROPERTY

If $f(x, y)$ is Fourier transformable, that is,

$$\mathscr{F}[f(x, y)] = F(p, q),$$

then

$$\mathscr{F}[f(x - x_0, y - y_0)] = F(p, q) \exp[-i(px_0 + qy_0)],$$

where x_0 and y_0 are arbitrary real constants. To show that this property holds, we write

$$\mathscr{F}[f(x - x_0, y - y_0)] = \int\int_{-\infty}^{\infty} f(x - x_0, y - y_0) \exp[-i(px + qy)] \, dx \, dy.$$

By substituting the variables $x' = x - x_0$ and $y' = y - y_0$ in the preceding equation, we can show that

$$\mathscr{F}[f(x - x_0, y - y_0)] = \exp[-i(px_0 + qy_0)] \int\int_{-\infty}^{\infty} f(x', y') \exp[-i(px' + qy')] \, dx \, dy'$$

$$= \exp[-i(px_0 + qy_0)]F(p, q).$$

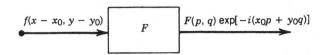

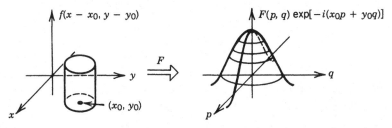

FIGURE 7.14 The spatially invariant property of Fourier transformation.

From this, we see that the translation of an object function $f(x, y)$ in the spatial domain causes a linear phase shift in the Fourier or spatial frequency domain. In other words, for a real-object function $f(x, y)$, the corresponding (magnitude) Fourier transform is spatially invariant in the spatial-frequency plane, as illustrated in Figure 7.14.

EXAMPLE 7.14

Find the Fourier transform of a rectangular pulse of duration d, as shown in Figure 7.15a. If the rectangular function is shifted from the origin to $x = d/2$, as depicted in Figure 7.15b, find the corresponding Fourier transform.

From the definition given by Eq. 7.9, we get the following Fourier transform of Figure 7.15a:

$$
\begin{aligned}
F(p) &= \int_{-\infty}^{\infty} \text{rect}\left(\frac{x}{d}\right) e^{-px} \, dx \\
&= \int_{-d/2}^{d/2} e^{-ipx} \, dx \\
&= \frac{1}{ip}\left[\exp\left(i\frac{pd}{2}\right) - \exp\left(-i\frac{pd}{2}\right)\right] \\
&= d\frac{\sin(pd/2)}{pd/2}.
\end{aligned}
$$

Figure 7.15c shows the distribution of $F(p)$, and the dotted line indicates the magnitude $|F(p)|$.

For the shifted rectangular function shown in Figure 7.15b, the Fourier transform can be shown as

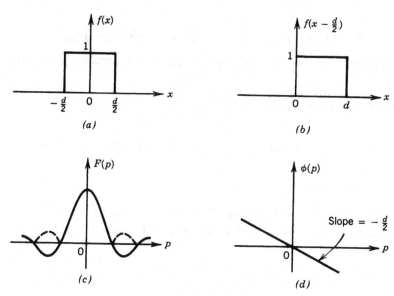

FIGURE 7.15

$$F'(p) = \int_{-\infty}^{\infty} \text{rect}\left(\frac{x - d/2}{d}\right) e^{-ipx}\, dx$$

$$= \int_{0}^{d} e^{-ipx}\, dx$$

$$= \frac{1}{ip}(1 - e^{-ipd})$$

$$= d\frac{\sin(pd/2)}{pd/2}\exp\left(-i\frac{pd}{2}\right) = F(p)\exp\left(-i\frac{pd}{2}\right).$$

From this equation we can see that the amplitude variation is essentially identical to that in Figure 7.15c. There is, however, a linear-phase factor added to this result, as sketched in Figure 7.15d.

7.5.2 RECIPROCAL TRANSLATION PROPERTY

In view of the definition of the Fourier transform given in Eqs 7.8 and 7.9, we note that the inverse Fourier transform is essentially the same as the direct Fourier transform, except with a positive kernel. In other words, if we reverse the positive and negative coordinate axes of (x, y), the inverse Fourier transform is the same as the direct Fourier transform. By virtue of Fourier translation, the reciprocal translation theorem can therefore be stated as follows. If $f(x, y)$ is Fourier-transformable, that is,

$$\mathscr{F}[f(x, y)] = F(p, q),$$

then

$$\mathscr{F}^{-1}[F(p - p_0, q - q_0)] = f(x, y) \exp[i(xp_o + yq_0)].$$

By substituting $p' = p - p_0$ and $q' = q - q_0$ in the preceding equation, we have

$$\mathscr{F}^{-1}[F(p - p_0, q - q_0)]$$

$$= \exp[i(p_0 x + q_0 y)] \frac{1}{4\pi^2} \int\limits_{-\infty}^{\infty} \int F(p', q') \exp[i(p'x + q'y)] \, dp' \, dq'$$

$$= \exp[i(p_0 x + q_0 y)] f(x, y).$$

Thus, we see that the translation of the Fourier spectrum $F(p, q)$ in the spatial-frequency plane would also introduce a linear-phase factor in the object function.

EXAMPLE 7.15

Using Example 7.14, let us shift the Fourier spectrum shown in Figure 7.15c to $p = p_0$, as shown in Figure 7.16a, and then evaluate the corresponding inverse Fourier transform.

The shifted Fourier spectrum can be written as

$$F(p - p_0) = d \frac{\sin[(p - p_0)d/2]}{(p - p_0)d/2}.$$

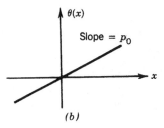

$|f(x)|$

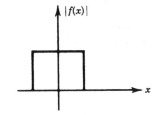

$F(p - p_0)$

(a)

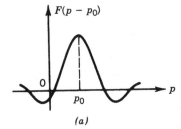

$\theta(x)$

Slope $= p_0$

(b)

FIGURE 7.16

The inverse Fourier transform can be obtained by

$$f(x) = \frac{d}{2\pi} \int_{-\infty}^{\infty} \frac{\sin[(p-p_0)d/2]}{(p-p_0)d/2} e^{ipx}\, dp.$$

By letting $p - p_0 = p'$, we have

$$f(x) = \frac{d}{2\pi} e^{ip_0 x} \int_{-\infty}^{\infty} \frac{\sin(p'd/2)}{p'd/2} e^{ip'x}\, dp'$$

$$= e^{ip_0 x}\, \text{rect}\!\left(\frac{x}{d}\right),$$

which is a rectangular function of finite duration d multiplexed with a linear-phase factor $\theta = p_0 x$. Sketches of $f(x)$ are shown in Figure 7.16b.

7.5.3 SCALE CHANGES OF FOURIER TRANSFORMS

If $f(x, y)$ is Fourier-transformable, that is,

$$\mathscr{F}[f(x, y)] = F(p, q),$$

then

$$\mathscr{F}[f(ax, by)] = \frac{1}{ab} F\!\left(\frac{p}{a}, \frac{q}{b}\right),$$

where a and b are arbitrary positive constants. By direct substitution of the direct Fourier transformation of Eq. 7.19, we have

$$\mathscr{F}[f(ax, by)] = \int\!\!\!\int_{-\infty}^{\infty} f(ax, by) \exp[-i(px + qy)]\, dx\, dy$$

$$= \frac{1}{ab} \int\!\!\!\int_{-\infty}^{\infty} f(ax, by) \exp\!\left[-i\!\left(\frac{p}{a}ax + \frac{q}{b}bx\right)\right] d(ax)\, d(by)$$

$$= \frac{1}{ab} F\!\left(\frac{p}{a}, \frac{q}{b}\right).$$

From this result we see that a scale reduction of $f(x, y)$ in the spatial domain will enlarge the Fourier transform $F(p, q)$ in the spatial frequency domain. The overall amplitude spectrum of $F(p, q)$ is also proportionally reduced by a factor of $1/ab$.

EXAMPLE 7.16

Given a set of rectangular functions with various durations, as depicted in the left-hand column of Figure 7.17, show that alternating the duration will also affect the height and width of the Fourier spectra.

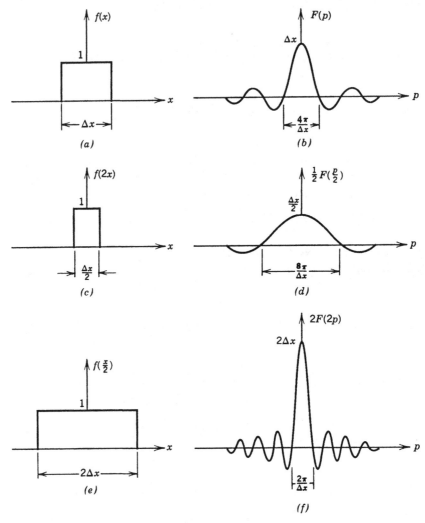

FIGURE 7.17

The Fourier transform of Figure 7.17a can be shown as

$$F(p) = \int_{-\infty}^{\infty} \text{rect}\left(\frac{x}{\Delta x}\right) e^{-ipx}\, dx$$

$$= \Delta x \frac{\sin[(p/2)\Delta x]}{(p/2)\Delta x},$$

which is sketched in Figure 7.17b.

Similarly, the Fourier transforms shown in Figures 7.17c and 7.17e respectively, can be written as

$$F_1(p) = \int_{-\infty}^{\infty} \text{rect}\left(\frac{x}{\Delta x/2}\right) e^{-ipx}\, dx$$

$$= \frac{\Delta x}{2} \frac{\sin[(p/4)\,\Delta x]}{(p/4)\,\Delta x} = \frac{1}{2} F\left(\frac{p}{2}\right)$$

and

$$F_2(p) = \int_{-\infty}^{\infty} \mathrm{rect}\left(\frac{x}{2\,\Delta x}\right) e^{-ipx}\, dx$$

$$= 2\,\Delta x \frac{\sin(p\,\Delta x)}{(p\,\Delta x)} = 2F(2p).$$

The sketches of $F_1(p)$ and $F_2(p)$ are shown in Figures 7.17d and 7.17f. By comparing these figures with those in the right-hand column, we see that the spectral width is inversely proportional to the duration, but that the height is proportional to the width of the rectangular phase.

7.5.4 CONVOLUTION PROPERTY

If $f_1(x, y)$ and $f_2(x, y)$ are Fourier-transformable, that is,

$$\mathcal{F}[f_1(x, y)] = F_1(p, q)$$

and

$$\mathcal{F}[f_2(x, y)] = F_2(p, q),$$

then

$$\mathcal{F}\left[\int\!\!\int_{-\infty}^{\infty} f_1(x, y) f_2(\alpha - x, \beta - y)\, dx\, dy\right] = F_1(p, q) F_2(p, q).$$

Since the convolution integral is a function of α and β variables,

$$C(\alpha, \beta) = \int\!\!\int_{-\infty}^{\infty} f_1(x, y) f_2(\alpha - x, \beta - y)\, dx\, dy,$$

the Fourier transform of $C(\alpha, \beta)$ can be written as

$$\mathcal{F}\left[\int\!\!\int_{-\infty}^{\infty} f_1(x, y) f_2(\alpha - x, \beta - y)\, dx\, dy\right] = \int\!\!\int_{-\infty}^{\infty} f_1(x, y)\mathcal{F}[f_2(\alpha - x, \beta - y)]\, dx\, dy.$$

By substituting the definition of the Fourier transform given in Eq. 7.9 and letting $\alpha - x = \alpha'$ and $\beta - y = \beta'$, we can write this equation as

$$\int\!\!\int_{-\infty}^{\infty} f_1(x, y) \left\{\int\!\!\int_{-\infty}^{\infty} f_2(\alpha - x, \beta - y) \exp[-i(\alpha p + \beta q)]\, d\alpha\, d\beta\right\} dx\, dy$$

$$= \int\!\!\int_{-\infty}^{\infty} f_1(x, y) \left\{\exp[-i(px + qy)] \int\!\!\int_{-\infty}^{\infty} f_2(\alpha', \beta') \exp[-i(\alpha'p + \beta'q)]\, d\alpha'\, d\beta'\right\} dx\, dy$$

$$= \int\!\!\int_{-\infty}^{\infty} f_1(x, y) \exp[-i(px + qy)]\, dx\, dy\, F_2(p, q)$$

$$= F_1(p, q) F_2(p, q).$$

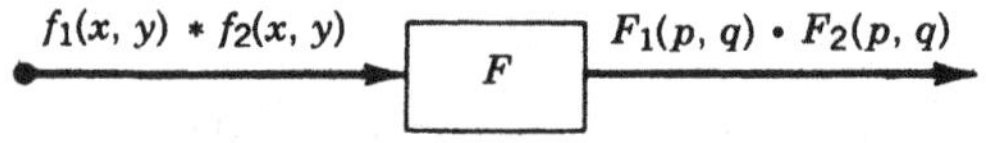

FIGURE 7.18 The convolution property of Fourier transformation.

In other words, the Fourier transform of $f_1(x, y)$ convolved with $f_2(x, y)$ equals the product of their Fourier transformations. For convenience of notation, we write the convolution integral as

$$\iint\limits_{-\infty}^{\infty} f_1(x, y) f_2(\alpha - x, \beta - y) \, dx \, dy = f_1(x, y) * f_2(x, y),$$

where the asterisk represents the convolution operation. According to the convolution property, the preceding equation can be written in the following simplified form,

$$\mathscr{F}[f_1(x, y) * f_2(x, y)] = F_1(p, q) F_2(p, q),$$

which is illustrated by the block diagram of Figure 7.18.

EXAMPLE 7.17

Given the following Fourier-transformable functions,

$$f_1(x) = \cos(p_0 x)$$

and

$$f_2(x) = \frac{p_1}{\pi} \frac{\sin(p_1 x)}{p_1 x}, \quad p_1 \gg p_0,$$

find the Fourier transform of $f_1(x)$ convolved with $f_2(x)$.

The Fourier transform of $f_1(x)$ can be obtained, using the definition given by Eq. 7.9:

$$F_1(p) = \int_{-\infty}^{\infty} \cos(p_0 x) e^{-ipx} \, dx.$$

Since

$$\cos p_0 x = \tfrac{1}{2}(e^{-ip_0 x} + e^{-ip_0 x}),$$

then

$$F_1(p) = \frac{1}{2} \int_{-\infty}^{\infty} (e^{ip_0 x} + e^{-ip_0 x}) e^{-ipx} \, dx$$
$$= \tfrac{1}{2}\delta(p - p_0) + \tfrac{1}{2}\delta(p - p_0).$$

Similarly,

$$F_2(p) = \frac{P_1}{\pi} \int_{-\infty}^{\infty} \frac{\sin(P_1 x)}{P_1 x} e^{-ipx} \, dx$$
$$= \text{rect}\left(\frac{p}{2p_1}\right).$$

Thus we have

$$\mathscr{F}[f_1(x) * f_2(x)] = \tfrac{1}{2}[\delta(p - p_0) + \delta(p + p_0)] \, \text{rect}\left(\frac{p}{2p_1}\right).$$

7.5.5 CROSS-CORRELATION PROPERTY

If $f_1(x, y)$ and $f_2(x, y)$ are Fourier-transformable, that is,

$$\mathscr{F}[f_1(x, y)] = F_1(p, q)$$

and

$$\mathscr{F}[f_2(x, y)] = F_2(p, q),$$

then

$$\mathscr{F}\left[\iint_{-\infty}^{\infty} f_1^*(x, y) f_2(x + \alpha, y + \beta) \, dx \, dy\right] = F_1^*(p, q) F_2(p, q),$$

where the asterisk represents the complex conjugate.

Again, we note that the cross-correlation integral is a function of the α and β variables,

$$R_{12}(\alpha, \beta) = \iint_{-\infty}^{\infty} f_1^*(x, y) f_2(x + \alpha, y + \beta) \, dx \, dy,$$

and that the Fourier transform of $R_{12}(\alpha, \beta)$ is operating with the α and β variables. It is therefore trivial that

$$\mathscr{F}\left[\iint_{-\infty}^{\infty} f_1^*(x, y) f_2(x + \alpha, y + \beta) \, dx \, dy\right]$$
$$= \iint_{-\infty}^{\infty} f_1^*(x, y) \mathscr{F}[f_2(x + \alpha, y + \beta)] \, dx \, dy.$$

If we let $\alpha' = x + \alpha$ and $\beta' = y + \beta$, then by the definition of Fourier transformation. we have

$$\int\limits_{-\infty}^{\infty}\!\!\int f_1^*(x,y)\left\{\int\limits_{-\infty}^{\infty}\!\!\int f_2(x+\alpha,y+\beta)\exp[-i(\alpha p+\beta q)]\,d\alpha\,d\beta\right\}dx\,dy$$

$$=\int\limits_{-\infty}^{\infty}\!\!\int f_1^*(x,y)\exp[i(px+qy)]$$

$$\times\left\{\int\limits_{-\infty}^{\infty}\!\!\int f_2(\alpha',\beta')\exp[-i(\alpha'p+\beta'q)]\,d\alpha'\,d\beta'\right\}dx\,dy$$

$$=\int\limits_{-\infty}^{\infty}\!\!\int f_1^*(x,y)\exp[i(px+qy)]\,dx\,dy\,F_2(p,q)$$

$$=F_1^*(p,q)F_2(p,q).$$

EXAMPLE 7.18

Given the Fourier-transformable functions

$$f_1(x)=\sin(p_0 x)$$

and

$$f_2(x)=\text{rect}\left(\frac{x-\Delta x}{\Delta x}\right),$$

find the Fourier transform of $f_1(x)$ cross-correlated with $f_2(x)$.
The Fourier transformations of $f_1(x)$ and $f_2(x)$ can be shown, respectively, as

$$F_1(p)=\int_{-\infty}^{\infty}\sin(p_0 x)e^{-ipx}\,dx$$

$$=\frac{1}{2i}\int_{-\infty}^{\infty}(e^{ip_0 x}-e^{-ip_0 x})e^{-ipx}\,dx$$

$$=\frac{1}{2i}[\delta(p-p_0)-\delta(p+p_0)]$$

and

$$F_2(p)=\Delta x\frac{\sin[(p/2)\Delta x]}{(p/2)\Delta x}e^{-ip\Delta x}.$$

The Fourier transform of their cross-correlation is therefore

$$\mathscr{F}[f_1(x)\circledast f_2(x)]=F_1^*(p)F_2(p)$$

$$=\tfrac{1}{2}[\delta(p-p_0)-\delta(p+p_0)]\,\Delta x\frac{\sin[(p/2)\,\Delta x]}{(p/2)\,\Delta x}\exp[-i(p\,\Delta x+\pi/2)].$$

7.5.6 AUTOCORRELATION PROPERTY

If $f(x, y)$ is Fourier-transformable, that is,

$$\mathscr{F}[f(x, y)] = F(p, q),$$

then

$$\mathscr{F}\left[\int\int_{-\infty}^{\infty} f^*(x, y)f(x + \alpha, y + \beta)\,dx\,dy\right] = |F(p, q)|^2,$$

where the asterisk represents the complex conjugate. This equation is written more simply as

$$\mathscr{F}[R_{11}(\alpha, \beta)] = \mathscr{F}[f^*(x, y) \circledast f(x, y)] = |F(p, q)|^2,$$

where $R_{11}(\alpha, \beta) = f^*(x, y) \circledast f(x, y)$ is known as the *autocorrelation function*, and $\circledast$ denotes the correlation operation. Conversely, we have

$$\mathscr{F}^{-1}[|F(p, q)|^2] = R_{11}(\alpha, \beta).$$

In other words, the autocorrelation function and the power spectral density are the Fourier transforms of each other. This result is also known as the *Wiener–Khinchine theorem*. The proof of this theorem is similar to that of the cross-correlation property, which can be shown as

$$\mathscr{F}\left[\int\int_{-\infty}^{\infty} f^*(x, y)f(x + \alpha, y + \beta)\,dx\,dy\right]$$

$$= \int\int_{-\infty}^{\infty} f^*(x, y)\mathscr{F}[f(x + \alpha, y + \beta)]\,dx\,dy$$

$$= \int\int_{-\infty}^{\infty} f^*(x, y)F(p, q)\exp[i(px + qy)]\,dx\,dy$$

$$= F^*(p, q)F(p, q) = |F(p, q)|^2.$$

EXAMPLE 7.19

Find the power spectral density and the autocorrelation function of a shifted rectangular pulse, as shown in Figure 7.19. The Fourier transform of $f(x)$ is

$$F(p) = \int_{-\infty}^{\infty} \text{rect}\left(\frac{x - x_0}{\Delta x}\right)e^{-ipx}\,dx$$

$$= \Delta x \frac{\sin[(p/2)\,\Delta x]}{(p/2)\,\Delta x}e^{-ipx_0}.$$

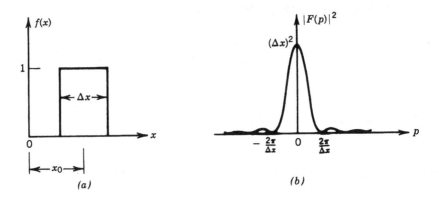

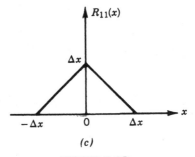

(c)

FIGURE 7.19

The power spectral density can be written as

$$|F(p)|^2 = F^*(p)F(p)$$
$$= \left\{ \Delta x \frac{\sin[(p/2)\,\Delta x]}{(p/2)\,\Delta x} \right\}^2,$$

which is sketched in Figure 7.19b.

The autocorrelation function can be found by applying the Wiener–Khinchine theorem,

$$R_{11}(x) = \mathcal{F}^{-1}[F(p,q)^2]$$
$$= \frac{(\Delta x)^2}{2\pi} \int_{-\infty}^{\infty} \left\{ \frac{\sin[(p/2)\,\Delta x]}{(p/2)\,\Delta x} \right\}^2 e^{-ipx}\,\mathrm{d}p$$
$$= \begin{cases} x + \Delta x, & -\Delta x \le x < 0 \\ -x + \Delta x, & 0 < x \le \Delta x \\ 0, & \text{otherwise,} \end{cases}$$

which is sketched in Figure 7.19c.

7.5.7 CONSERVATION PROPERTY

If $f(x, y)$ is Fourier transformable, that is,

$$\mathscr{F}[f(x, y)] = F(p, q),$$

then

$$\int\!\!\int_{-\infty}^{\infty} |f(x, y)|^2 \, dx \, dy = \frac{1}{4\pi^2} \int\!\!\int_{-\infty}^{\infty} |F(p, q)|^2 \, dp \, dq.$$

This result is also known as *Parseval's theorem*. In othe words, Parseval's theorem implies the conservation of energy. To show that this property holds, we write

$$\int\!\!\int_{-\infty}^{\infty} |f(x, y)|^2 \, dx \, dy = \int\!\!\int_{-\infty}^{\infty} f(x, y) f^*(x, y) \, dx \, dy$$

$$= \int\!\!\int_{-\infty}^{\infty} dx \, dy \left\{ \frac{1}{4\pi^2} \int\!\!\int_{-\infty}^{\infty} F(p', q') \exp[i(xp' + yq')] \, dp' \, dq' \right\}$$

$$\times \left\{ \frac{1}{4\pi^2} \int\!\!\int_{-\infty}^{\infty} F^*(p'', q'') \exp[-i(xp'' + yq'')] \, dp'' \, dq'' \right\}$$

$$= \frac{1}{4\pi^2} \int\!\!\int_{-\infty}^{\infty} F(p', q') \, dp' \, dq' \int\!\!\int_{-\infty}^{\infty} F^*(p'', q'') \, dp'' \, dq''$$

$$\times \left\{ \frac{1}{4\pi^2} \int\!\!\int_{-\infty}^{\infty} \exp\{i[x(p' - p'') + y(q' - q'')]\} \, dx \, dy \right\}$$

$$= \frac{1}{4\pi^2} \int\!\!\int_{-\infty}^{\infty} F(p', q') \, dp' \, dq' \int\!\!\int_{-\infty}^{\infty} F^*(p'', q'') \, dp'' \, dq'' \delta(p' - p'', q' - q'')$$

$$= \frac{1}{4\pi^2} \int\!\!\int_{-\infty}^{\infty} |F(p, q)|^2 \, dp \, dq.$$

EXAMPLE 7.20

By referring to function $f(x)$ and its Fourier transform $F(p)$ in Example 7.19, we can show that

$$\int_{-\infty}^{\infty} |f(x)|^2 \, dx = \int_{-\infty}^{\infty} \left| \text{rect}\left(\frac{x - x_0}{\Delta x}\right) \right|^2 \, dx$$

$$= (\Delta x)^2$$

and

$$\frac{1}{2\pi}\int_{-\infty}^{\infty}|F(p)|^2\,dp = \frac{(\Delta x)^2}{2\pi}\int_{-\infty}^{\infty}\frac{\sin^2[(p/2)\,\Delta x]}{[(p/2)\,\Delta x]^2}\,dp$$
$$= (\Delta x)^2.$$

7.5.8 SYMMETRIC PROPERTIES

Assume that $f(x,y)$ is Fourier transformable, that is,

$$\mathcal{F}[f(x,y)] = F(p,q).$$

(1) If $f(x,y)$ is a real function over (x,y) domain,

$$f^*(x,y) = f(x,y),$$

then

$$F^*(-p,-q) = F(p,q).$$

To show that this property holds, we take the conjugate of the Fourier transform,

$$F^*(p',q') = \iint\limits_{-\infty}^{\infty} f^*(x,y)\exp[i(p'x + q'y)]\,dx\,dy.$$

If we let $p' = -p$ and $q' = -q$, we have

$$F^*(-p,-q) = \iint\limits_{-\infty}^{\infty} f^*(x,y)\exp[-i(px + qy)]\,dx\,dy.$$

Since $f(x,y)$ is real, that is, $f^*(x,y) = f(x,y)$, we see that
$$F^*(-p,-q) = F(p,q).$$

(2) If $f(x,y)$ is a real and even function,
$$f^*(x,y) = f(x,y) = f(-x,-y),$$

then $F(p,q)$ is also a real and even function,
$$F(p,q) = F^*(p,q) = F^*(-p,-q).$$

To prove that this property holds, we show that

$$F(p,q) = \iint\limits_{-\infty}^{\infty} f(x,y)\exp[-i(px + qy)]\,dx\,dy$$

$$= \iint\limits_{-\infty}^{\infty} f(-x,-y)\exp[-i(px + qy)]\,dx\,dy.$$

By letting $x = -x'$ and $y = -y'$, we can write the preceding equation as

$$F(p,q) = \frac{1}{4\pi^2} \int\limits_{-\infty}^{\infty}\int f(x',y') \exp[i(px' + qy')]\,dx'\,dy'.$$

Since $f(x',y')$ is real, that is, $f(x',y') = f^*(x',y')$, we see that

$$F(p,q) = F^*(p,q) = F^*(-p,-q).$$

(3) If $f(x,y)$ is a real and odd function,

$$f(x,y) = f^*(x,y) = -f(-x,-y),$$

then

$$F(p,q) = -F(-p,-q) = F^*(-p,-q).$$

The proof of this property can be obtained in

$$F(p,q) = \frac{1}{4\pi^2} \int\limits_{-\infty}^{\infty}\int f(x,y) \exp[-i(px + qy)]\,dx\,dy$$

$$= \frac{1}{4\pi^2} \int\int -f(-x,-y) \exp[-i(px + qy)]\,dx\,dy.$$

By letting $x' = -x$ and $y' = -y$, we have

$$F(p,q) = -\frac{1}{4\pi^2} \int\limits_{-\infty}^{\infty}\int f(x',y') \exp[+i(px' + qy')]\,dx\,dy$$

$$= -F(-p,-q).$$

And assuming that $f(x,y)$ is real, we can see that

$$F(p,q) = -F(-p,-q) = F^*(-p,-q).$$

REFERENCES

1. A. PAPOULIS, *The Fourier Integral and Its Applications*, McGraw-Hil, New York, 1962.
2. D. K. CHENG, *Analysis of Linear System*, Addison-Wesley, Reading, MA, 1959.
3. F. T. S. YU, *Optical Information Processing*, Wiley-Interscience, New York, 1983, Chapter 1.

PROBLEMS

7.1   If the transfer function of a linear spatially invariant system is

$$H(p) = \frac{ip}{1 + ip},$$

determine the spatial impulse response.

7.2 Show that the additivity and homogeneity properties in Problem 7.1 hold.

7.3 If the arbitrary input excitation in Problem 7.1 is translated to $x = x_0$, that is, $f(x - x_0)$, show that the output excitation is $g(x - x_0)$.

7.4 If for a second-order nonlinear phase transfer function of a system

$$H(p) = Ae^{-ip^2},$$

an input excitation is given by

$$f(x) = \cos(p_1 x) + \cos(p_2 x).$$

Calculate the output response.

7.5 Assume an ideal linear-phase system for which the transfer function is

$$H(p) = \begin{cases} e^{-i\alpha p}, & |p| \le |p_c| \\ 0, & |p| > |p_c|, \end{cases}$$

where α is an arbitrary constant and p_c is the angular cut-off spatial frequency. If the input excitation is

$$f(x) = \delta(x) + \delta(x - x_0),$$

calculate the output response $g(x)$.

7.6 We assume that the input excitation in Problem 7.5 is a rectangular function of finite duration Δx. Sketch the output responses for such cases as

$$\Delta x \ll \frac{\pi}{p_c}, \qquad \Delta x = \frac{\pi}{p_c}, \qquad \text{and} \quad \Delta x \gg \frac{\pi}{p_c}.$$

7.7 Determine the Fourier transforms of the following functions.

(a) $f(x) = \sin(p_1 x) + \cos(p_2 x)$,

where p_1 and p_2 are arbitrary angular spatial frequencies.

(b) $f(x) = e^{ip_0 x} + e^{-ip_0 x}$.

(c) $f(x) = \sum_{n=-\infty}^{\infty} \delta(x - n\alpha)$,

where α is an arbitrary constant.

7.8 Determine the Fourier transforms of

$$f_1(x) = \cos(p_1 x)$$

and

$$f_2(x) = \sin(p_1 x).$$

Using the results, show that the Fourier transformation is a linear transformation.

7.9   Find the Fourier transform of $f(x)$ defined by

$$f(x) = \begin{cases} e^{-\alpha x}, & x > 0 \\ 0, & x < 0, \end{cases}$$

where $\alpha > 0$.

7.10   Using Problem 7.9, show that the following Fourier transform is shift invariant.

$$|\mathscr{F}[f(x = x_0)]| = |\mathscr{F}[f(x)]|.$$

7.11   Show that multiplying $f(x)$ by $\cos(p_o x)$ translates its spectrum to $\pm p_0$ in the angular spatial frequency domain p.

7.12   Given a band-limited signal $f(x)$, in which the upper angular spatial-frequency limit is p_m. If $f(x)$ is sampled by a periodic impulse function of period d, show that a periodic sequence of the spectra will be generated in the spatial frequency domain of p. Then determine what sampling interval d is required so that the spectra do not overlap.

7.13   We assume that the spatial impulse response of a linear spatially invariant system is given by

$$h(x) = \frac{p_c}{\pi} \frac{\sin(p_c x)}{p_c x}.$$

Find the output response, with an implicit integral form produced by rectangular excitation, such as

$$f(x) = \text{rect}\left(\frac{x}{\Delta x}\right) = \begin{cases} 1, & |x| \leq \dfrac{\Delta x}{2} \\ 0, & \text{otherwise}, \end{cases}$$

where Δx is the pulse duration.

     If the pulse duration of the input excitation is very large compared to the pulse width of the impulse response, that is, if

$$\Delta x \gg \frac{2\pi}{p_c},$$

determine the asymptotic output response.

7.14   Show that the inverse Fourier transform of $F^*(p)$ is $f^*(-x)$, or

$$\mathscr{F}^{-1}[F^*(p)] = f^*(-x).$$

7.15   Let us assume that the transfer function of a linear spatially invariant system is given by

$$H(p) = F^*(p),$$

which is equal to the complex conjugate of the input Fourier transform.

(a) Find the corresponding output response $g(x)$.

    (b) If the input excitation is an arbitrary function $u(x)$, where $u(x) \neq f(x)$, evaluate the output response.

    (c) State the functional operations of parts **a** and **b** and the significance of the Wiener–Khinchine theorem.

7.16   Find the Fourier transform of a rectangular pulse train of duration Δx and period T.

DIFFRACTION $\boxed{8}$

Suppose that a point source of light is illuminating an opaque object, casting a shadow of the object on an observation screen. If we examine the sharpness of the shadow, we see that the edge of the shadow fades gradually over a short distance rather than changing abruptly. Furthermore, if the point source of light is monochromatic, there will be narrow bands of light, called *fringes*, parallel to the edges of the geometrical shadow. It is obvious that the light that passes the edges of the object deviates from a straight-line propagation. This phenomenon is called the *diffraction* of light.

Historically, it was the observation of diffraction that led to general acceptance of the wave theory of light. Wave theory shows that the magnitude of the diffraction effect, that is, the angle of deviation from straight-line propagation, is directly proportional to the wavelength. Thus, with a sound wave we are usually not conscious of a shadow at all, because it has a long wavelength and the diffraction angle is large. Even with sound, however, shadows can be demonstrated if supersonic frequencies are used. Since the wavelengths of visible light are extremely short, the diffraction angle is also small. Therefore, the straight-line propagation assumed in geometrical optics is only the limit approached as the wavelength approaches zero.

8.1   Fraunhofer and Fresnel Diffractions

Above, we have spoken of the diffraction of light as it passes the edge of an obstacle. Diffraction may be treated more simply, however, if we consider the light as passing through one or more small apertures in a diffraction screen.

It is customary to divide diffraction into two types, each of which has been named after one of the early investigators of diffraction. If the source and viewing point are located at effectively infinite distances relative to the

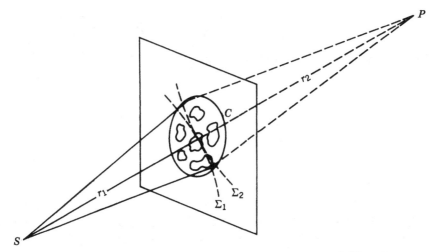

FIGURE 8.1 Geometry for defining Fraunhofer and Fresnel diffraction.

aperture in the diffracting screen, the diffraction is known as *Fraunhofer diffraction*. On the other hand, if either the source or the viewing screen is located at a finite distance relative to the aperture, the diffraction is known as *Fresnel diffraction*. The boundary between these two types is somewhat arbitrary and depends on the accuracy desired. In most instances the Fraunhofer method can be used if the difference in the distances does not exceed one-twentieth of a wavelength. Of course, Fraunhofer diffraction may be achieved without the source being at a great physical distance, if a collimating lens is used to make the rays of light from the source nearly parallel.

Figure 8.1 illustrates the preceding definitions. A point source of monochromatic light is at S, the viewing point is at P, and between them is an opaque screen with a finite number of apertures. A circle C, which is as small as possible while still enclosing all the apertures, has been drawn on the plane of the screen. Circle C is the base of cones that have S and P as vertices. Spherical surfaces Σ_1 and Σ_2, with S and P as their centers, are at the bases of radii r_1 and r_2, which are the shortest distances from S and P to the circle C. If the longest distance from C to Σ_1 and from C to Σ_2 is not more than one-twentieth of the wavelength of the light used, the diffraction is Fraunhofer, and the light falling on the observing screen forms a *Fraunhofer diffraction pattern*.

On the other hand, because of the large size of C, or the shortness of the distance to S or P, the distance between C and Σ_1 or Σ_2 is greater than one-twentieth of the wavelength, the diffraction is Fresnel, and a *Fresnel diffraction pattern* is produced.

The radius of the circle C is denoted by ρ in Figure 8.2, l is the shortest distance from S to the screen, and Δl is the greatest separation between

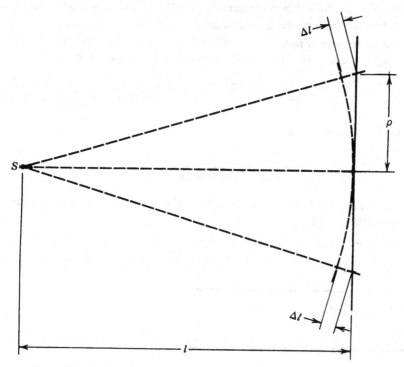

FIGURE 8.2 Determining the distance for Fraunhofer diffraction.

sphere and screen. According to the definition of Fraunhofer diffraction, Δl must be a small fraction of the wavelength. However, ρ may be many wavelengths long (as it can also be in Fresnel diffraction). In the right triangle in Figure 8.2, we have

$$(l + \Delta l)^2 = l^2 + \rho^2, \tag{8.1}$$

and, because of the small size of $(\Delta l)^2$ in comparison with the other quantities, we may make the approximation

$$l \simeq \frac{\rho^2}{2\,\Delta l}. \tag{8.2}$$

EXAMPLE 8.1

Let us assume that a diffraction screen, which contains a number of open apertures, is illuminated by a monochromatic point source of visible green light of wavelength 5×10^{-5} cm. If the open apertures in the diffraction screen are encompassed within a circle of diameter 2 cm, calculate the length of the separation between the light source and diffraction screen (as well as the distance between the diffraction screen and the viewing screen) that is necessary to achieve the Fraunhofer diffraction condition.

To calculate the Fraunhofer diffraction condition, we determine that $\rho = 1\,\text{cm}$ and $\lambda = 5 \times 10^{-5}\,\text{cm}$. By letting $\Delta l = \lambda/20$, we have

$$\Delta l = \frac{5 \times 10^{-5}}{20} = 2.5 \times 10^{-6}\,\text{cm}.$$

Applying Eq. 8.2, we obtain

$$l \simeq \frac{\rho^2}{2\,\Delta l} = \frac{1}{2(2.5 \times 10^{-6})} = 2\,\text{km}.$$

The light source should be about 2 km from the diffraction screen to achieve the Fraunhofer diffraction condition.

8.2 The Fresnel–Kirchhoff Integral

The Fresnel–Kirchhoff integral can generally be derived from the *scalar wave theory* by the application of Green's theorem. However, this approach is rather tedious and mathematically involved. So in this section we shall approach the Fresnel–Kirchhoff integral by *simple linear system theory*.

According to Huygens' principle in Section 1.2, the complex amplitude observed from a point p' of the coordinate system $\sigma(\alpha, \beta, \gamma)$, caused by a monochromatic light source located in another coordinate system $\rho(x, y, z)$, as shown in Figure 8.3, may be calculated by assuming that each point of the light source is an infinitesimal spherical radiator. Thus, the complex light amplitude $h_l(x, y, z)$ contributed by a point p in the ρ coordinate system must come from an unpolarized monochromatic point source, for which

$$h_l(x, y, z) = -\frac{i}{\lambda r} e^{i(kr - \omega t)}, \tag{8.3}$$

where λ is the wavelength, $k = 2\pi/\lambda$ is the wavenumber, and ω is the angular time frequency of the point source p, respectively, and r is the distance between the point source p and the point of observation p', which can be written as

$$r = [(l + \gamma - z)^2 + (\alpha - x)^2 + (\beta - y)^2]^{1/2}. \tag{8.4}$$

If the separation l of the two coordinate systems is assumed to be large compared with the regions of interest in the ρ and σ coordinate systems, then the r in the denominator of Eq. 8.3 can be approximated by l, and that in the exponent is replaced by

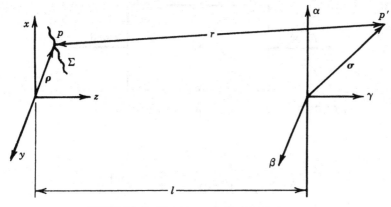

FIGURE 8.3 The Fresnel–Kirchhoff theory.

$$r \simeq (l + \gamma - z) + \frac{(\alpha - x)^2}{2l} + \frac{(\beta - y)^2}{2l}. \tag{8.5}$$

Therefore, Eq. 8.3 can be written as

$$h_l(\alpha - x, \beta - y, \gamma - z) \simeq -\frac{i}{\lambda l} \exp\left\{ ik \left[l + \gamma - z + \frac{(\alpha - x)^2}{2l} + \frac{(\beta - y)^2}{2l} \right] \right\}, \tag{8.6}$$

where the time-dependent part of the exponent, ωt, has been dropped for convenience. Since Eq. 8.6 represents the free-space radiation from a mono-chromatic point source, it is known as the free-space or *spatial impulse response*. In other words, the complex amplitude produced at the σ coordinate system by a monochromatic radiating surface located in the ρ coordinate system can be written in the following abbreviated form:

$$g(\sigma) = \iint_\Sigma f(\rho) h_l(\sigma - \rho; k) \, d\Sigma, \tag{8.7}$$

which is essentially a two-dimensional convolution integral, where $f(\rho)$ is the complex light field of the monochromatic radiating surface, Σ denotes the surface integral, and $d\Sigma$ is the incremental surface element. We note that the convolution integral given by Eq. 8.7 is called the Fresnel–Kirchhoff integral.

For simplicity, we assume that a complex monochromatic radiating field $f(x, y)$ is distributed over the x, y plane. The complex light disturbances at the α, β coordinate plane can be obtained using the following convolution integral:

$$g(\alpha, \beta) = \iint_{-\infty}^{\infty} f(x, y) h_l(\alpha - x, \beta - y) \, dx \, dy, \tag{8.8}$$

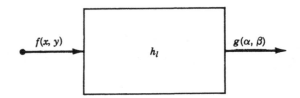

FIGURE 8.4 A linear system representation of Eq. 8.8.

where

$$h_l(x, y) = C \exp\left[i\frac{k}{2l}(x^2 + y^2)\right] \tag{8.9}$$

is the spatial impulse response between the spatial coordinate systems (x, y) and (α, β), and

$$C = -\frac{i}{\lambda l}e^{(ikl)}$$

is a complex constant. Equation 8.8 can be represented by the block diagram shown in Figure 8.4.

In addition, if the complex light disturbances at (α, β) are known, the monochromatic radiating field of $f(x, y)$ can be determined in a similar manner,

$$f(x, y) = \int\int_{-\infty}^{\infty} g(\alpha, \beta)h_l^*(x - \alpha, y - \beta)\, d\alpha\, d\beta, \tag{8.10}$$

where the superscript asterisk denotes the complex conjugate,

$$h_l^*(\alpha, \beta) = C^* \exp\left[-i\frac{k}{2l}(x^2 + y^2)\right], \tag{8.11}$$

and

$$C^* = \frac{i}{\lambda l}e^{-ikl}.$$

Notice that Eq. 8.11 represents a convergent spherical wavefront instead of a divergent wavefront as described by Eq. 8.9.

EXAMPLE 8.2

Given a monochromatic point source located at the origin of the (x, y) coordinate system, as shown in Figure 8.5, use the Fresnel–Kirchhoff theory to calculate the complex light field arriving at the α, β plane.

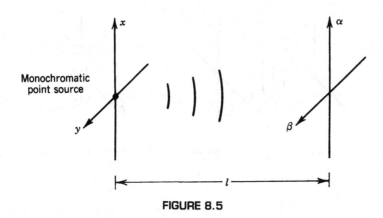

FIGURE 8.5

Since the distance between the x, y and α, β planes is assumed linear and spatially invariant, the spatial impulse response can be written as

$$h_l(x, y) = C \exp\left[i\frac{k}{2l}(x^2 + y^2)\right],$$

where C is a proportionality complex constant. The complex light distributed over the α, β plane can be computed by

$$g(\alpha, \beta) = \delta(x, y) * h_l(x, y)$$

$$= \int\int_{-\infty}^{\infty} \delta(x, y)h_l(\alpha - x, \beta - y)\, dx\, dy$$

$$= C \exp\left[i\frac{k}{2l}(\alpha^2 + \beta^2)\right],$$

which is a spherical wavefront.

EXAMPLE 8.3

Referring to the final result of Example 8.2, find the complex light field at distance $l' \neq l$ in front of the α, β plane, as shown in Figure 8.6a.

Let us first draw a block diagram for this problem, as shown in Figure 8.6b. We note that the spatial impulse response is written in conjugate form, which represents a convergent spherical wavefront. The complex light distribution over the x', y' plane can be obtained using

$$f(x', y') = g(\alpha, \beta) * h_{l'}^*(\alpha, \beta),$$

where

$$h_{l'}^*(\alpha, \beta) = C^* \exp\left[-i\frac{k}{2l'}(\alpha^2 + \beta^2)\right].$$

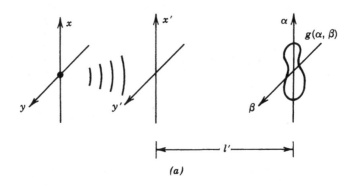

(a)

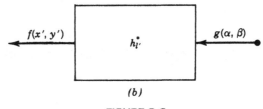

(b)

FIGURE 8.6

Thus, we have

$$f(x',y') = \int\!\!\int_{-\infty}^{\infty} g(\alpha,\beta)h_{l'}^{*}(x'-\alpha,y'-\beta)\,d\alpha\,d\beta$$

$$= C\exp\left[i\frac{k}{2l'}(x'^{2}+y'^{2})\right].$$

8.3 Fourier Transform in Fraunhofer Diffraction

We assume a diffraction screen that is illuminated by a monochromatic point source, as shown in Figure 8.7. To calculate the complex light distribution over the α,β plane, we first draw an analog system diagram to represent this problem, as shown in Figure 8.8. Referring to this diagram, we see that the complex light distribution over the output plane can be obtained by

$$g(\alpha,\beta) = [\delta(\xi,\eta) * h_{l_1}(\xi,\eta)]f(x,y) * h_{l_2}(x,y), \tag{8.12}$$

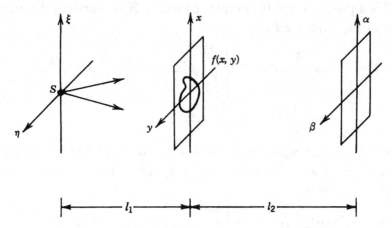

FIGURE 8.7 Fourier transform in Fraunhofer diffraction.

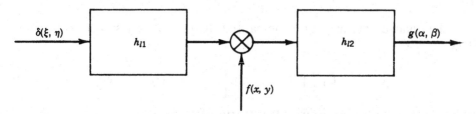

FIGURE 8.8 An analog system diagram of Figure 8.7.

where the asterisks represent the convolution operations, and

$$h_l(\xi, \eta) = C \exp\left[\frac{ik}{2l}(\xi^2 + \eta^2)\right] \tag{8.13}$$

is the spatial impulse response.

In order to achieve a Fraunhofer diffraction, we first let l_1 approach infinity. Thus, we see that

$$\lim_{l_1 \to \infty} \delta(\xi, \eta) * h_{l_1}(\xi, \eta) = \lim_{l_1 \to \infty} C \int\limits_{-\infty}^{\infty}\!\!\int \delta(\xi, \eta) \exp\left\{\frac{ik}{2l_1}[(\alpha - x)^2 + (\beta - y)^2]\right\} d\xi\, d\eta$$

$$= C \int\limits_{-\infty}^{\infty}\!\!\int \delta(\xi, \eta) = C, \tag{8.14}$$

which is the plane wavefront. Thus, as $l_1 \to \infty$, Eq. 8.12 becomes

$$g(\alpha, \beta) = Cf(x, y) * h_{l_2}(x, y), \tag{8.15}$$

which can be written as

$$g(\alpha, \beta) = C' \int\limits_{-\infty}^{\infty}\!\!\int f(x, y) \exp\left\{\frac{ik}{2l_2}[(\alpha - x)^2 + (\beta - y)^2]\right\} dx\, dy, \tag{8.16}$$

where C' is a proportionality complex constant. By expanding the quadratic exponent of Eq. 8.16, we have

$$g(\alpha, \beta) = C' \exp\left[\frac{ik}{2l_2}(\alpha^2 + \beta^2)\right] \int\int_{-\infty}^{\infty} f(x, y) \exp\left[-\frac{ik}{l_2}(\alpha x + \beta y)\right]$$

$$\times \exp\left[i\frac{k}{2l_2}(x^2 + y^2)\right] dx\,dy. \tag{8.17}$$

To achieve the Fraunhofer diffraction at (α, β), we should let l_2 be sufficiently large, as compared with the region of $f(x, y)$. Thus, Eq. 8.17 can be reduced to the following form:

$$g(\alpha, \beta) = C' \exp\left[\frac{ik}{2l_2}(\alpha^2 + \beta^2)\right] \int\int f(x, y) \exp\left[-\frac{ik}{l_2}(\alpha x + \beta y)\right] dx\,dy. \tag{8.18}$$

If the quadratic phase variation caused by $\alpha^2 + \beta^2$ is assumed to be small, then

$$g(\alpha, \beta) \simeq C_1 \int\int f(x, y) \exp\left[-\frac{ik}{l_2}(\alpha x + \beta y)\right] dx\,dy, \tag{8.19}$$

which is essentially the Fourier transform of $f(x, y)$.

In addition, the corresponding intensity distribution at the output plane can be written as

$$I(\alpha, \beta) = g(\alpha, \beta)g^*(\alpha, \beta)$$

$$= |g(\alpha, \beta)|^2$$

$$= K|F(p, q)|^2, \tag{8.20}$$

which is proportional to the power spectral distribution of $f(x, y)$, where K is a proportionality constant, and $p = (2\pi/\lambda l_2)\alpha$ and $q = (2\pi/\lambda l_2)\beta$ are the angular spatial-frequency axes. Thus, we see that the Fraunhofer diffraction is indeed approaching the Fourier transformation.

EXAMPLE 8.4

Assuming that a diffraction screen with a square aperture of dimension W is illuminated by a monochromatic plane wave, calculate the far-field Fraunhofer diffraction.

The transmission function of the diffraction screen can be written as

$$f(x, y) = \begin{cases} 1, & |x| \leq \dfrac{W}{2} \quad \text{and} \quad |y| \leq \dfrac{W}{2}, \\ 0, & \text{otherwise.} \end{cases}$$

According to the Fresnel–Kirchhoff theory, the complex light distribution behind the diffraction screen can be written as

$$g(\alpha, \beta) = \int\!\!\int_{-\infty}^{\infty} f(x,y) h_l(\alpha - x, \beta - y)\, dx\, dy,$$

where l is the distance behind the screen and $h_l(x,y)$ is the spatial impulse response. Since we assumed that distance l is sufficiently large in comparison with the size of the diffraction aperture, the complex light field can be written as (see Eq. 8.19)

$$g(\alpha, \beta) = C \int_{-W/2}^{W/2} \int_{-W/2}^{W/2} \exp\left[-\frac{ik}{l}(\alpha x + \beta y)\right] dx\, dy,$$

where C is a proportionality complex constant. Making use of the separable nature of this equation, we have

$$g(\alpha, \beta) = C \int_{-W/2}^{W/2} \exp\left(-\frac{ik}{l}\, dx\right) dx \int_{-W/2}^{W/2} \exp\left(-\frac{ik}{l}\beta y\right) dy$$

$$= C \left[\frac{\sin\left(\dfrac{\pi W \alpha}{\lambda l}\right)}{\dfrac{\pi W \alpha}{\lambda l}}\right] \left[\frac{\sin\left(\dfrac{\pi W \beta}{\lambda l}\right)}{\dfrac{\pi W \beta}{\lambda l}}\right],$$

which is the Fourier transform of the square aperture.

8.4 The Fresnel Zone Plate

A diffraction screen contains concentric narrow circular slits, as shown in Figure 8.9. The screen is normally illuminated by a monochromatic plane

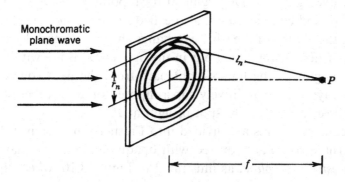

FIGURE 8.9 The focusing effect of a Fresnel zone plate.

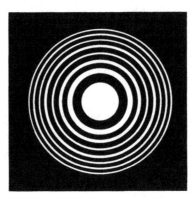

FIGURE 8.10 Fresnel zone plate with the center zone open.

wave of wavelength λ. At a point P behind the diffraction screen, the optical disturbances produced by the individual circular slits are in phase. We are to determine the distances between the narrow slits.

In order for all the optical disturbances to be in phase at distance f behind the screen, the path lengths from each of the circular slits to point P have to be of the order of a full wavelength:

$$l_n - f = n\lambda, \qquad n = 0, 1, 2, \ldots, \infty. \tag{8.21}$$

The radii of the concentric circles can now be computed:

$$r_n = \sqrt{l_n^2 - f^2}, \qquad n = 0, 1, 2, \ldots, \infty. \tag{8.22}$$

By substituting Eq. 8.21 into Eq. 8.22, we get

$$r_n = [(n\lambda + f)^2 - f^2]^{1/2}$$
$$= [(n\lambda)^2 + 2fn\lambda]^{1/2}, \qquad n = 0, 1, 2, \ldots, \infty. \tag{8.23}$$

Thus, we see that these concentric circular slits are capable of focusing a monochromatic plane wave into a very small region with very high intensity.

Suppose that the widths of the concentric slits are enlarged to include all the constructuve light rays converging at focal point P. Then the slits would become half-period circular zones, also called *Fresnel zones*. In other words, by enlarging the widths of the slits, we change the optical path lengths of the diffracted light rays. If the variations in the optical paths are within the limit of half-wavelengths of the line-of-sight path from the light source, the diffracted light rays will be additively superimposed on each other at point P. Thus, we will see a very bright spot of light at P.

If the circular apertures are divided into Fresnel zones, as just described, and the alternate zones are covered with opaque material, we have what is called the *Fresnel zone plate*, as illustrated in Figure 8.10, where the central zone is shown open.

Notice that the focusing effect will be the same if we start with a closed zone at the center. Furthermore, if the transmittance of a Fresnel zone plate varies sinusoidally, it is known as a *Fresnel zone lens*.

EXAMPLE 8.5

We are given an amplitude transmittance function of a one-dimensional Fresnel zone lens with infinite length,

$$T(x) = \frac{1}{2} + \frac{1}{2}\cos\left(\frac{\pi}{\lambda R}x^2\right),$$

where λ is an arbitrary wavelength and R is a positive constant.

If this Fresnel zone lens is illuminated by a monochromatic plane wave of wavelength λ, calculate the focal distance and the complex amplitude distribution of the convergent diffracted light rays.

Using the Fresnel–Kirchhoff theory, we can compute the complex light distribution behind the zone lens as follows,

$$g(\alpha) = \int_{-\infty}^{\infty} T(x)h_l(\alpha - x)\,dx,$$

where

$$h_l(x) = C\exp\left(\frac{i\pi x^2}{\lambda l}\right),$$

and l is the distance behind the zone lens. Since

$$\cos\left(\frac{\pi}{\lambda R}x^2\right) = \frac{1}{2}\left[\exp\left(-i\frac{\pi}{\lambda R}x^2\right) + \exp\left(i\frac{\pi}{\lambda R}x^2\right)\right],$$

we have

$$g(\alpha) = C\int_{-\infty}^{\infty}\left\{\frac{1}{2} + \frac{1}{4}\left[\exp\left(-i\frac{\pi}{\lambda R}x^2\right) + \exp\left(i\frac{\pi}{\lambda R}x^2\right)\right]\right\}\exp\left[\frac{i\pi}{\lambda l}(\alpha - x)^2\right]dx,$$

the terms in which can be integrated.

Let us now evaluate the second term, which is the convergent wave field,

$$g_2(\alpha) = C'\int_{-\infty}^{\infty}\exp\left(-\frac{i\pi}{\lambda R}x^2\right)\exp\left[\frac{i\pi}{\lambda l}(\alpha - x)^2\right]dx,$$

where C' is a proportionality complex constant. By expanding the quadratic factor, we can write this integral as

$$g_2(\alpha) = C'\exp\left(\frac{i\pi}{\lambda l}\alpha^2\right)\int_{-\infty}^{\infty}\exp\left[i\frac{\pi}{\lambda}\left(\frac{1}{l} - \frac{1}{R}\right)x^2\right]\exp\left[-\frac{i2\pi}{\lambda l}(\alpha x)\right]dx.$$

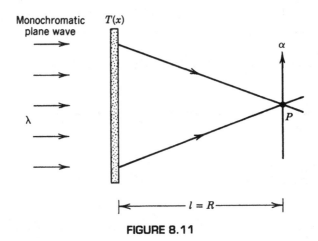

Monochromatic plane wave

$T(x)$

α

λ

P

$l = R$

FIGURE 8.11

To eliminate the quadratic phase factor of x^2, we would let, $l = R$, which is the focal distance of the zone lens. Thus, at $l = R$, the integral reduces to

$$g_2(\alpha) = C' \exp\left(\frac{i\pi}{\lambda R}\alpha^2\right) \int_{-\infty}^{\infty} \exp\left[-\frac{i2\pi}{\lambda R}(\alpha x)\right] dx,$$

which is the Fourier transform of a unit constant function. This integral can therefore be reduced to the following result,

$$g_2(\alpha) = C''\delta(\alpha),$$

where C'' is a proportionality complex constant. From this result we see that the diffracted rays (i.e., those caused by the second term) from the Fresnel zone lens will be focused to a point P at a distance $l = R$ behind the zone lens, as sketched in Figure 8.11.

EXAMPLE 8.6

If the Fresnel zone lens in Example 8.5 is illuminated by a monochromatic plane wave of wavelength λ_1, that is, $\lambda_1 \neq \lambda$, calculate the distance of the focused diffracted light rays and discuss the effect under white-light illumination. Since the zone lens is illuminated by a monochromatic plane wave with different wavelengths, the Fresnel–Kirchhoff integral should be written in terms of λ_1,

$$g(\alpha; \lambda_1) = \int_{-\infty}^{\infty} T(x; \lambda)h_l(\alpha - x; \lambda_1) dx,$$

where

$$h_l(x; \lambda_1) = C \exp\left(i\frac{\pi}{\lambda_1 l}x^2\right).$$

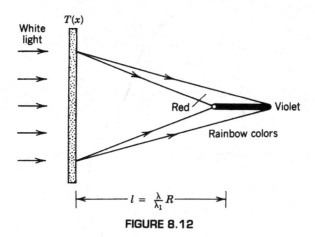

FIGURE 8.12

With substitutions for $T(x; \lambda)$ and $h_l(x; \lambda_l)$, the preceding equation becomes

$$g(\alpha; \lambda_1) = C \int_{-\infty}^{\infty} \left\{ \frac{1}{2} + \frac{1}{4} \left[\exp\left(-i\frac{\pi}{\lambda R}x^2\right) + \exp\left(i\frac{\pi}{\lambda R}x^2\right) \right] \right\} \exp\left[i\frac{\pi}{\lambda_1 l}(\alpha - x)^2 \right] dx.$$

The second integral is

$$g_2(\alpha; \lambda_1) = C' \int_{-\infty}^{\infty} \exp\left(-i\frac{\pi}{\lambda R}x^2\right) \exp\left[\frac{i\pi}{\lambda_1 l}(\alpha - x)^2 \right] dx,$$

which can be written as

$$g_2(\alpha; \lambda_1) = C' \exp\left(i\frac{\pi}{\lambda_1 l}\alpha^2\right) \int_{-\infty}^{\infty} \exp\left[i\pi\left(\frac{1}{\lambda_1 l} - \frac{1}{\lambda R}\right)x^2 \right] \exp\left[-i\frac{2\pi}{\lambda_1 l}(\alpha x) \right] dx.$$

To eliminate the quadratic phase factor x^2, we see that

$$l = \frac{\lambda}{\lambda_1}R,$$

which is the focal distance of the covergent light rays,

$$g_2(\alpha; \lambda_1) = C'' \delta(\alpha; \lambda_1), \qquad l = \frac{\lambda}{\lambda_1}R.$$

Since the focal distance is inversely proportional to the illuminating wavelength λ_1, the focal point will be closer to the zone lens for illumination of longer wavelengths and farther from the zone lens for illumination of shorter wavelengths.

It is apparent that if the zone lens is illuminated by a white-light plane wave, the focal point will smear into rainbow colors. For example, the red color focal point will be located closer to the zone lens, and the violet color will be focused at a greater distance, as illustrated in Figure 8.12.

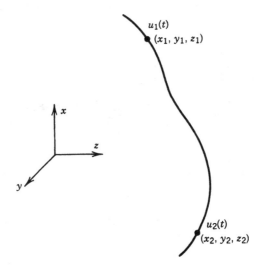

FIGURE 8.13 An electromagnetic wavefront in space.

8.5 Partial Coherence

The widespread application of partially coherent light makes it appropriate to include a discussion of the principles of coherence in radiation. If the radiations from two point sources maintain a fixed phase relation between them, they are said to be *mutually coherent*. An extended source is coherent if all points of the source have fixed phase differences between them. Here we discuss some general aspects of partial coherence.

In the classical theory of electromagnetic radiation, as in the development of Maxwell's equations, it is usually assumed that the electric and magnetic fields are always measurable at any position. Thus, in these situations there is no need to take into account partial coherence theory. There are situations, however, for which this assumption of known fields cannot be made and for which it is often helpful to apply partial coherence theory. For example, if we want to determine the diffraction pattern caused by the radiation from several sources, we cannot obtain an exact result unless the degree of coherence from the separate sources is taken into account. In such a situation, however, it is desirable to obtain an ensemble average to represent the *statistically* most likely result from any such combination of sources. It may thus be more useful to provide a statistical description than to follow the dynamical behavior of a wave field in detail.

Our treatment of partial coherence uses such ensemble averages. Let us assume an electromagnetic wave field propagating in space, as depicted in Figure 8.13, where $u_1(t)$ and $u_2(t)$ denote the instantaneous wave

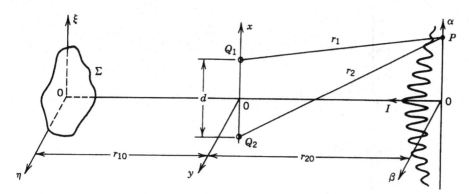

FIGURE 8.14 Young's experiment. Σ is an extended but nearly monochromatic source.

disturbances at positions 1 and 2, respectively. We choose the second-order moment as the quantity to be averaged, which is why it is called the *mutual coherence function*. This function is defined in the following,

$$\Gamma_{12}(\tau) \triangleq \langle u_1(t+\tau)u_2^*(t)\rangle, \qquad (8.24)$$

where the asterisk denotes the complex conjugate, and $\Gamma_{12}(\tau)$ is the mutual coherence function between these points for a time delay τ. The angle brackets $\langle\ \rangle$ denote a time average, which can be written as

$$\Gamma_{12}(\tau) = \lim_{T\to\infty} \frac{1}{T}\int_0^T u_1(t+\tau)u_2^*(t)\,dt. \qquad (8.25)$$

It is apparent that the mutual coherence function defined by Eq. 8.25 is essentially the temporal *cross-correlation function* between the complex disturbances of $u_1(t)$ and $u_2(t)$.

The *normalized mutual coherence function* can therefore be defined as

$$\gamma_{12}(\tau) \triangleq \frac{\Gamma_{12}(\tau)}{[\Gamma_{11}(0)\Gamma_{22}(0)]^{1/2}}, \qquad (8.26)$$

where $\Gamma_{11}(\tau)$ and $\Gamma_{22}(\tau)$ are the *self-coherence functions* of $u_1(t)$ and $u_2(t)$, respectively. The function $\gamma_{12}(\tau)$ may also be called the *complex degree of coherence* or the *degree of correlation*.

We now demonstrate that the normalized mutual coherence function $\gamma_{12}(\tau)$ can be measured by applying Young's experiment on interference. In Figure 8.14, Σ represents an extended source of light, which is assumed to be incoherent but nearly monochromatic; that is, its spectrum is of finite width. The light from this source falls upon a screen at a distance r_{10} from the source, and upon two small apertures (pinholes) in this screen, Q_1 and Q_2, separated by a distance d. On an observing screen located r_{20} away from the diffracting screen, an interference pattern is formed by the light passing

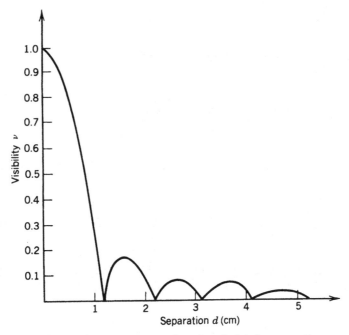

FIGURE 8.15 Visibility as a function of pinhole separation.

through Q_1 and Q_2. Now let us suppose that the changing characteristics of the interference fringes are observed as the parameters in Figure 8.14 are changed. As a measurable quantity, let us adopt *Michelson's visibility* of the fringes, which is defined as

$$\nu \triangleq \frac{I_{max} - I_{min}}{I_{max} + I_{min}}, \tag{8.27}$$

where I_{max} and I_{min} are the maximum and minimum intensities of the fringes.

For the visibility to be measurable, the conditions of the experiment, such as the narrowness of the spectrum and the closeness of optical path lengths, must be such that they permit I_{max} and I_{min} to be clearly defined. Let us assume that these ideal conditions exist.

As we begin, our parameters change. First, we find that the average visibility of the fringes increases as the size of the source Σ is made smaller. Next, as the distance between pinholes Q_1 and Q_2 is increased, while Σ is held constant (and circular in form), the visibility shifts in the manner shown in Figure 8.15. When Q_1 and Q_2 are very close together, the intensity between the fringes falls to zero, and the visibility becomes unity. As d is increased, the visibility falls rapidly and reaches zero as I_{max} and I_{min} become equal. An additional increase in d causes the fringes to reappear, although they are shifted on the screen by half a fringe; that is, the areas that were previously light are now dark, and vice versa. As the distance d between pinholes is made even greater, there are the repeated fluctuations in visibility shown in the figure. A curve similar to that shown in Figure 8.15 is obtained when the pinhole spacing is kept constant while the size of Σ is changed. These effects can be predicted from the Van Cittert–Zernike theorem (see Section 8.6.1). The visibility versus pinhole separation curve is sometimes used as a measure of *spatial coherence*, as discussed in Section 8.6.1. Screen separations r_{10} and r_{20} are both assumed to be large compared with the aperture spacing d, and with the dimensions of the source. Beyond this limitation, changes in r_{10} or r_{20} shift the scale of effects, as is shown in Figure 8.15, but without altering their general character.

When the point of observation P is moved away from the center of the observing screen, visibility decreases as the path difference $\Delta r = r_2 - r_1$ increases, until it eventually becomes zero. The effect also depends on how nearly monochromatic the source is. The visibility of the fringes has been found to be appreciable only for a path difference of

$$\Delta r \simeq \frac{2\pi c}{\Delta \omega}, \tag{8.28}$$

where c is the velocity of light and $\Delta \omega$ is the spectral width of the source. Equation 8.28 is often used to define the *coherence length* of the source, which is the distance that its light beam is longitudinally coherent.

The preceding discussion has indicated that it is not necessary to have completely coherent light to produce an interference pattern, but that under the right conditions such a pattern may be obtained from an incoherent source. This effect is called *partial coherence*.

It will be helpful to develop the preceding equations further. Thus, $u_1(t)$ and $u_2(t)$ in Eq. 8.24, the complex wave fields at points Q_1 and Q_2, are subject to the scalar wave equation in free space,

$$\nabla^2 u = \frac{1}{c^2} \frac{\partial^2 u}{\partial t^2}, \tag{8.29}$$

where c is the velocity of light. This is a linear equation, and the wave field at point P on the observing screen is a sum of the fields from Q_1 and Q_2,

$$u_p(t) = c_1 u_1 \left(t - \frac{r_1}{c} \right) + c_2 u_2 \left(t - \frac{r_2}{c} \right), \tag{8.30}$$

where c_1 and c_2 are the appropriate complex constants. The corresponding irradiance at P may be written as

$$I_p = \langle u_p(t) u_p^*(t) \rangle = I_1 + I_2 + 2 \operatorname{Re} \left\langle c_1 u_1 \left(t - \frac{r1}{c} \right) c_2^* u_2^* \left(t - \frac{r_2}{c} \right) \right\rangle, \tag{8.31}$$

where I_1 and I_2 are proportional to the squares of the magnitudes of $u_1(t)$ and $u_2(t)$. We now define the variables

$$t_1 = \frac{r_1}{c} \quad \text{and} \quad t_2 = \frac{r_2}{c}. \tag{8.32}$$

Then Eq. 8.31 can be written as

$$I_p = I_1 + I_2 + 2c_1 c_2^* \, \mathrm{Re}\langle u_1(t - t_1) u_2^*(t - t_2)\rangle. \tag{8.33}$$

Thus, we see that the averaged quantity in Eq. 8.33 is the cross-correlation of the two complex wave fields.

If we make $t_2 - t_1 = \tau$, Eq. 8.33 can be written as

$$I_p = I_1 + I_2 + 2c_1 c_2^* \, \mathrm{Re}\langle u_1(t + \tau) u_2^*(t)\rangle,$$

and combining this with Eq. 8.24, we obtain

$$I_p = I_1 + I_2 + 2c_1 c_2^* \, \mathrm{Re}[\Gamma_{12}(\tau)]. \tag{8.34}$$

The *self-coherence functions* (i.e., the autocorrelations) of the radiations from the two pinholes can therefore be defined as

$$\Gamma_{11}(0) = \langle u_1(t) u_1^*(t)\rangle \quad \text{and} \quad \Gamma_{22}(0) = \langle u_2(t) u_2^*(t)\rangle. \tag{8.35}$$

If we let the following relations hold,

$$|c_1|^2 \Gamma_{11}(0) = I_1 \quad \text{and} \quad |c_2|^2 \Gamma_{22}(0) = I_2,$$

the intensity at P can then be written in terms of the degree of complex coherence, as given in Eq. 8.26:

$$I_p = I_1 + I_2 + 2(I_1 I_2)^{1/2} \, \mathrm{Re}[\gamma_{12}(\tau)]. \tag{8.36}$$

Let us write $\gamma_{12}(\tau)$ in the form

$$\gamma_{12}(\tau) = |\gamma_{12}(\tau)| \exp[i\phi_{12}(\tau)], \tag{8.37}$$

and assume also that $I_1 = I_2 = I$, which can be called the *best condition*. Then Eq. 8.36 becomes

$$I_p = 2I[1 + |\gamma_{12}(\tau)| \cos \phi_{12}(\tau)]. \tag{8.38}$$

Thus we see that the maximum value of I_p is $2I[1 + |\gamma_{12}(\tau)|]$ and the minimum value is $2I[1 - |\gamma_{12}(\tau)|]$. By substituting these values into the visibility equation (Eq. 8.27), we have

$$\nu = |\gamma_{12}(\tau)|. \tag{8.39}$$

That is, under the best condition, the visibility of the fringes is a measure of the degree of coherence.

EXAMPLE 8.7

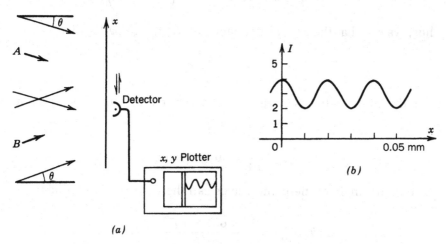

FIGURE 8.16

Assume that two monochromatic plane waves of equal intensity are falling obliquely over an observation plane, as shown in Figure 8.16a. The fringe intensity distribution is as plotted in Figure 8.16b.

(a) Calculate the degree of coherence between these two beams of light.
(b) Determine the spatial frequency of the fringe pattern and the oblique angle of the plane waves. The wavelength of illumination is assumed to be $\lambda = 0.633\,\mu m$.
(c) If the intensity ratio is equal to 2 (i.e., $I_A/I_B = 2$), calculate the visibility of the pattern of interference fringes.

Answers
(a) Referring to Figure 8.16b, we note that

$$I_{max} = 4,$$
$$I_{min} = 2.$$

From the relation given in Eq. 8.39 and the definition given by Eq. 8.27, we have

$$\gamma = \nu = \frac{I_{max} - I_{min}}{I_{max} + I_{min}} = \frac{4 - 2}{4 + 2} = \frac{1}{3}.$$

(b) The intensity distribution over x can be written as

$$I = I_A + I_B + 2\sqrt{I_A I_B}|\gamma|\cos(\phi_A - \phi_B),$$

where $\phi_A = -kx \sin\theta$ and $\phi_B = kx \sin\theta$. By substituting ϕ_A and ϕ_B, we have

$$I = I_A + I_B + 2\sqrt{I_A I_B}|\gamma| \cos\left(\frac{4\pi}{\lambda} x \sin\theta\right).$$

Thus, we see that the spatial frequency of the fringe pattern is

$$f_x = \frac{2}{\lambda} \sin\theta.$$

From the intensity traces in Figure 8.16b, the spatial frequency is computed as

$$f_x = \frac{5}{0.1} = 50 \, \text{cycles/mm}.$$

The oblique angle of the plane waves can therefore be determined:

$$\sin\theta = \frac{f_x \lambda}{2} = \frac{50 \times 633 \times 10^{-6}}{2} = 0.01583$$
$$\theta = 0.01583 \, \text{rad}.$$

(c) Using Eq. 8.27, we can write the visibility as

$$\nu = \frac{I_{max} - I_{min}}{I_{max} + I_{min}} = \frac{(I_A + I_B + 2\sqrt{I_A I_B}|\gamma|) - (I_A + I_B - 2\sqrt{I_A I_B}|\gamma|)}{(I_A + I_B + 2\sqrt{I_A I_B}|\gamma|) + (I_A + I_B - 2\sqrt{I_A I_B}|\gamma|)}$$
$$= \frac{2\sqrt{I_A I_B}|\gamma|}{I_A + I_B}$$
$$= \frac{2\sqrt{I_A I_B}|\gamma|}{1 + I_A/I_B}.$$

By substituting $I_A/I_B = 2$, and $|\gamma| = \frac{1}{3}$, we have

$$\nu = \frac{(2\sqrt{2})(\frac{1}{3})}{1 + 2} = 0.314.$$

8.6 Spatial and Temporal Coherence

The term *spatial coherence* is applied to effects that are due to the size (in space) of the source of radiation. If we consider a point source, and look at two points that are at equal light path distances from the source, the radiations reaching these points will be exactly the same. This mutual coherence will be equal to the self-coherence at either point. That is, if the points are Q_1 and Q_2, then

$$\Gamma_{12}(Q_1, Q_2, \tau) = \langle u(Q_1, t + \tau) u^*(Q_2, t) \rangle = \Gamma_{11}(\tau). \qquad (8.40)$$

As the source is made larger, we can no longer claim an equality of mutual coherence and self-coherence. This lack of complete coherence is a *spatial* effect. *Temporal coherence* is an effect that is due to the finite spectral width of the source. The coherence is complete for strictly monochromatic radiation but becomes only partial as other wavelengths are added, giving a finite spectral width to the source. It is never possible to separate completely the two effects (i.e., spatial and temporal coherence), but it is well to name them and point out their significance, as we do in the following discussion.

8.6.1 SPATIAL COHERENCE

We utilize Young's experiment to determine the angular size of a spatially incoherent source (e.g., an extended source) as it relates to spatial coherence. Thus, we see that spatial coherence depends on the size of a light source. Let us now consider the paths of two light rays from a point on a linearly extended source ΔS, as they pass through the two narrow slits Q_1 and Q_2 depicted in Figure 8.17. If we let $r_{10} \gg d$, the intensity distribution at observation screen P_3 can, as a consequence of Eq. 8.36, be written as

$$I_\rho(\alpha) = I_1 + I_2 + 2\sqrt{I_1 I_2} \cos\left[k\left(\frac{d}{r_{10}}\xi + \frac{d}{r_{20}}\alpha\right)\right]. \qquad (8.41)$$

Here I_1 and I_2 are the corresponding intensity distributions that are produced by the light rays passing through Q_1 and Q_2, respectively, and $k = 2\pi/\lambda$ is the wavenumber.

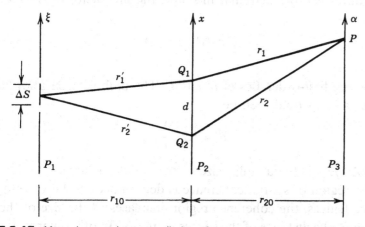

FIGURE 8.17 Young's experiment applied to the measurement of spatial coherence.

Equation 8.41 describes a set of parallel fringes along the α axis that come from a single radiation point on the extended source, ΔS. Thus, we see that the other points on the extended source cause the superimposition of the fringes on the observation screen P_3. For a uniformly one-dimensional extended source of size ΔS, the resulting fringe pattern can be written as

$$I'_p(\alpha) = I_1\,\Delta S + I_2\,\Delta S + 2\sqrt{I_1 I_2}\int_{-\Delta S/2}^{\Delta S/2}\cos\left[k\left(\frac{d}{r_{10}}\xi + \frac{d}{r_{20}}\alpha\right)\right]d\xi$$

$$= \Delta S\left\{I_1 + I_2 + 2\sqrt{I_1 I_2}\,\mathrm{sinc}\left[\left(\frac{\pi d}{\lambda r_{10}}\right)\Delta S\right]\cos\left(k\frac{d}{r_{20}}\alpha\right)\right\}, \qquad (8.42)$$

where

$$\mathrm{sinc}\,\pi\chi \triangleq \frac{\sin \pi\chi}{\pi\chi}.$$

The sinc factor in Eq. 8.42 indicates that the fringes vanish at

$$d = \frac{\lambda r_{10}}{\Delta S}. \qquad (8.43)$$

Similarly, we can write Eq. 8.43 in terms of the angular size of the source θ_s,

$$d = \frac{\lambda}{\theta_s}, \qquad (8.44)$$

where $\theta_s \triangleq \Delta S/r_{10}$. We note that the sinc factor in Eq. 8.42 is primarily due to the incoherent addition of the point radiators over the extended source ΔS. Equation 8.42 presents a useful example for the discussion of coherent and incoherent illumination. For completely coherent illumination there are interference fringes, but for completely incoherent illumination the interference fringes vanish. Moreover, Eq. 8.42 shows that the interference pattern (i.e., the cosine factor) is weighted by a broad sinc factor, whereby the interference fringes go to zero and reappear as a function of d, ΔS, or $1/r_{10}$. If the extended source had been circular in shape, the sinc factor in Eq. 8.42 would have been

$$\frac{J_1(\pi\,\Delta S\,d/\lambda r_{10})}{\pi\,\Delta S\,d/\lambda r_{10}}, \qquad (8.45)$$

where J_1 is the first-order Bessel function. The pinhole separation at which the fringes vanish would then be

$$d = 1.22\frac{\lambda}{\theta_s}, \qquad (8.46)$$

whre $\theta_s \triangleq \Delta S/r_{10}$ is the angular size of the circular source.

Since the region of spatial coherence is determined by the visibility of the interference fringes, the coherence region increases as the size of the source decreases or as the distance of the source increases. In other words, the degree of spatial coherence increases as the distance of the propagated wave increases.

In addition, there is a relation between the intensity distribution of the source and the degree of spatial coherence. This relation is described in the *Van Cittert–Zernike theorem*. The theorem essentially states that the normalized complex degree of coherence $\gamma_{12}(0)$ between two points on a plane that come from an extended incoherent source at another plane is proportional to the normalized inverse Fourier transform of the intensity distribution of the source:

$$\gamma_{12}(0) = \frac{\int_{\Delta S} I(\xi) \exp\left[i \frac{k}{r_{10}}(x_1 - x_2)\xi\right] d\xi}{\int_{\Delta S} I(\xi) \, d\xi}, \tag{8.47}$$

Here the subscripts 1 and 2 denote the two points on the diffraction screen P_2; x_1 and x_2 are the corresponding position vectors; ξ is the position vector at the source plane, $k = 2\pi/\lambda$; $I(\xi)$ is the intensity distribution of the source; and the integration is over the source size. Thus, it is apparent that if the intensity distribution of the source is uniform over a circular disc, the mutual coherence function is described by a rotational symmetric Bessel function J_1. On the other hand, if the source intensity is uniform over a rectangular slit, the mutual coherence function is described by a two-dimensional sinc factor.

8.6.2 TEMPORAL COHERENCE

It is possible for us to split a light wave into two paths, to delay one of them, and then to recombine them to form an interference fringe pattern. In this way, we can measure the temporal coherence of the light wave. The degree of temporal coherence is the measure of the cross-correlation of a wave field at one time with respect to another wave field at a later time. Thus, the definition of temporal coherence can also refer to *longitudinal coherence* as opposed to transverse (i.e., spatial) coherence. The maximum difference in the optical path lengths of two waves derived from a source is known as the *coherent length* of that source. Since spatial coherence is determined by the wave field in the transverse direction, and temporal coherence is measured along the longitudinal direction, spatial and temporal coherences would describe the degree of coherence of a wave field within a volume of space.

The *Michelson interferometer* is one of the instruments most commonly used to measure the temporal coherence of a light source (see Figure 8.18). Looking at this figure, we can see that the beam splitter BS divides the light beam into two paths, one of which goes to mirror M_1 and the other to mirror M_2. By varying the path length (i.e., τ) to one of the mirrors, we can observe

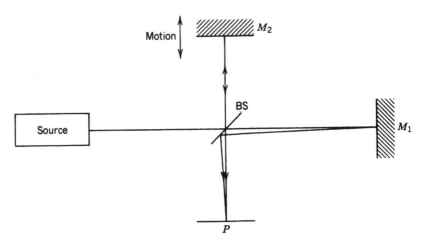

FIGURE 8.18 The Michelson interferometer for the measurement of temporal coherence. BS, beam splitter; M_1, M_2, mirrors; P, observation screen.

the variation of the interference fringes at observation screen P. Thus, the measurement of the visibility of the interference fringes corresponds to the measurement of the degree of temporal coherence $\gamma_{11}(\tau)$ at a time delay τ, where subscript 11 signifies the interference derived from the same point in space. Therefore, when the difference in path lengths is zero, the measurement of visibility corresponds to $\gamma_{11}(0)$.

Since the nature of the source affects coherence, the spectral width of the source affects the temporal coherence of the source. The time interval Δt, during which the wave is coherent, can be approximated by

$$\Delta t \simeq \frac{2\pi}{\Delta\omega}, \tag{8.48}$$

where $\Delta\omega$ is the spectral bandwidth of the source. Thus, the coherence length of the light source can be defined, using Eq. 8.28, as

$$\Delta r = \Delta t c \simeq \frac{2\pi c}{\Delta\omega}, \tag{8.49}$$

where $c = 3 \times 10^8$ m/s is the velocity of light.

By substituting the relations

$$c = \frac{\omega\lambda}{2\pi} \quad \text{and} \quad \frac{\omega}{\Delta\omega} = \frac{\lambda}{\Delta\lambda},$$

we can also write Eq. 8.49 as

$$\Delta r \simeq \frac{\lambda^2}{\Delta\lambda}, \tag{8.50}$$

where λ is the center wavelength and $\Delta\lambda$ is the spectral width of the light source.

EXAMPLE 8.8

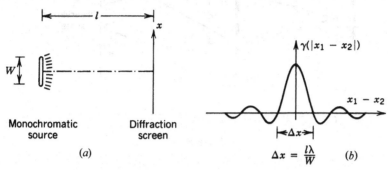

FIGURE 8.19

The intensity distribution of a one-dimensional monochromatic source is given by

$$I(\xi) = \begin{cases} 1, & |\xi| \le W/2, \\ 0, & \text{otherwise.} \end{cases}$$

If the light source illuminates a diffraction screen (as shown in Figure 8.19a), calculate the complex degree of coherence (i.e., the normalized spatial coherence) at the screen.

Using the Van Cittert–Zernike theorem as given in Eq. 8.47, we can determine the complex degree of coherence:

$$\gamma_{12}(0) = \gamma_{12}(|x_1 - x_2|) = \frac{\displaystyle\int_{-W/2}^{W/2} \exp(ik|x_1 - x_2|\xi)\, d\xi}{\displaystyle\int_{-W/2}^{W/2} d\xi}$$

$$= \frac{\sin\left(\dfrac{\pi W}{l\lambda}|x_1 - x_2|\right)}{\dfrac{\pi W}{l\lambda}|x_1 - x_2|}$$

$$= \text{sinc}\left(\frac{\pi W}{l\lambda}|x_1 - x_2|\right).$$

A sketch of the complex degree of coherence across the diffraction screen is given in Figure 8.19b. We see that the *coherence distance* can be written as $\Delta x = l\lambda/W$, which is equal to the width of the sinc factor. In other words, any two points within distance Δx (i.e., $|x_2 - x_1| \le \Delta x$) on the diffraction screen are highly coherent.

In order to gain a sense of magnitude, we let $W = 1\,\text{mm}$, $l = 0.2\,\text{m}$, and $\lambda = 600 \times 10^{-9}\,\text{m}$. Then the coherence distance is

$$\Delta x = \frac{0.2 \times 600 \times 10^{-9}}{10^{-3}} = 0.12\,\text{mm}.$$

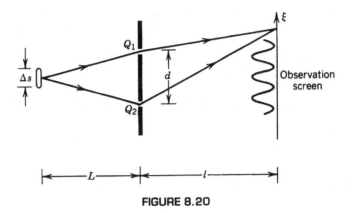

FIGURE 8.20

EXAMPLE 8.9

Refer to the depiction of Young's experiment in Figure 8.20, where Δs is a linear extended monochromatic source, $d = 1\,\text{mm}$, $\lambda = 6330\,\text{Å}$, and $L = 1\,\text{m}$. Using these parameters, calculate the degrees of coherence between Q_1 and Q_2 for $\Delta s = 0.1\,\text{mm}$ and $0.5\,\text{mm}$, respectively.

Using the Van Cittert–Zernike theorem given by Eq. 8.47 or as applied in Example 8.8, we obtain

$$\gamma_{12}(\Delta s) = \text{sinc}\left(\frac{\pi \Delta s}{\lambda L} d\right).$$

Substituting $d = 1\,\text{mm}$, $\lambda = 6330\,\text{Å}$, $L = 1\,\text{m}$, and $\Delta s = 0.1\,\text{mm}$, or

$$\frac{d}{\lambda L} = \frac{1}{6330 \times 10^{-7} \times 10^3} = 1.58\,\text{mm}^{-1},$$

we get the following degree of coherence,

$$\gamma_{12}(0.1\,\text{mm}) = \frac{\sin(\pi \times 0.1 \times 1.58)}{\pi \times 0.1 \times 1.58} = 0.959.$$

Similarly, for $\Delta s = 0.5\,\text{mm}$, the degree of coherence is

$$\gamma_{12}(0.5\,\text{mm}) = \frac{\sin(\pi \times 0.5 \times 1.58)}{\pi \times 0.5 \times 1.58} = 0.247.$$

8.7 Coherence Measurement

Although in the 1930s Pieter H. Van Cittert and Frits Zernike predicted a profound relationship between the spatial coherence and the intensity distribution of a light source, it was Brian J. Thompson and Emil Wolf who demonstrated a two-beam interference technique to measure the degree of partial coherence. They showed that, under quasi-monochromatic illumination, the degree of spatial coherence depends on the size of the source and the distance between two arbitrary points. The degree of temporal coherence, however, depends on the spectral bandwidth of the light source. Thompson and Wolf also illustrated several coherence measurements that are very consistent with the Van Cittert–Zernike predictions.

In this section we illustrate some results obtained by interferometric techniques, as described in the previous section, for the measurement of spatial and temporal coherence.

Figure 8.21 shows the sequence of interference fringe patterns obtained with the dual-beam technique, which uses a circular extended source. This set of fringe patterns is obtained by increasing the separation d of the two pinholes. From this set of fringe patterns we see that the visibility (i.e., the degree of spatial coherence) decreases from Figure 8.21a to Figure 8.21d, then increases in Figure 8.21e, and decreases again in Figure 8.21f, in accord with the plotting in Figure 8.15. We also note that the phase of the fringe patterns is shifted a few times from Figures 8.21a through f, which is due to the bipolar nature of the first-order Bessel function, as described in Eq. 8.45. In other words, Figures 8.21a–c give the fringe patterns recorded in the first lobe in Figure 8.15, Figures 8.21d and 8.21e give those in the second lobe, and Figure 8.21f gives the fringe patterns in the third lobe. It is also apparent from these six figures that the spatial frequency of the interference fringes is proportional to the separation d.

Figure 8.22 shows a set of interference fringe patterns produced when the separation d is held constant while the diameter of the circular source is increased. From this set of figures we see that the visibility decreases as the source size increases, and that the spatial frequency of the fringe patterns remains unchanged.

For the measurement of temporal coherence, however, we use the Michelson interferometer, as described earlier. Figure 8.23 shows a sequence of interference fringe patterns that we have obtained with this technique. In Figures 8.23a–c we see that the visibility of the fringe patterns decreases as the difference in the path lengths (i.e., τ) of the two waves increases. Thus, we see that the coherence length of a temporal, partially coherent source can be measured with this technique.

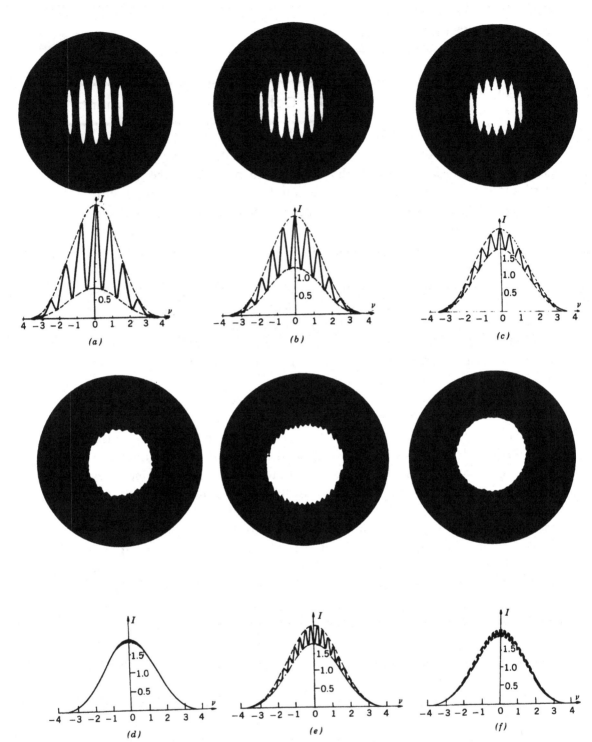

FIGURE 8.21 Measurement of spatial coherence as a function of pinhole separation. d, Separation between pinholes; γ, degree of coherence.

(a) $d = 0.6\,\text{cm}$, $|\gamma_{12}| = 0.593$;   (b) $d = 0.8\,\text{cm}$, $|\gamma_{12}| = 0.361$;
(c) $d = 1\,\text{cm}$, $|\gamma_{12}| = 0.236$;   (d) $d = 1.2\,\text{cm}$, $|\gamma_{12}| = 0.015$;
(e) $d = 1.7\,\text{cm}$, $|\gamma_{12}| = 0.123$;   (f) $d = 2.3\,\text{cm}$, $|\gamma_{12}| = 0.035$.
(By permission of Brian J. Thompson).

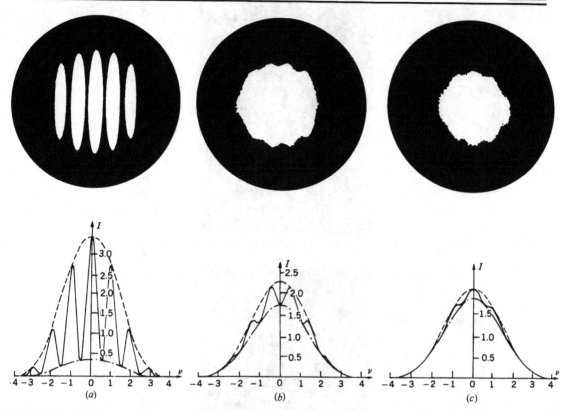

FIGURE 8.22 Measurement of spatial coherence as a function of source size. ϕ, Phase shift; d, separation between pinholes; γ, degree of coherence.
(a) $\phi_{12} = 0$, $d = 0.5\,\text{cm}$, $|\gamma_{12}| = 0.703$;
(b) $\phi_{12} = \pi$, $d = 0.5\,\text{cm}$, $|\gamma_{12}| = 0.132$;
(c) $\phi_{12} = 0$, $d = 0.5\,\text{cm}$, $|\gamma_{12}| = 0.062$;
(By permission of Brian J. Thompson.)

EXAMPLE 8.10

If the extended source in Example 8.9 is replaced by two point sources with a spacing b, calculate the degree of coherence for $b = 0.1$ and $0.5\,\text{mm}$. Since the two point sources can be described by two delta functions,

$$I(\xi) = \delta\left(\xi + \frac{b}{2}\right) + \delta\left(\xi - \frac{b}{2}\right),$$

by substituting into Eq. 8.47, we can show that

$$\gamma_{12} = \frac{\int \left[\delta\left(\xi + \frac{b}{2}\right) + \delta\left(\xi - \frac{b}{2}\right)\right] e^{-ik\xi d}\, \mathrm{d}\xi}{\int \left[\delta\left(\xi + \frac{b}{2}\right) + \delta\left(\xi - \frac{b}{2}\right)\right] \mathrm{d}\xi}$$

$$= \cos\left(\pi b \frac{d}{\lambda L}\right).$$

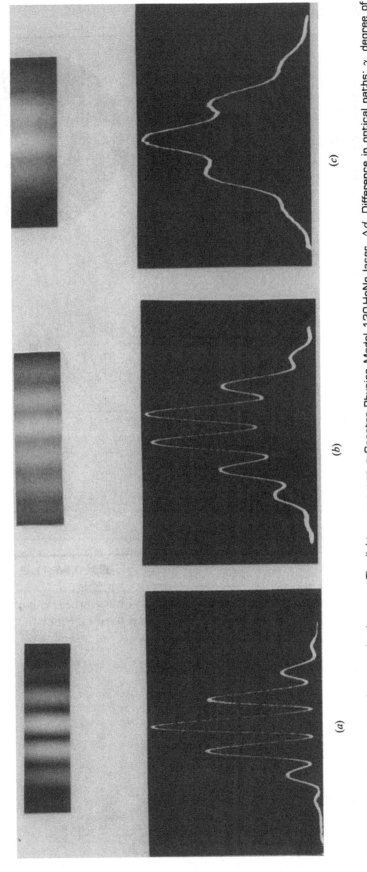

FIGURE 8.23 Measurement of temporal coherence. The light source was a Spectra-Physics Model 120 HeNe laser. Δd, Difference in optical paths; γ, degree of coherence.

(a) $\Delta d \cong 0$, $\gamma = 0.94$;
(b) $\Delta d = 25$ cm, $\gamma = 0.67$;
(c) $\Delta d = 43$ cm, $\gamma = 0.12$.

Thus, for the separation for $b = 0.1\,\text{mm}$, the degree of coherence between Q_1 and Q_2 is

$$\gamma_{12}(0.1\,\text{mm}) = \cos(\pi \times 0.1 \times 1.58) = 0.879.$$

Similarly for $b = 0.5\,\text{mm}$, we have

$$|\gamma_{12}(0.5\,\text{mm})| = |\cos(\pi \times 0.5 \times 1.58)| = 0.79.$$

Note that $0 \leq |\gamma_{12}| \leq 1$.

EXAMPLE 8.11

Given a partially coherent source, a central wavelength λ that is assumed equal to $4880\,\text{Å}$, and a spectral width $\Delta\lambda$ of about $0.03\,\text{Å}$, calculate the coherence length of the light source.

Using Eq. 8.49, we can compute the coherence length:

$$\Delta\gamma = \frac{2\pi c}{\Delta\omega} = \frac{\lambda^2}{\Delta\lambda} = \frac{(4880 \times 10^{-8})^2}{0.03 \times 10^{-8}} = 7.9\,\text{cm}.$$

We note that

$$c = \frac{\omega\lambda}{2\pi} \quad \text{and} \quad \frac{\omega}{\Delta\omega} = \frac{\lambda}{\Delta\lambda}.$$

REFERENCES

1. M. Born and E. Wolf, *Principles of Optics*, second edition, Pergamon Press, New York, 1964.
2. J. M. Stone, *Radiation and Optics*, McGraw-Hill, New York, 1963.
3. M. J. Beran and G. B. Parrent, Jr., *Theory of Partial Coherence*, Prentice-Hall, Englewood Cliffs, NJ, 1964.
4. F. T. S. Yu, *Optical Information Processing*, Wiley-Interscience, New York, 1983.
5. P. H. Van Cittert, Die Wahrscheinliche Schwingungs verteilung in einer von einer lichtquelle direkt Oden Mittels einer linse, *Physica*, 1 (1934) 201.
6. F. Zernike, The concept of degree of coherence and its application to optical problems, *Physica*, 5 (1938) 785.
7. B. J. Thompson and E. Wolf, Two-beam interference and partially coherent light, *Journal of the Optical Society of America*, 47 (1957) 895.
8. B. J. Thompson, Illustration of the phase change in two-beam interference with partially coherent light, *Journal of the Optical Society of America*, 48 (1958) 95.

PROBLEMS

8.1   Consider a diffraction screen that contains an open circular aperture which is 5 mm in diameter. If this screen is normally illuminated by a red light plane wave of $\lambda = 6.3 \times 10^{-5}$ cm, calculate the separation between the diffraction screen and the observation plane to achieve the far-field Fraunhofer diffraction.

8.2   In Section 8.3 we showed the Fourier transform property of Fraunhofer diffraction. Compute the complex light distribution of a circular aperture of diameter D under monochromatic plane-wave illumination. You may use the following identities:

$$J_0(z) = \frac{1}{2\pi} \int_0^{2\pi} \exp[-iz \cos(\theta - \phi)] \, d\theta,$$

and

$$z J_1(z) = \int_0^Z \alpha J_0(\alpha) \, d\alpha,$$

where J_0 and J_1 are the zero-order and first-order Bessel functions of the first kind, respectively.

8.3   Assume that the expression of the complex wave field of an open aperture caused by Fraunhofer diffraction is given as $g(x, y)$. If two identical apertures with separation d are located in the diffraction screen, as shown in Figure 8.24, calculate the resultant intensity distribution, in terms of $g(x, y)$, at the observation screen. (*Hint*: use the Fourier transform property of Fraunhofer diffraction.)

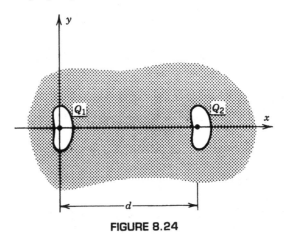

FIGURE 8.24

8.4   An image transparency, with an amplitude transmittance function $f(x, y)$, is normally illuminated by a monochromatic plane wave of λ, as shown in Figure 8.25.

(a) Draw an analog system diagram to evaluate the complex light field over the α, β plane.

(b) Calculate the corresponding light distribution $g(\alpha, \beta)$.

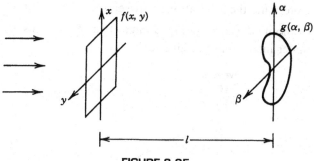

FIGURE 8.25

8.5 Use the image transparency in Problem 8.4, which is now illuminated by a monochromatic light field $u(x, y)$ instead of a plane wave.

(a) Draw an analog system diagram to represent this problem.

(b) Evaluate the corresponding complex light distribution over the α, β plane.

8.6 Use the result obtained in Problem 8.4b, that is, $g(\alpha, \beta)$.

(a) Draw an analog system diagram to evaluate the light field that emerges from the transparency.

(b) Show that the complex light field that emerges from the transparency is indeed equal to $f(x, y)$.

8.7 Using the optical imaging system shown in Figure 8.26:

(a) Draw an analog system diagram to represent this optical system.

(b) If a monochromatic point source is located at the front focal length of the optical system, calculate the complex light field at the output plane.

Remember that the phase transform of a positive lens is

$$T(x, y) = C \exp\left[-i\frac{k}{2f}(x^2 + y^2)\right].$$

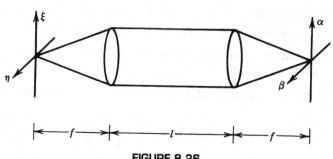

FIGURE 8.26

8.8 A diffraction screen contains five narrow slits, as shown in Figure 8.27. If a monochromatic plane wave of $\lambda = 6000\,\text{Å}$ is normally incident on the screen, and if the optical disturbances produced by the individual slits, at a point P lying 1 m behind the screen, are assumed to be in phase:

(a) Determine the separations between the narrow slits.

(b) Compute the corresponding intensity at P, in terms of the irradiances under incoherent illumination.

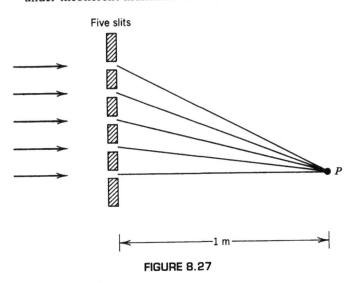

Five slits

$\vdash$————1 m————$\dashv$

FIGURE 8.27

8.9 Referring to Eq. 8.23:

(a) Sketch the transmittance as a function of x for the Fresnel zone plate shown in Figure 8.10.

(b) Sketch the transmittance as a function of x^2.

(c) Expand the transmittance function sketched in (b) into a one-dimensional Fourier series.

8.10 The transmittance function of a Fresnel zone plate can be written in the following Fourier series expansion,

$$T(x,y) = A_0 + \sum_{n=1}^{\infty} A_n \cos\left[\frac{n\pi}{\lambda R}(x^2 + y^2)\right],$$

where A_0, A_n, and R are arbitrary constants. If this Fresnel zone plate is normally illuminated by a monochromatic plane wave of λ:

(a) Calculate the focal distances of the zone plate.

(b) Sketch the locations of the convergent focal points. Recall that $\cos\theta = \frac{1}{2}(e^{i\theta} + e^{-i\theta})$.

8.11 Assume that the transmittance function of a Fresnel zone lens is

$$T(x,y) = K_1 + K_2 \cos\left[\frac{2\pi}{\lambda}\left(\frac{x^2 + y^2}{2R} - x\sin\theta\right)\right].$$

If a monochromatic plane wave of λ is normally incident on this zone lens:

(a) Compute the focal distance of the zone lens.

(b) Sketch the location of the convergent focal point.

8.12 If the zone lens in Problem 8.11 is normally illuminated by a white-light plane wave:

(a) Show that the focal point would be smeared into rainbow colors.

(b) Sketch the location of the smeared focal point.

8.13 Two *mutually coherent* monochromatic light beams illuminate an observation screen, as shown in Figure 8.28. If the complex amplitude distributions produced by these two beams are represented by $U_1(x,y)$ and $U_2(x,y)$, compute the resultant intensity distribution $I(x,y)$.

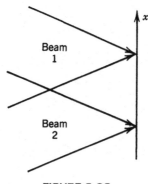

FIGURE 8.28

8.14 If two light beams such as those in Problem 8.13 are instead *mutually incoherent*:

(a) Calculate the resultant intensity distribution.

(b) Compare the results obtained in part a in Problem 8.13. State your observation.

8.15 Two monochromatic plane waves of equal intensity illuminate an observation screen, as shown in Figure 8.29a. Use the fringe intensity distribution as plotted in Figure 8.29b.

(a) Calculate the degree of coherence between these two beams of light.

(b) If the wavelength of both of these partially coherent beams is $\lambda = 500\,\text{nm}$, compute the spatial frequency of the fringes.

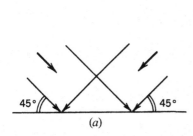 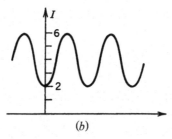

FIGURE 8.29

8.16 Refer to Young's experiment, which was discussed in Example 8.9. Assume that the light source is a point source (i.e., $\Delta s = 0$) that produces two spectral lines, $\lambda_1 = 5770\,\text{Å}$ and $\lambda_2 = 5790\,\text{Å}$, of equal intensity.

(a) Calculate the visibility of the interference fringes at the observation plane.

(b) Sketch the corresponding interference fringes calculated in (a).

8.17 Assuming that the spectral line width of an argon laser is $\Delta\lambda = 0.02\,\text{Å}$ and that its central wavelength is $\lambda = 4880\,\text{Å}$, compute the coherent length of the light source.

INTERFERENCE

When two light beams originating from a point source are superimposed on an observation screen by suitable optical elements, the intensity in the overlapped region is found to vary from point to point. The maximum intensity can exceed the sum of the intensities of the two beams, whereas the minimum intensity may be zero. This phenomenon is known as *interference*, which is one of the most important properties of the wave nature of light. Historically, the observation of interference fringes by Young in 1801 provided the evidence for the wave theory of light. Today, *interferometry* offers viable measurement techniques that can be applied in many areas, such as metrology and spectroscopy. In this chapter, the theory of interference will be introduced. Several interferometers and their applications will also be discussed.

9.1 Condition for Interference

The phenomenon of interference is illustrated schematically in Figure 9.1. Two plane waves are projected onto a screen. In the region where these two waves meet, the observed intensity is not simply the sum of that of the two waves. Instead, interference fringes are observed. However, such a phenomenon is not common in our daily life. For example, when we illuminate our classroom with several light bulbs, no interference fringes are observed on the wall. In other words, interference occurs only under certain circumstances.

To understand the condition for interference, we start with the *principle of superposition*, which is a basic property of all kinds of waves, be they mechanic, acoustic, optic, or electromagnetic waves. When two waves are superimposed, the resulting amplitude distribution is the addition of the instantaneous amplitude of these two waves. Now let us consider two monochromatic waves from two point sources S_1 and S_2, as shown in Figure 9.2. Assume that P is a point within the superimposed region and that the

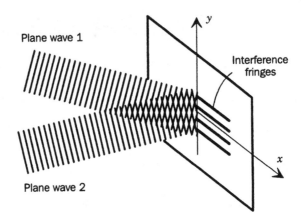

FIGURE 9.1 Interference of two plane waves.

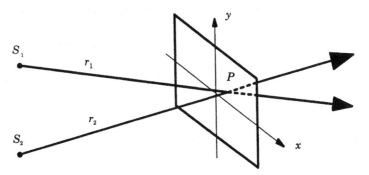

FIGURE 9.2 Superposition of two optical waves at point P.

distances from P to S_1 and S_2 are r_1 and r_2, respectively. If these two optical waves have the same polarization (i.e., vibration of electric vectors), their electric fields at point P can be written as

$$E_1 = A_1 \cos\left(\omega_1 t - \frac{2\pi r_1}{\lambda_1} + \phi_1\right) \tag{9.1}$$

$$E_2 = A_2 \cos\left(\omega_2 t - \frac{2\pi r_2}{\lambda_2} + \phi_2\right), \tag{9.2}$$

where A_1 and A_2 denote the amplitude of the optical fields, ω_1 and ω_2 are the angular frequencies, λ_1 and λ_2 are the wavelengths, and ϕ_1 and ϕ_2 are the initial phases of the two waves. The superimposed amplitude can be shown to be

$$E_1 + E_2 = A_1 \cos(\omega_1 t + \alpha_1) + A_2 \cos(\omega_2 t + \alpha_2), \tag{9.3}$$

where we have combined the second and third terms in the parenthesis in Eqs. 9.1 and 9.2 into a single constant term α_1 or α_2. The corresponding intensity of an optical wave is defined as the square of amplitude averaged over the response time of detectors, such as

$$I = \left\langle |E_1 + E_2|^2 \right\rangle$$

$$= \left\langle A_1^2 \cos(\omega_1 t + \alpha_1) + A_2^2 \cos(\omega_2 t + \alpha_2) \right.$$

$$+ A_1 A_2 \cos[(\omega_1 + \omega_2)t + (\alpha_1 + \alpha_2)]$$

$$\left. + A_1 A_2 \cos[(\omega_1 - \omega_2)t + (\alpha_1 - \alpha_2)] \right\rangle. \tag{9.4}$$

The frequency of visible light is of the order of 10^{14} Hz. At present, there is no physical detector that is able to follow instantaneously a vibration of such high frequency. Instead, the detected intensity is always the average of many cycles of the light waves. The resulting intensity is therefore given by

$$I = \frac{1}{2}A_1^2 + \frac{1}{2}A_2^2 = I_1 + I_2, \tag{9.5}$$

where I_1 and I_2 are the intensities of the two light waves. In other words, the total intensity is the addition of two optical fields between which no interference is observed.

However, if the two light waves oscillate at same frequency, $\omega_1 = \omega_2 = \omega$, Eq. 9.4 can be written as

$$I = \left\langle A_1^2 \cos(\omega t + \alpha_1) + A_2^2 \cos(\omega t + \alpha_2) \right.$$

$$\left. + A_1 A_2 \cos[2\omega t + (\alpha_1 + \alpha_2)] + A_1 A_2 \cos(\alpha_1 - \alpha_2) \right\rangle$$

$$= I_1 + I_2 + 2\sqrt{I_1 I_2} \cos(\delta), \tag{9.6}$$

where

$$\delta = \alpha_1 - \alpha_2 = \frac{2\pi(r_1 - r_2)}{\lambda} + (\phi_1 - \phi_2) \tag{9.7}$$

is the phase difference between the two light waves at point P. In order to obtain a stable intensity distribution, the phase difference at any given point must be the same. This means that the difference in initial phase shift between the two light waves $(\phi_1 - \phi_2)$ should be constant. At various points in space, the resultant intensity can be greater than, less than, or equal to $I_1 + I_2$, depending on the value of δ. It is apparent that maximum intensity occurs at $\cos \delta = 1$, such that

$$I_{max} = I_1 + I_2 + 2\sqrt{I_1 I_2}, \tag{9.8}$$

when

$$\delta = 0, \pm 2\pi, \pm 4\pi, \ldots. \tag{9.9}$$

In this case the phase difference between the two waves is an integer multiple of 2π, and the two light waves are said to oscillate *in phase*. The minimum intensity results when the waves are $180°$ out of phase ($\cos \delta = -1$), such that

$$I_{min} = I_1 + I_2 - 2\sqrt{I_1 I_2}, \qquad (9.10)$$

when

$$\delta = \pm\pi, \pm3\pi, \pm5\pi, \ldots \qquad (9.11)$$

When the resultant intensity is greater than the sum intensity of two beams, i.e., when $I > I_1 + I_2$, we have *constructive interference*, and when $I < I_1 + I_2$ we have *destructive interference*. The alternation of constructive and destructive interferences forms fringes on a screen, as shown in Figure 9.1.

EXAMPLE 9.1

If the two plane waves shown in Figure 9.1 are described by

$$E_1 = A \cos\left[\omega t - \frac{2\pi(z \cos\theta + y \sin\theta)}{\lambda}\right]$$

and

$$E_2 = A \cos\left[\omega t - \frac{2\pi(z \cos\theta - y \sin\theta)}{\lambda}\right],$$

what is the period of the interference fringes?

The phase difference between the plane waves is

$$\delta = \frac{4\pi y \sin\theta}{\lambda}.$$

Let the phase difference equal 2π. Then the period of the interference fringes is

$$\Lambda = \frac{\lambda}{2\sin\theta}.$$

If the wavelength is 0.633 nm and the oblique angle is $1°$, the period of fringes is about 18 µm, which corresponds to about 55 fringes/mm.

Recalling all the assumptions we made to derive Eq. 9.6, the following conditions must be met for two waves to interfere:

(1) *Same polarization.* Two orthogonal polarization states cannot interfere. They generally lead to a resulting wave that is in a different polarization state.

(2) *Same frequency*. Two waves must oscillate at the same frequency to cancel the frequency-dependent term in the argument of the cosine function in Eq. 9.4.

(3) *Constant phase relationship*. The phase difference as represented by Eq. 9.7 must be a constant at any given point in the superimposed region. Otherwise, no stable interference fringes can be observed.

Light waves meeting those conditions are called *coherent waves*. They are usually derived from the same light source and then divided by some optical elements such as mirrors, beam splitters, or slits. The methods of beam splitting are generally categorized into two classes: wavefront splitting and amplitude splitting. Interferometers based on each type of splitting method will be discussed in the following sections.

9.2 Young's Experiment

The first experiment demonstrating the interference of light was conducted by Young in 1801. The experimental set-up is illustrated schematically in Figure 9.3. It is a typical wavefront-splitting interferometer, in which the wavefront originating from a monochromatic light source S passes through two slits in a screen.

From the discussion on diffraction in the preceding chapter, we have learned that when a wave passes through an aperture it is always spread to some extent into the region which is not directly exposed to the oncoming wave. To explain such a bending of light, Huygens proposed the rule that each point on a wavefront may be regarded as a new source of waves, which is called the *Huygens principle*. Based on this principle, the slits in the screen will produce new waves with cylindrical wavefront, as shown in Figure 9.3.

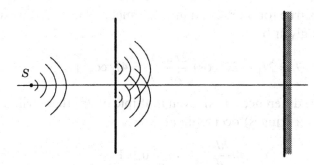

FIGURE 9.3 Experimental set-up for Young's double slit experiment.

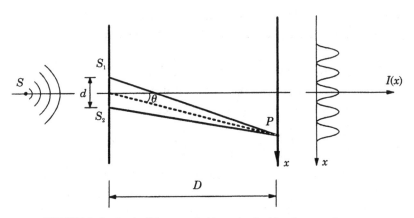

FIGURE 9.4 Path difference in Young's double slit experiment.

These two waves, originating from the same light source S, have exactly the same polarization, frequency, and fixed phase relationship. When they meet in space, interference occurs.

The geometry of Young's double slit experiment is illustrated in Figure 9.4. A monochromatic point source, which itself may be a slit or a pinhole, illuminates the diffraction screen containing two slits, S_1 and S_2, separated by a small distance d. For simplicity, and without loss of generality, all slits are assumed to be vanishingly thin. The distance d is typically of the order of millimeters. Since the slits are taken to be equidistant from the light source S, the waves start out from S_1 and S_2 in the same phase. Furthermore, the amplitudes are practically the same if S_1 and S_2 are of equal width and very close (as is usually the case). For an observation point P, the paths from the two slits have different lengths. As a result, the phase difference δ varies from point to point, as does the fringe intensity as given by Eq. 9.7.

The phase difference at point P can be written as

$$\delta = \frac{2\pi}{\lambda}(S_1P - S_2P) \approx \frac{2\pi dx}{\lambda D}. \qquad (9.12)$$

In the derivation of Eq. 9.12 we have assumed that D is much larger than d, which is always true for a practical interferometer. The corresponding intensity is therefore given by

$$I = 2I_1 + 2I_1 \cos\left(\frac{2\pi dx}{\lambda D}\right) = 4I_1 \cos^2\left(\frac{\pi dx}{\lambda D}\right). \qquad (9.13)$$

When the phase difference δ is an even multiple of 2π, there will be maximum intensities (bright fringes) occurring at

$$x = m\frac{\lambda D}{d} \qquad m = 0, \pm 1, \pm 2, \ldots. \qquad (9.14)$$

When d is an odd number of π, the intensity at P will be zero (dark fringes), which occurs at

$$x = \left(m + \frac{1}{2}\right)\frac{\lambda D}{d} \qquad m = 0, \pm 1, \pm 2, \ldots. \qquad (9.14)$$

The number m, which characterizes a specific bright fringe, is known as the order of interference. Hence the fringes with $m = 0, 1, 2, \ldots$, are called the zero, first, second, etc., orders.

The spacing between two adjacent bright (or dark) fringes can be shown to be

$$\Lambda = \frac{\lambda D}{d} \approx \frac{\lambda}{\theta}, \qquad (9.15)$$

which is directly proportional to the distance between the diffraction screen and the observation screen D; the wavelength of the light source λ is inversely proportional to the separation of slit d. Equation 9.15 is trivial when the configuration of the Young's interferometer is known, and the measured fringe spacing can be used to determine the wavelength of the light source.

EXAMPLE 9.2

Two narrow parallel slits separated by a distance of 1 mm are illuminated by a diode laser. The plane of slit is located at a distance of 800 mm from the observation screen. If the spacing between the zero and tenth-order fringes is measured to be 5.44 mm, determine the wavelength of the diode laser.

From the given data, we have $d = 1$ mm, $D = 800$ mm, and $\Lambda = 0.544$ mm. Then the wavelength of the laser diode would be

$$\lambda = \frac{\Lambda d}{D} = \frac{0.544 \text{ mm} \times 1 \text{ mm}}{800 \text{ mm}} = 0.68 \text{ μm}.$$

9.3 Visibility of Fringes

Equation 9.13 represents an ideal condition where sharp fringes can be observed no matter how large the phase difference is. In practice, however, the sharpness (i.e., the contrast) of the fringes decreases as the phase difference between the two beams increases. When the phase difference is beyond a certain value, the fringes eventually vanish. This is a result of the limited

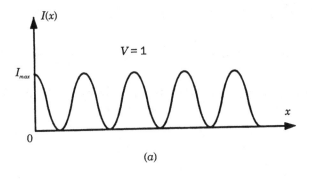

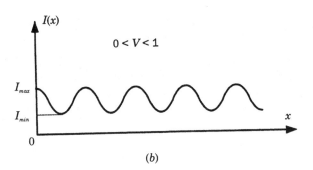

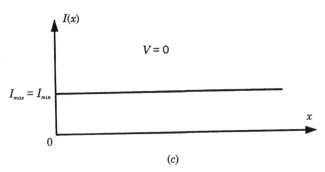

FIGURE 9.5 Fringes with different visibility values.

coherence of light sources. The contrast of the fringes can be described quantitatively as the *visibility*, which is defined as

$$V = \frac{I_{\max} - I_{\min}}{I_{\max} + I_{\min}} \tag{9.16}$$

The value of this visibility is between 0 and 1. When the fringes are at the highest possible intensity value in the bright areas and zero intensity in the dark areas, the visibility is equal to 1, as shown in Figure 9.5a. As the phase difference increases, the coherence between the light waves decreases and the visibility is reduced, as shown in Figure 9.5b. Finally, as the coherence between the two light waves vanishes, I_{max} and I_{min} converge and the visibility V goes to zero. No fringe is observed, as illustrated in Figure 9.5c.

From Eq. 9.6, the visibility of the fringes resulting from two beams can be expressed as

$$V = \frac{2\sqrt{I_1 I_2}}{I_1 + I_2} = \frac{2\sqrt{I_1/I_2}}{1 + I_1/I_2}. \qquad (9.17)$$

The closer the intensities of two waves, the higher is the visibility of the fringes. When $I_1 = I_2$, we have $V = 1$. Therefore, in an interferometer the intensity of the two beams should be kept as close as possible.

We have so far considered purely monochromatic light sources. In practice, however, the spectrum of a light source always has a finite bandwidth. If a light source with finite spectral bandwidth is used, the visibility is smaller than 1 even though the two light waves have very little phase difference. The visibility is, in fact, a measure of the temporal coherence of the light source, which is determined by the spectral bandwidth of the source. The maximum optical path difference between two beams that leads to visible fringes (i.e., $V > 0$) is called the *coherence length*. For a light source with spectral bandwidth of $\Delta\lambda$, the coherence length is given by

$$\Delta r = \frac{\lambda^2}{\Delta\lambda}. \qquad (9.18)$$

A more detailed discussion of coherence and the measurement of the coherence length can be found in Chapter 8 (Sections 8.5–8.7).

9.4 Michelson Interferometer

The Michelson interferometer, developed by Albert Michelson in the 1880s, played a premier role in the development of modern physics, particularly in the theory of relativity. It is a highly accurate measuring instrument and is widely used in many scientific and engineering applications.

The arrangement of a Michelson interferometer is illustrated schematically in Figure 9.6. As an example of amplitude-splitting interferometers, the two interfering beams are obtained using a half-reflective and half-transmissive mirror, the so-called *beam splitter*. These two beams are then sent in different directions against two highly polished plane mirrors, whence they are reflected and combined again by the beam splitter to form interference fringes. A plate made of the same material as the substrate of the beam splitter is set in one of the two arms as a compensation plate. This ensures

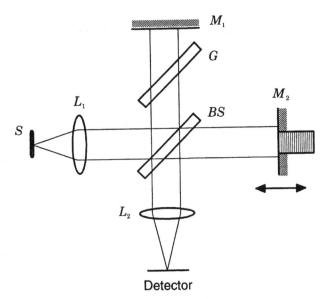

M_1

G

L_1

M_2

S

BS

L_2

Detector

FIGURE 9.6 Michelson interferometer. S, light source; BS beam splitter; M_1, mirror; M_2, movable mirror; G, compensation glass; L, lenses.

that the optical path in each arm is rendered equivalent by passing through the same thickness of the same material. When wide-band light sources are used, the compensation plate is essential to compensate for the dispersion of the different spectral components that occurs within the beam splitter substrate. Mirror M_2 is equipped with a micrometer screw so that it may be moved along the optical axis, thereby adjusting the optical path difference between the two arms.

If a point light source is placed at the front focal point of lens L_1, the input light to the interferometer is a collimated beam. The output beam is then focused by lens L_2 onto the detector. Although no fringes are observable, the detected intensity does change while the mirror M_2 moves. It changes from I_{max} to I_{min}, and then back to I_{max} again, when M_2 moves a distance of $\lambda/2$. If the mirror moves a distance of l, the number of cycles in the intensity change is

$$N = \frac{2l}{\lambda}. \tag{9.19}$$

The change in output intensity can also be explained by the *Doppler effect*. When a light source moves towards (or away from) a detector, the apparent frequency felt by the detector is shifted from original frequency and is given by

$$\omega = \omega_0 \left(1 \pm \frac{V}{c}\right), \tag{9.20}$$

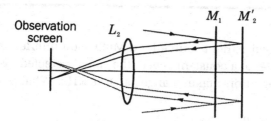

FIGURE 9.7 Formation of circular fringes in a Michelson interferometer.

where ω_0 is the original frequency, V is the velocity of the light source (or the detector), and the plus or minus sign is determined by the direction of movement. Assume that the M_1 stands still and M_2 moves toward the beam splitter at a velocity of V (resulting in an effective velocity of the light source of $2V$). The light reflected from M_1 and M_2 will have a frequency difference as given by

$$\Delta\omega = \omega_0 \frac{2V}{c} = \frac{4\pi V}{\lambda}. \tag{9.21}$$

When these two frequencies are detected simultaneously, a *beat* is seen. If the detected intensity changes N times within a period T, then

$$N = \frac{1}{2\omega} \int_0^T \Delta\omega \, dt = \frac{2l}{\lambda}, \tag{9.22}$$

where l is the movement of M_2 during T. It can be seen that Eq. 9.22 is identical to Eq. 9.19.

To obtain fringes in the observation plane, an extend source should be used and the lens L_1 may be removed. The image of the light source, reflected from M_1 and M_2, respectively, are superimposed at the output plane to produce interference fringes. When M_2 is adjusted exactly perpendicular to M_1, concentric circular fringes are formed. This process is illustrated in Figure 9.7, in which M_2' corresponds to the image of M_2 in the beam splitter.

If the mirrors are slightly tilted, the center of the circular fringes moves from the center of the observing plane to a position well removed from the center. The interference fringes then appear as almost straight lines in the observation plane. Moving the mirror M_2 will cause the fringes to sweep through the field. Each movement of M_2 by $\lambda/2$ will effect a path difference of λ, and will move the bright fringe at the center of the field one pitch and replace it with the next adjacent bright fringe. By counting the fringes swept from the center of the observing field, the movement of the mirror can be measured. Alternatively, if the movement of M_2 is accurately measured, the wavelength of the light source can be determined.

EXAMPLE 9.3

A HeNe laser is used as light source for a Michelson interferometer. When one of the mirrors moves at a constant speed, the signal detected by a detector at the center of the observation plane changes at 1.58 KHz. What is the speed of the moving mirror?

From the given conditions, we have

$$\lambda = 0.633\,\mu m$$

and

$$\frac{N}{t} = 1580\ 1/s.$$

The velocity of the moving mirror is:

$$V = \frac{l}{t} = \frac{\lambda}{2}\frac{N}{t} = 0.5\,mm/s.$$

In contrast to monochromatic light sources (e.g., lasers), white-light sources have a very short coherent length. As an example, a light source with a uniform spectrum ranging from 400 to 780 nm has a coherent length as short as about 1 μm. Because of the extremely short coherent length, the two arms in the Michelson interferometer must be adjusted to have equal optical paths so that visible interference fringes can be obtained. If the two arms are set strictly equal, all wavelengths interfere constructively and a white, bright fringe is observed at the center of the observation field. As the length of one arm varies with respect to the other, some wavelengths interfere constructively, while others interfere destructively. Consequently, the fringes are of different colors. Table 9.1 lists the colors observed as a

TABLE 9.1 Color of fringes from a white-light Michelson interferometer

Path difference (μm)	Color of fringes
0	White
0.158	Brown-white
0.259	Bright red
0.332	Blue
0.565	Green
0.664	Orange
0.747	Red
0.866	Violet
1.101	Green
1.376	Violet

function of the path difference between the two arms. While mirror M_2 keeps moving, the intensity of these color fringes will fade and the field eventually reverts to a white-light field with no visible fringes. Since the coherent length of white light is typically of the order of 1 μm, one can usually observe only five to eight fringes.

The white-light Michelson interferometer can provide a zero reference for length measurement. If both a white-light source and a laser are used, each of them will generate interference fringes. Whenever the white, bright fringe appears, the counter for laser fringes resets to zero. As the mirror moves, the counter counts the laser fringes swept through the center of the observation which determines the moving distance of the mirror with respect to the position where white-light fringes were obtained. Such a system has been used in the calibration of step-motor-driven stages.

9.5 Mach–Zehnder Interferometer

The Mach–Zehnder interferometer is another amplitude-splitting interferometer. As shown in Figure 9.8, it consists of two beam splitters and two totally reflecting mirrors. The two waves within the apparatus travel along different paths. If the incoming light is a collimated beam, straight fringes with uniform fringe spacing can be obtained by slightly tilting one beam splitter. Since the two paths are separated, the insertion of some samples into one of the arms produces a path difference, which generates interference fringes in the observing plane.

A common application of the Mach–Zehnder interferometer is in the observation of the density variation in gas flow chambers such as wind

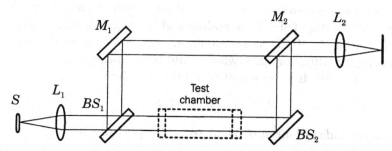

FIGURE 9.8 The Mach–Zehnder interferometer. S, Extended light source; BS_1, BS_2, beam splitter; L_1, L_2, lenses; M_1, M_2, mirrors.

tunnels and shock tubes. One beam passes through the test chamber (sealed with optically flat windows). Since some wind tunnels are several meters in diameter, such interferometers are usually very large. To compensate for the dispersion caused by the windows of the test chamber, some optical plates may be inserted in the other arm. It has been found experimentally that the value of $(\eta - 1)$ is directly proportional to the air pressure at a given temperature, such that

$$(\eta - 1) = \frac{2}{3} P, \qquad (9.23)$$

where η is the refractive index of the gas and P is the pressure. The variation in air density in space thus causes nonuniform distribution of refractive index. The beam within the test chamber propagates through regions having a spatially varying index of refraction. The resulting distortion in the wavefront generates the fringe contours at the observation plane.

Another application of the Mach–Zehnder interferometer is to measure the refractive index of certain types of gas at different temperatures and pressures. Two similar evacuated tubes T_1 and T_2 are placed in the two arms of the interferometer. While T_1 is kept unchanged, gas is slowly admitted into T_2. The number of fringes swept through the center of the observation plane is counted while the gas in T_2 reaches the desired temperature and pressure. If the length of the tubes is l and total number of fringes counted is N, the refractive index can be determined by

$$(\eta - 1)l = N\lambda. \qquad (9.24)$$

Since the tube can be very long, a small change in refractive index can be measured accurately.

EXAMPLE 9.4

A HeNe laser is used in a Mach–Zehnder interferometer. Two 1 m long tubes filled with air are placed in the two arms of the interferometer. Tube T_1 is kept at 0°C and 76 mmHg. Tube T_2 is maintained at the same temperature, but is filled with more air. If 15 fringes sweep across the center of the observation plane during the filling process, what would be the pressure in T_2?

Equation 9.24 needs to be modified for this problem, that is

$$(\eta_2 - \eta_1)l = N\lambda.$$

The refractive index of air at 0°C and 76 mmHg pressure is

$$\eta_1 = 1.000292$$

and the refractive index of the air in T_2 can be calculated as,

$$\eta_2 = \eta_1 + \frac{N\lambda}{l} = 1.000292 + \frac{15 \times 0.633 \times 10^{-6}\,\text{m}}{1\,\text{m}} = 1.0003015$$

According to Eq. 9.23, the pressure is calculated as:

$$P_2 = \frac{\eta_2 - 1}{\eta_1 - 1} P_1 = \frac{0.0003015}{0.000292} \times 76 = 78.47\,\text{mmHg}$$

It can be seen that a subtle change in pressure can cause a shift of 15 fringes.

The Mach–Zehnder interferometer can also be used to measure the rotation of an object. If the interferometer is rotating clockwise, the light passing through the first arm (reflected by M_1) will have a longer time of flight than the light traveling through the other arm. This time delay creates an optical path difference, and results in a shift of fringes at the observation plane. This phenomenon is called the *Sagnac effect*. The fringe shift is a measure of the angular velocity of the interferometer as well as the object on which the interferometer is mounted. To increase the path difference caused by the rotation and thereby to improve the sensitivity of the measurement, the interferometer is often modified as shown in Figure 9.9. The beam splitter BS_2 is replaced by another mirror. BS_1 is used for both beam splitting and combining. The optical beams traveling clockwise and counterclockwise interfere at the output plane. If the observed fringe shift is N, the angular velocity of the interferometer is given by

$$\Omega = \frac{c\lambda N}{4A}, \tag{9.25}$$

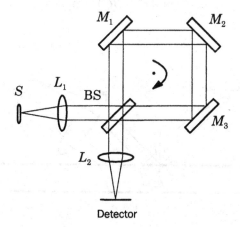

FIGURE 9.9 A modified Mach–Zehnder interferometer used to measure rotation.

where A is the area of the interferometer, c is the velocity of light, and λ is the wavelength of the light source. It is noted that the fringe shift N is not necessarily an integer. This type of interferometer (implemented with optical fibers) is now being used as a laser (or optical fiber) *gyroscope* in inertial guidance systems.

9.6 Two-beam Interference with Dielectric Plates

Interference resulting from dielectric plates or thin films is commonly observed and has many applications. By a dielectric plate we mean a layer of transparent material composed of two surfaces. The thickness of a dielectric plate can be from a fraction of a wavelength to several centimeters. Common examples of interference from dielectric plates are the spectacular color patterns arising from oil slicks or soap films.

First, let us consider a transparent parallel plate of thickness d illuminated by a monochromatic point source S, as shown in Figure 9.10. Since the reflectance of an uncoated interface is usually very low (about 4–5%), only the reflected beams from the top and the bottom interfaces (both having undergone only one reflection) need be considered; the multiple reflections can be neglected. Thus the transparent plate serves as an amplitude-splitting element. For any incident light ray, the two reflected rays are parallel on leaving the plate and are brought to a point P on the focal plane of the lens L (or the retina of the eye when it is focused at infinity). From Figure 9.10, the optical path difference for the two reflected rays is given by

$$\Delta = \eta_p(AB + BC) - \eta_1 AD. \tag{9.26}$$

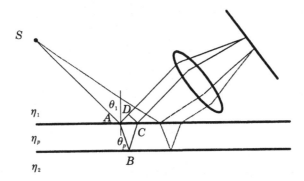

FIGURE 9.10 A parallel plate illuminated by a point source.

It is trivial to see that

$$AB = BC = \frac{d}{\cos \theta_p},$$ (9.27)

and

$$AD = AC \sin \theta_1 = (2d \tan \theta_p)\left(\frac{\eta_p}{\eta_1} \sin \theta_p\right).$$ (9.28)

By substituting Eqs 9.27 and 9.28 into Eq. 9.26, we have

$$\Delta = 2\eta_p d \cos \theta_p.$$ (9.29)

If the plate is immersed in a single medium (e.g., air), the index of refraction can be simply written as $\eta_1 = \eta_2 = \eta$. Regardless of the polarization of the incoming light, the two reflected light rays (one internally and the other externally reflected) will experience a relative phase shift of π. Hence the phase difference between these two reflected light rays can be expressed as

$$\delta = \frac{4\pi\eta_p}{\lambda} d \cos \theta_p + \pi.$$ (9.30)

Constructive interference (bright fringe) occurs when $\delta = 2m\pi$ and destructive interference (dark fringe) occurs when $\delta = (2m + 1)\pi$. From Eq. 9.30, the conditions for interference maxima and minima can be written as

$$\text{Maxima:} \quad d \cos \theta_p = \left(m - \frac{1}{2}\right)\frac{\lambda}{2}, \quad m = 0, 1, 2, \ldots,$$ (9.31)

$$\text{Minima:} \quad d \cos \theta_p = m\frac{\lambda}{2}, \quad m = 0, 1, 2, \ldots.$$ (9.32)

In other words, for a given thickness of the plate d, the interference intensity is determined by the refractive angle θ_p or, equivalently, the incident angle θ_1. These interference fringes are called *fringes of equal inclination*. Since the interference intensity depends only on the incident angle, extended light sources can be used. Keep in mind that each point on an extended source is incoherent with respect to other points. Since the resulting fringes are incoherently (i.e., intensity) added, the fringe pattern would be very bright and, therefore, easy to observe.

If the thickness of the plate is not uniform, its variation should be taken into consideration. In practice, there is a type of interference for which the optical thickness $\eta_p d$ is the dominant parameter (rather than θ_1). These are referred to as *fringes of equal thickness*. Under white-light illumination, the iridescence of soap bubbles and oil slicks are examples of equal thickness fringes. Interference fringes of this kind are analogous to the constant height contour lines of a topographic map. Each fringe is the locus of all points in the plate with the same optical thickness. Consequently, they are very useful for determining the surface quality of optical elements. Figure 9.11 shows an

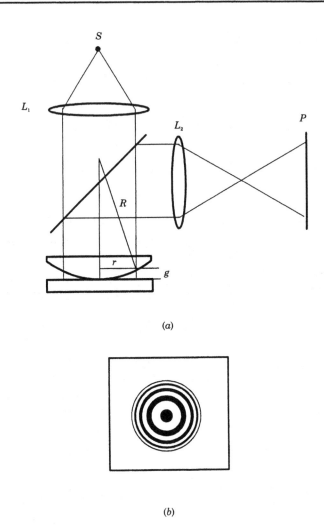

(a)

(b)

FIGURE 9.11 Newton's rings: (a) experimental arrangement; (b) Newton's ring obtained with monochromatic light.

apparatus used in optical shops to evaluate the surface curvature of optical lenses. The lens to be tested is put into contact with an optical flat glass. The air in the space between the two surfaces generates circular interference fringes. As this phenomenon was first studied by Newton, the ring-shaped fringes depicted in Figure 9.11b are also known as *Newton's rings*.

Light from a monochromatic point source is collimated and then projected onto the optical flat at the normal incident angle. The outcoming light is then reflected by the beam splitter and is imaged on the observation plane by the lens L_2. The path difference between the two light rays reflected from the bottom surface of the lens to be examined and the top surface of the flat glass is given by

$$\Delta = 2g = 2\left(R - \sqrt{R^2 - r^2}\right),\tag{9.33}$$

where g is the thickness of the air gap, R is the nominal radius of the lens surface, and r is the radial distance from the central contact point, as shown in Figure 9.11a. Thus the phase difference between the two reflected light rays can be written as

$$\delta = \frac{4\pi}{\lambda}\left(R - \sqrt{R^2 - r^2}\right) + \pi. \tag{9.34}$$

Constructive interference (bright fringe) occurs when $\delta = 2m\pi$ and destructive interference (dark fringe) occurs when $\delta = (2m + 1)\pi$. Ideally, the fringes resulted from a perfect lens surface should be concentric circular patterns. Any deviation from circular fringes indicates irregular variations in the lens surface, or defects. The spacing of the rings can be measured to determine the radius of the lens surface. Precise evaluation of the surface curvature is usually conducted using an accurate test plate that closely matches the surface under examination. This has the effect of widely spacing the rings, allowing a much more accurate measurement of the fringe spacing.

EXAMPLE 9.5

Consider that a HeNe laser is used to illuminate the experimental set-up shown in Figure 9.11. The diameters of the fifth and the fifteenth bright ring fringes are measured as 3.375 and 6.058 mm, respectively. Calculate the radius of curvature of the lens surface.

From the optical path difference equation (Eq. 9.34), we see that

$$\delta_5 = \frac{4\pi}{\lambda}\left(R - \sqrt{R^2 - r_5^2}\right) + \pi = 5 \times 2\pi$$

$$\delta_{15} = \frac{4\pi}{\lambda}\left(R - \sqrt{R^2 - r_{15}^2}\right) + \pi = 15 \times 2\pi$$

By subtracting the above two equations, we have

$$\sqrt{R^2 - r_5^2} - \sqrt{R^2 - r_{15}^2} = 5\lambda.$$

Substituting the wavelength of the laser (0.6328 μm) and the radii of the fifth and fifteenth bright fringe rings, the curvature of the lens surface can be calculated as

$$R = 500 \text{ mm.}$$

TABLE 9.2 Intensity of reflected and transmitted beams from a glass plate

	No. of reflections				
	1	2	3	4	5
Reflected beam	0.80	0.032	0.021	0.013	0.005
Transmitted beam	0.040	0.026	0.017	0.010	0.007

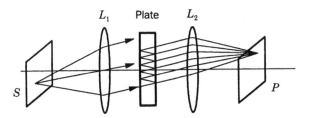

FIGURE 9.12 Multiple beam interference from a parallel plate.

9.7 Multiple Beam Interference

In the discussion in the preceding section, we assumed that the reflectance coefficients at the interfaces of a plate are small, such that the high-order reflections are so weak that they can be neglected. However, if for any reason the reflectance of the interfaces is not negligibly small, the multiple reflections should be taken into account. The relative intensities of the first five reflected and transmitted beams for a glass plate coated with thin films of 80% reflectance on both sides are listed in Table 9.2. It can be seen that, except for the first reflected beam, the intensities of the reflected and transmitted beams are very similar. When these beams meet in space, multiple beam interference takes place.

The interference between the multiple transmitted beams from a parallel plate is shown schematically in Figure 9.12. The light beam from any point of a monochromatic light source S is collimated by the lens L_1 and then obliquely projected onto the transparent plate at an incident angle θ_1, which corresponds to the refractive angle θ_p within the plate. Assuming that the amplitude reflection coefficients for both interfaces is r, then the phase shift between each pair of adjacent transmitted beams can be shown to be

$$\delta = \frac{4\pi\eta_p}{\lambda} d \cos\theta_p + \pi. \tag{9.35}$$

The corresponding complex amplitudes of the transmitted light beams can be written as

$$A_1 = (1 - r^2)A_0,$$
$$A_2 = r^2(1 - r^2)A_0 e^{-i\delta},$$
$$\vdots$$
$$A_n = r^{2(n-1)}(1 - r^2)A_0 e^{-i(n-1)\delta}, \tag{9.36}$$

where A_0 is the complex amplitude of the incident beam, and $A_1, A_2, A_3, \ldots, A_n$ denote the amplitudes of the transmitted beams. The resulting amplitude of these superimposed beams is then given by

$$A = \sum_{n=1}^{\infty} A_n = \frac{(1 - r^2)}{1 - r^2 e^{-i\delta}} A_0. \tag{9.37}$$

and the corresponding intensity distribution is

$$I_T = |A|^2 = \frac{(1 - R^2)I_0}{(1 - R^2) + 4R \sin^2\left(\dfrac{\delta}{2}\right)}, \tag{9.38}$$

where $R = r^2$ is the reflectance at the transparent plate surfaces, as discussed in Chapter 1 (Section 1.7).

Similarly, the interference intensity from the reflected light beams can be shown to be

$$I_R = \frac{4R \sin^2\left(\dfrac{\delta}{2}\right) I_0}{(1 - R^2) + 4R \sin^2\left(\dfrac{\delta}{2}\right)}. \tag{9.39}$$

It should be noted that $I_R + I_T = 1$, for which the law of energy conservation holds. The form of Eqs 9.38 and 9.39 suggests that we introduce a new quantity known as the *coefficient of finesse*, such that

$$F = \frac{4R}{(1 - R)^2}. \tag{9.40}$$

Then the relative interference intensity distribution can be written as

$$\frac{I_T}{I_0} = \frac{1}{1 + F \sin^2\left(\dfrac{\delta}{2}\right)}, \tag{9.41}$$

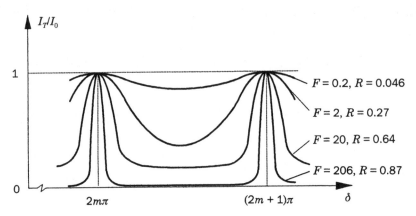

FIGURE 9.13 Relative intensities of transmitted multiple beam interference fringes.

and

$$\frac{I_R}{I_0} = \frac{F \sin^2\left(\frac{\delta}{2}\right) I_0}{1 + F \sin^2\left(\frac{\delta}{2}\right)}. \tag{9.42}$$

The relative transmitted interference intensities from the plates with different reflectance values are plotted in Figure 9.13, from which it can be seen that sharp bright fringes can be obtained with high reflectance.

The multiple beam interference from a Fabry–Perot etalon has many applications in spectroscopy, laser, and atomic physics. A Fabry–Perot etalon consists of two exactly parallel surfaces with high reflectance. In practice, the reflectance of the two surfaces can be as high as 90–99.9% and the separation between the two plates can vary from a fraction of millimeter to several centimeters. If a Fabry–Perot etalon is used in the position of the parallel plate, as shown in Figure 9.12, especially if the separation between the two plates is adjustable, the arrangement is called a *Fabry–Perot interferometer*. Although both reflected and transmitted beams interfere with each other, the Fabry–Perot interferometer is usually used in the transmissive mode. We shall therefore focus our discussion in the rest of this section on the interference of multiple transmitted beams.

Like the two-beam interference, constructive interference occurs whenever the phase difference δ is $2m\pi$, and destructive interference occurs when $\delta = (2m + 1)\pi$. By substituting these conditions into Eq. 9.38, the maximum and minimum intensities can be derived as

$$I_{Tmax} = I_0, \tag{9.43}$$

$$I_{Tmin} = \left(\frac{1-R}{1+R}\right)^2 I_0. \tag{9.44}$$

It can be seen that the maximum intensity remains constant, whereas the minimum intensity reduces as the surface reflectance increases. For a large reflectance R, a small deviation of phase difference δ from $2m\pi$ will lead to a significant reduction in intensity. Consequently, very narrow bright fringes are produced.

The sharpness of the fringes can be measured by the full width at half-intensity (FWHI), $\Delta\delta$. For the mth order fringe, the phase differences corresponding to the half-intensity value are given by

$$\delta = 2m\pi \pm \frac{\Delta\delta}{2}. \tag{9.45}$$

By substituting this in Eq. 9.41, we obtain

$$\frac{I_T}{I_0} = \frac{1}{1 + F\sin^2\left(\dfrac{\Delta\delta}{4}\right)} = \frac{1}{2}. \tag{9.46}$$

Since the surface reflectance of a Fabry–Perot etalon is always very high, the width of the bright fringes is very small, as can be seen from Figure 9.13. Hence, the small angle approximation can be made, as given by

$$\sin\left(\frac{\Delta\delta}{4}\right) \approx \frac{\Delta\delta}{4}. \tag{9.47}$$

The width of the bright fringes can therefore be written as

$$\Delta\delta = \frac{4}{\sqrt{F}} = \frac{2(1-R)}{R}. \tag{9.48}$$

Once again we have shown that the larger the reflectance R, the sharper the transmitted interference fringes.

Another quantity of special interest is the ratio of the separation between adjacent maxima to the half-intensity width. This is known as the *finesse*, and is defined as

$$S = \frac{2\pi}{\Delta\delta} = \frac{\pi\sqrt{F}}{2}, \tag{9.49}$$

where F is the coefficient of finesse as given in Eq. 9.40. If the reflectance of the plate surface is 80%, the finesse of the transmitted interference fringes is about 14. In other words, the fringes are about 14 times as sharp as the fringes obtained by two-beam interference. For a 99.5% reflectance Fabry–Perot etalon, the finesse can be higher than 600.

EXAMPLE 9.6

A Fabry–Perot interferometer is used to measure the interval between two sodium spectral lines. The thickness of the air gap in the Fabry–Perot etalon is 60 μm. The average wavelength of these two lines is known to be 0.5893 nm. Assume that the Fabry–Perot etalon is illuminated by collimated light from a sodium lamp at normal incidence angle. If the shift between the two sets of fringes is about 21% of the fringe spacing, calculate the sodium spectral line interval.

By referring to Eq. 9.35, we have

$$\delta_1 = \frac{4\pi d}{\lambda_1} + \pi = 2m\pi,$$

and

$$\delta_2 = \frac{4\pi d}{\lambda_2} + \pi = 2(m + \Delta m)\pi.$$

Subtracting these two quantities, we obtain

$$2d\left(\frac{1}{\lambda_2} - \frac{1}{\lambda_1}\right) = \frac{2d\,\Delta\lambda}{\lambda^2} = \Delta m.$$

Thus the spectral line interval is calculated as

$$\Delta\lambda = \frac{\lambda^2\,\Delta m}{2d} = \frac{0.5893\,\text{nm} \times 0.5893\,\text{nm} \times 0.21}{2 \times 60\,\mu\text{m}} = 0.6\,\text{nm}.$$

9.8   Optical Thin Film

The reflection at the surfaces of optical elements not only reduces the light efficiency of the optical system, but also generates output noise. It is therefore desirable to eliminate or reduce such reflections. In practice, this can be achieved by depositing a thin film, called *antireflection film*, on the surface of the optical elements.

In Section 1.7, we learned that the reflectance at the boundary surface between air and glass is given by

$$R = \left(\frac{\eta_g - \eta_0}{\eta_g + \eta_0}\right)^2, \tag{9.50}$$

where η_0 and η_g are the refractive indices of air and glass, respectively. The typical value of η_g is 1.5–1.7, which corresponds to a reflectance of 4–6.7%. If

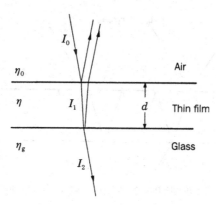

FIGURE 9.14 Two-beam interference with an optical thin film.

an optical system consists of many elements, the total reflection from all surfaces is significant.

If a thin film with refractive index of η is deposited on the surface of a plate of glass, two interfaces are obtained at the air–film and at the film–glass boundaries, as illustrated in Figure 9.14. First let us consider the interference due to two beams. Assume that the incident intensity is I_0, the intensities within the thin film and within the glass can be written as follows:

$$I_1 = \left[1 - \left(\frac{\eta - \eta_0}{\eta + \eta_0}\right)^2\right] I_0, \tag{9.51}$$

and

$$I_2 = \left[1 - \left(\frac{\eta_g - \eta}{\eta_g + \eta}\right)^2\right] I_1 \approx \left[1 - \left(\frac{\eta_g - \eta}{\eta_g + \eta}\right)^2\right] I_0. \tag{9.52}$$

Under the condition of no reflection, the transmitted intensity is equal to the incident intensity, that is

$$I_1 = I_2. \tag{9.53}$$

Thus we have

$$\frac{\eta - \eta_0}{\eta + \eta_0} = \frac{\eta_g - \eta}{\eta_g + \eta}, \tag{9.54}$$

by which we can show that

$$\eta = \sqrt{\eta_0 \eta_g}. \tag{9.55}$$

The thickness of the thin film should be such that the interference of the reflected beams from the two interfaces has the minimum intensity, for which the phase difference can be shown as

$$\delta = \frac{4\pi \eta d}{\lambda} = (2m + 1)\pi, \qquad m = 0, 1, 2, \ldots. \tag{9.56}$$

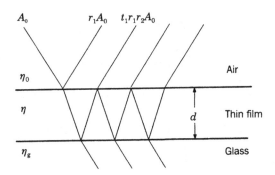

FIGURE 9.15 Multiple beam interference with an optical thin film.

This leads to the following result:

$$\eta d = (2m + 1)\frac{\lambda}{4}, \qquad m = 0, 1, 2, \ldots.. \qquad (9.57)$$

Thus we can see that the required optical thickness of the coated thin film should be $\lambda/4$, $3\lambda/4$, $5\lambda/4$, etc. Equations 9.55 and 9.57 are the criteria for thin film design. It should be noted that, in practice, the selection of the refractive index η is limited by the available materials.

In certain applications, a high reflectance for a given wavelength is needed. In the analysis of high reflectance thin films, multiple beam interferences must be taken into consideration. Multiple reflection and transmission at the two interfaces of a thin film is illustrated in Figure 9.15. Similar to the analysis in Section 9.7, the amplitudes of the reflected beams are given by:

$$A_1 = r_1 A_0,$$
$$A_2 = r_2 t_1^2 A_0 e^{-i\delta},$$
$$A_3 = -r_2^2 r_1 t_1^2 A_0 e^{-i\delta},$$
$$\vdots$$
$$A_n = (-1)^{n-1} t_1^2 r_2^{(n-1)} r_1^{n-2} A_0 e^{-i(n-1)\delta}, \qquad (9.58)$$

where A_0 is the amplitude of the incident beam, $\delta = (4\pi\eta d/\lambda)\cos\theta_p$ is the phase difference between the two adjacent beams, and the $(-1)^{(n-1)}$ term is due to the π phase shift when light is reflected from the interface back to the medium with lower refractive index. The resulting complex amplitude from these superimposed beams is given by:

$$A = \sum_{n=1}^{\infty} A_n = \frac{r_1 + r_2 e^{-i\delta}}{1 - r_1 r_2 e^{-i\delta}} A_0. \qquad (9.59)$$

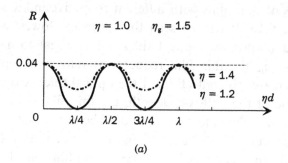

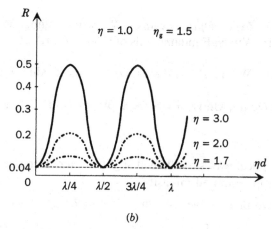

FIGURE 9.16 Reflectance for a single layer optical thin film.

The overall reflectance is obtained as:

$$R = \left(\frac{A}{A_0}\right)^2 = \frac{r_1^2 + r_2^2 + 2r_1r_2\cos\delta}{1 + r_1^2 + r_2^2 + 2r_1r_2\cos\delta}. \quad (9.60)$$

For normal incidence the reflectance can be written as

$$R = \frac{(\eta_0 - \eta_g)^2 \cos^2\left(\frac{\delta}{2}\right) + \left(\frac{\eta_0\eta_g}{\eta} - \eta\right)^2 \sin^2\left(\frac{\delta}{2}\right)}{(\eta_0 + \eta_g)^2 \cos^2\left(\frac{\delta}{2}\right) + \left(\frac{\eta_0\eta_g}{\eta} + \eta\right)^2 \sin^2\left(\frac{\delta}{2}\right)} \quad (9.61)$$

The constructive and destructive interference intensities are given by:

$$R_{max} = \left(\frac{\eta_0\eta_g - \eta^2}{\eta_0\eta_g + \eta^2}\right)^2, \qquad \eta d = \frac{\lambda}{4}, \frac{3\lambda}{4}, \frac{5\lambda}{4}, \ldots, \quad (9.62)$$

$$R_{min} = \left(\frac{\eta_0 - \eta_g}{\eta_0 + \eta_g}\right)^2, \qquad \eta d = \frac{\lambda}{2}, \lambda, \frac{3\lambda}{2}, \ldots. \quad (9.63)$$

The reflectances of thin films with different refractive index values are illustrated in Figure 9.16. It can be seen that both the lowest and the highest reflectance occurs when the optical thickness is equal to a quarter of the wavelenth. If $\eta < \eta_g$ the thin film reduces the reflection, whereas is $\eta > \eta_g$ the thin film increases the reflection. If the optical thickness (ηd) of the thin film is a multiple of the half-wavelength, the thin film does not change the reflectance of the air–glass interface. It should be noted that the antireflection effect of a single layer thin film is wavelength dependent. If antireflection over a broad spectral range is desired, multilayer thin films must be deposited.

REFERENCES

1. U. Hecht and A. Zajac, *Optics*, Addison-Wesley, Reading, MA, 1973.
2. F. Jenkins and H. White, Fundamentals of Optics, fourth edition, McGraw-Hill, New York, 1976.
3. M. Born and E. Wolf, *Principles of Optics*, sixth edition, Pergamon Press, Oxford, 1980.
4. K. D. Möller, *Optics*, University Science Books, Mill Valley, CA, 1988.

PROBLEMS

9.1   If two optical waves oscillate at slightly different frequencies, what would be the superimposed intensity?

9.2   If the two plane waves shown in Figure 9.1 are described by

$$E_1 = A \cos\left[\omega t - \frac{2\pi}{\lambda}\left(x\cos\alpha + y\cos\alpha + z\sqrt{1 - 2\cos^2\alpha}\right)\right]$$

and

$$E_2 = A \cos\left[\omega t - \frac{2\pi}{\lambda}\left(x\cos\alpha + y\cos\alpha - z\sqrt{1 - 2\cos^2\alpha}\right)\right],$$

calculate the period of the interference fringes.

9.3   Two plane waves traveling in the z direction are given by

$$E_1 = \mathbf{x}A \cos\left(\omega t - \frac{2\pi}{\lambda}z + \frac{\pi}{2}\right)$$

and

$$E_2 = \mathbf{y}A \cos\left(\omega t - \frac{2\pi}{\lambda}z\right),$$

where $\mathbf{x}$ and $\mathbf{y}$ are unit vectors in the x and y directions. Calculate the intensity of the superimposed optical field.

9.4   An HeNe laser illuminates an area of a VLSI mask with two transparent slits. If a screen is placed 1 m away from the VLSI mask and the distance from the first and fifteenth fringes is measured to be 9.34 mm, determine the distance between the two slits.

9.5 When a thin film of transparent plastic is placed over one of the slits in Young's double slit experimental set-up, the central bright fringe is displaced by 4.5 fringes. Assume that the refractive index of the plastic is 1.480 and the wavelength of the light source is 514.5 nm.

(a) Calculate the optical path increase due to the plastic film.

(b) What is the thickness of the plastic film?

9.6 Given that two point sources, one oscillating at 540 nm and the other at 630 nm, illuminate the two slits of a Young's interferometer, will any of the maxima of the two interference patterns coincide? If so, explain the circumstances.

9.7 If the two slits in Young's experimental set-up move away from each other at a velocity V, how will the interference fringes change?

9.8 Calculate the visibility of the fringes obtained from two coherent beams with intensity ratios I_1/I_2 of 1/1, 1/2, 1/3, 1/5, 1/10, and 1/100.

9.9 A broad-spectrum light source has a central wavelength of 550 nm and a bandwidth of 200 nm. What is its coherent length? If it illuminates a Young's double slit interferometer, how many fringes will be observed?

9.10 An HeNe laser is used as the light source in a Michelson interferometer. When one of the mirrors moves at a speed of 1 mm/s, what is the frequency of the signal detected by a detector at the center of the observation plane?

9.11 Assume that a light source of 500 m wavelength is used in a Michelson interferometer. Calculate the angular radius of the tenth bright fringe, when the central path difference $2d$ is:

(a) 1.5 mm.

(b) 15 mm.

9.12 The moving mirror in a Michelson interferometer is moved through 247 µm, and 853 fringes are shifted across the center of the detector. Determine the wavelength of the light source.

9.13 A sodium lamp having two spectral lines at 0.5890 and 0.5896 µm is used as the light source in a Michelson interferometer. One of the two mirrors moves from the maximum fringe visibility position to the zero visibility location. Calculate the distance it has traveled.

9.14 An HeNe laser is used in a Mach–Zehnder interferometer. Two 1-m long tubes filled with air are placed in the two arms of the interferometer. Tube T_1 is kept at 0°C and 76 mmHg, and the T_2 is kept at the same temperature, but is filled with more air until its pressure reaches 80 mmHg. How many fringes have swept across the center of the observation plane during the filling process?

9.15 If the enclosed area of the interferometer shown in Figure 9.9 is 1 m^2 and the wavelength of the light source is 850 nm, what is the angular velocity corresponding to one fringe shift? If the optical path is implemented by optical fibers and 1000 turns of fiber are wound on a circular drum with an enclosed area of $1 \, m^2$, what is the angular velocity corresponding to one fringe shift?

9.16 A convex lens surface having a radius of 148 mm is tested against an optical flat glass. If the wavelength of the light source is 550 nm, calculate the diameter of the first dark fringe.

9.17 If the convex lens in Problem 9.16 is tested against a concave surface of radius equal to 150 mm. Calculate the diameter of the first dark fringe.

9.18 A transparent film has a thickness of 32.5 μm and a refractive index of 1.40. Find

(a) The order of interference m at $\theta = 0°$;

(b) the first three angles at which the light from an He–Ne laser forms bright fringes.

9.19 The amplitude reflection coefficient of a Fabry–Perot etalon is $r = 0.8944$. Determine:

(a) The coefficient of finesse F.

(b) The width of the bright fringes $\Delta\delta$.

(c) The finesse S.

9.20 A Fabry–Perot interferometer is used to measure the interval between the twin spectral lines. Assume that the average wavelength is 589.3 nm and the reflectance of the Fabry–Perot etalon is 90%. If the wavelength interval is 0.6 nm, what is the minimum distance between the two mirrors of the etalon?

9.21 A soap bubble of 0.5 μm thick is viewed in white light at normal incidence. What is the color of the transmitted light?

9.22 An antireflective film is to be deposited on a glass lens with a refractive index of 1.563. Determine

(a) The refractive index of the thin film material.

(b) The thickness of the film if 0% reflectance at 514.5 nm is required.

9.23 A reflection of 50% is desired for an optical surface under the normal incidence of argon ion laser illumination (wavelength 514.5 nm). Assume that the refractive index of the glass is 1.55.

(a) Determine the refractive index of the thin film material.

(b) Determine the thickness of the film.

(c) What is the reflectance for 45° incidence?

HOLOGRAPHY 10

The concept of holography, also known as wavefront reconstruction, was first introduced by Dennis Gabor in 1948. At that time he had encountered two major difficulties, namely that a high-intensity coherent source of light was not available and that virtual and real images could not be separated. Nevertheless, he set a basic foundation of holographic theory. In fact, the word "holography" was first coined by Gabor, by combining parts of two Greek words: *holos*, meaning "whole", and *graphein*, meaning "to write". Thus, *holography* means the recording of a complete message. In this chapter we discuss the fundamental concept of holography.

A *hologram* is a recording on a light-sensitive medium, such as a photographic plate, of interference patterns formed between two beams of coherent light coming from the same laser. In the process of holography, the laser beam is first divided into two beams by a beam splitter and then broadened. One beam goes directly to the photographic plate. The second beam of light is directed onto the three-dimensional object under observation, each point of which reflects the part of the beam reaching it toward the plate. As each point of the object scatters the light, it acts as a point source of spherical waves. At the photographic plate the innumerable spherical waves from the object combine with the light wave of the first beam. Because they are from the same laser, the sets of light waves are coherent and are in a condition to interfere. They form interference fringes on the plane of the photographic plate and are thereby recorded. These interference fringes are a series of zone plate-like rings, but these rings are also superimposed, making an incredibly complex pattern of lines and swirls. The developed negative of these interference fringes is the hologram.

In an ordinary photograph taken in noncoherent light, the film has recorded the intensity of light reaching it at every point. In holography the interference of the two beams allows the photographic plate to record both the intensity and the relative phase of the light waves at each point.

The complicated interference fringe pattern in the hologram would seem to bear no relation whatsoever to an image of the original object, but it does indeed contain all the information needed to reconstruct the wavefronts and reconstitute a three-dimensional image of great vividness and verisimilitude. For this reconstitution the hologram is illuminated by a parallel beam of light from the laser. Most of the light passes straight through, but the complex of fine fringes acts as an elaborate diffraction grating. Light is diffracted at a fairly wide angle. The diffracted rays form two images: a reconstituted virtual image of the object behind the transparency and a reconstituted real image on the eye side. Since the light rays pass through the point where the real image is, it can be photographed.

The vivid, three-dimensional virtual image of the hologram is for viewing. Observers can move to different positions and look around the image to the same extent that they would be able to were they looking directly at the real object. And the moving viewers observe parallax, the apparent displacement of nearer and more distant parts of the image.

10.1 On-axis Holography

Since the physical spaces involved in holographic construction and the reconstruction processes are generally assumed to be linear and spatially invariant, holographic processes can easily be evaluated by the simple object point approach. Thus, in holographic recording we assume that an object point, to which a beam of coherent light is directed from a monochromatic light source, is located at a distance l_1 from the photographic plane. A coherent reference plane wave, derived from the same coherent source, is superimposed over the recording medium with the spherical beam diffracted from the object, as depicted in Figure 10.1. If the amplitude transmittance of the recording medium is assumed to be linearly proportional to the intensity distribution of the recording, an analog system diagram of the holographic construction, like the one shown in Figure 10.2, can be drawn.

The complex light field coming from the object and the plane wave derived from the coherent source, which is called the *reference beam*, are represented by

$$u(x, y) = \delta(\xi, \eta) * h_{l1}(\xi, \eta)$$

$$= C \exp\left[i\frac{\pi}{\lambda_1 l_1}(x^2 + y^2)\right], \tag{10.1}$$

and

$$v(x, y) = C', \tag{10.2}$$

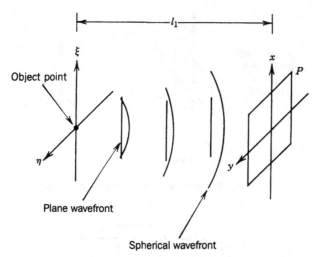

FIGURE 10.1   An on-axis object point holographic construction. P, Photographic plate.

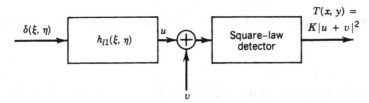

FIGURE 10.2   An analog system diagram of the holographic construction shown in Figure 10.1.

as they interfere over the recording plate. Here the asterisk denotes the convolution operation, and the C terms are the proportionality complex constants. The resultant intensity distribution over the recording plate is

$$I(x, y; \lambda_1) = |u(x, y) + v(x, y)|^2$$

$$= |C|^2 + |C'|^2 + 2|C||C'| \cos\left[\frac{\pi}{\lambda_1 l_1}(x^2 + y^2)\right]. \qquad (10.3)$$

Since the amplitude transmittance of the photographic plate is assumed to be linear in exposure, that is, the intensity multiplies the exposure time, we have

$$T(x, y; \lambda_1) = K_1 + K_2 \cos\left[\frac{\pi}{\lambda_1 l_1}(x^2 + y^2)\right], \qquad (10.4)$$

which can be written as

$$T(x, y; \lambda_1) = K_1 + \frac{K_2}{2}\exp\left[i\frac{\pi}{\lambda_1 l_1}(x^2 + y^2)\right] + \frac{K_2}{2}\exp\left[-i\frac{\pi}{\lambda_1 l_1}(x^2 + y^2)\right],$$

$$(10.5)$$

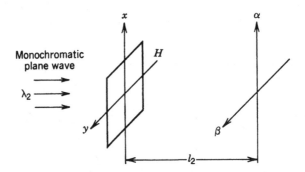

FIGURE 10.3 The geometry for hologram image reconstruction. H, designates the hologram.

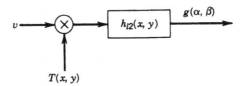

FIGURE 10.4 An analog system diagram of the hologram image reconstruction shown in Figure 10.3.

where K_1 and K_2 are proportionality constants. We note further that the amplitude transmittance function of an object point hologram (i.e., Eq. 10.4) is identical to a *Fresnel zone lens* which is composed of a glass plate and a negative and a positive lens (see Section 10.4).

In addition, if the object point hologram is normally illuminated by a monochromatic plane wave of λ_2, as shown in Figure 10.3, the holographic reconstruction process can be evaluated, with the help of the analog system diagram shown in Figure 10.4. The complex light distribution over the α, β plane can be evaluated as

$$g(\alpha, \beta; \lambda_2) = CT(x, y; \lambda_1) * h_{l2}(x, y; \lambda_2)$$
$$= C \iint_S T(x, y; \lambda_1) h_{l2}(\alpha - x, \beta - y; \lambda_2) \, dx \, dy, \qquad (10.6)$$

where C is a proportionality complex constant, S denotes the surface integration over the hologram, and

$$h_{l2}(x, y) = \exp\left[i\frac{\pi}{\lambda_2 l_2}(x^2 + y^2)\right] \qquad (10.7)$$

is the spatial impulse response between the hologram and the α, β plane.

The terms in Eq. 10.6 can be evaluated as

$$g_0(\alpha, \beta; \lambda_2) = CK_1 \iint\limits_S h_{l2}(\alpha - x, \beta - y; \lambda_2) \, dx \, dy, \qquad (10.8)$$

$$g_v(\alpha, \beta; \lambda_2) = \frac{CK_2}{2} \iint\limits_S \exp\left[i\frac{\pi}{\lambda_1 l_1}(x^2 + y^2)\right] h_{l2}(\alpha - x, \beta - y; \lambda_2) \, dx \, dy, \quad (10.9)$$

$$g_r(\alpha, \beta; \lambda_2) = \frac{CK_2}{2} \iint\limits_S \exp\left[-i\frac{\pi}{\lambda_1 l_1}(x^2 + y^2)\right] h_{l2}(\alpha - x, \beta - y; \lambda_2) \, dx \, dy,$$

$$(10.10)$$

and

$$g(\alpha, \beta; \lambda_2) = g_0(\alpha, \beta; \lambda_2) + g_v(\alpha, \beta; \lambda_2) + g_r(\alpha, \beta; \lambda_2), \qquad (10.11)$$

where the subscripts 0, v, and r denote the zero-order and the first-order virtual- and real-image diffractions, respectively. To simplify the analysis, we assume that the size of the hologram is infinitely extended and that the integral of Eq. 10.8 converges to a complex constant:

$$g_0(\alpha, \beta; \lambda_2) = C_1. \qquad (10.12)$$

The integral of Eq. 10.9 can be evaluated so that

$$g_v(\alpha, \beta; \lambda_2) = C_2 \int\limits_{-\infty}^{\infty}\!\!\int \exp\left[i\frac{\pi}{\lambda_1 l_1}(x^2 + y^2)\right]$$

$$\times \exp\left\{i\frac{\pi}{\lambda_2 l_2}[(\alpha - x)^2 + (\beta - y)^2]\right\} dx \, dy, \qquad (10.13)$$

which can be written as

$$g_v(\alpha, \beta; \lambda_2) = C_2 \exp\left[i\frac{\pi}{\lambda_2 l_2}(\alpha^2 + \beta^2)\right] \int\limits_{-\infty}^{\infty}\!\!\int \exp\left[i\pi\left(\frac{1}{\lambda_1 l_1} + \frac{1}{\lambda_2 l_2}\right)(x^2 + y^2)\right]$$

$$\times \exp\left[-i\frac{2\pi}{\lambda_2 l_2}(\alpha x + \beta y)\right] dx \, dy. \qquad (10.14)$$

In view of the preceding equation, a hologram image of the object point can be reconstructed if, and only if, the quadratic phase factor can be eliminated:

$$\frac{1}{\lambda_1 l_1} + \frac{1}{\lambda_2 l_2} = 0.$$

Thus, the longitudinal location of the hologram image would be

$$l_2 = -\frac{\lambda_1}{\lambda_2} l_1, \qquad (10.15)$$

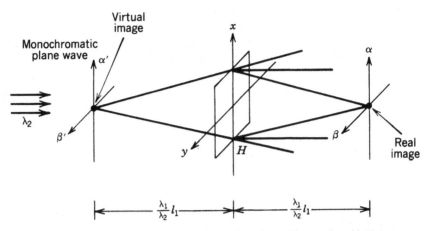

FIGURE 10.5 Hologram image reconstruction of an object point. *H*, Hologram.

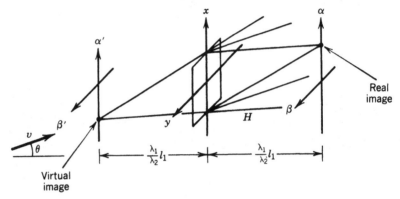

FIGURE 10.6 Oblique wavefront reconstruction of an on-axis object point hologram. *v*, Oblique plane wave.

where λ_1 and λ_2 are the construction and reconstruction wavelengths, respectively, and l_1 is the separation between the object and the hologram. Since λ_1, λ_2, and l_1 are positive quantities, the minus sign in Eq. 10.15 means that the image is located at the front of the hologram. For $l_2 = -l_1\lambda_1/\lambda_2$ the virtual image is reconstructed on the optical axis:

$$g_v(\alpha', \beta'; \lambda_2)\bigg|_{l_2=-l_1\lambda_1/\lambda_2} = C_2 \exp\left[i\frac{\pi}{\lambda_2 l_2}(\alpha^2 + \beta^2)\right]$$

$$\times \int\!\!\!\int_{-\infty}^{\infty} \exp\left[-i\frac{\pi}{l_2\lambda_2}(\alpha x + \beta y)\right] dx\,dy$$

$$= C_2\delta(\alpha', \beta'). \tag{10.16}$$

We can also show that the real image is reconstructed behind the holographic aperture:

$$g_r(\alpha, \beta; \lambda_2)\Big|_{l_2 = l_1\lambda_1/\lambda_2} = C_3\delta(\alpha, \beta). \tag{10.17}$$

From the results of Eqs 10.12, 10.16, and 10.17, we see that the holographic reconstruction process yields the zero-order, the virtual-image, and the real-image diffractions, as depicted in Figure 10.5. From this figure we see that the three orders of diffraction overlap, which causes spurious image distortion. We also notice that the overlapping diffractions cannot be separated, even under oblique illumination, as shown in Figure 10.6. For an oblique illumination the diffracted light field can be evaluated as follows:

$$g(\alpha, \beta; \lambda_2) = T(x, y; \lambda_1) \exp\left(i\frac{2\pi}{\lambda_2}x\sin\theta\right) * h_{l2}(x, y; \lambda_2)$$

$$= \iint_S T(x, y; \lambda_1) \exp\left(i\frac{2\pi}{\lambda_2}x\sin\theta\right) h_{l2}(\alpha - x, \beta - y; \lambda_2)\,dx\,dy.$$

$$\tag{10.18}$$

Again, by substituting the amplitude transmittance function of Eq. 10.5 in Eq. 10.18, we obtain

$$g_0(\alpha, \beta; \lambda_2) = C_1 \exp\left(i\frac{2\pi}{\lambda_2}\alpha\sin\theta\right), \tag{10.19}$$

$$g_v(\alpha', \beta'; \lambda_2) = C_2\delta\left(\alpha' + \frac{\lambda_1}{\lambda_2}l_1\sin\theta, \beta'\right), \tag{10.20}$$

and

$$g_r(\alpha, \beta; \lambda_2) = C_3\delta\left(\alpha - \frac{\lambda_1}{\lambda_2}l_1\sin\theta, \beta\right). \tag{10.21}$$

The image reconstructions are illustrated in Figure 10.6.

EXAMPLE 10.1

Assume that the on-axis object point hologram of Eq. 10.4 is constructed with a monochromatic red light of 600 nm and that the separation between the object and the holographic aperture is 50 cm. Calculate the locations of the hologram images if the hologram is illuminated by a normally incident plane wave of $\lambda = 500$ nm.

Let the construction wavelength be $\lambda_1 = 600$ nm, the separation be $l_1 = 0.5$ m, and the reconstruction wavelength be $\lambda_2 = 500$ nm. By substituting these parameters into Eq. 10.15, we have

$$l_2 = \frac{\lambda_1}{\lambda_2} l_1 = \frac{600}{500}(0.5) = 0.6 \, \text{m}.$$

Thus, the virtual and the real hologram images are located 0.6 m from the front and behind the hologram, respectively.

EXAMPLE 10.2

If the object point hologram in Example 10.1 is illuminated by a normally incident plane wave of white light, calculate the smearing length of the reconstructed image.

Since the spectral lines of visible light vary from 350 to 700 nm, a white-light source would have a spectral width of $\Delta\lambda = 700 - 350 = 350$ nm. By virtue of Eq. 10.15, the smearing length of real and virtual images can be calculated as

$$\Delta l = \frac{600}{350}(0.5) - \frac{600}{700}(0.5)$$
$$= 0.8571 - 0.4287 = 0.4284 \, \text{m}.$$

Thus, we see that the hologram images smear into rainbow colors; the red image is located close to the holographic plate and the violet image is situated farther away.

10.2 Off-axis Holography

An oblique reference beam should be used in the construction process to separate the hologram image diffractions, as illustrated in Figure 10.7. An analog system diagram representing the off-axis (also called the carrier spatial frequency) holographic construction is depicted in Figure 10.8. Thus, the amplitude transmittance function of the constructed hologram is

$$T(x, y; \lambda_1) = K_1 K_2 \cos\left[\frac{2\pi}{\lambda_1}\left(\frac{x^2 + y^2}{2l_1} - x \sin\theta\right)\right], \qquad (10.22)$$

where the K terms are proportionality constants.

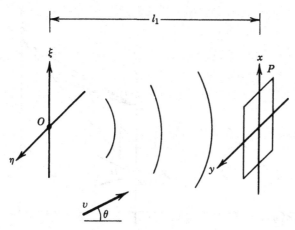

FIGURE 10.7 An off-axis or carrier spatial-frequency holographic construction. O, Object point; P, photographic plate; v, oblique plane wave.

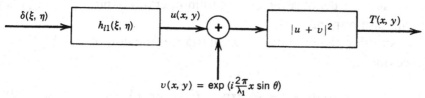

FIGURE 10.8 An analog system diagram of the holographic construction shown in Figure 10.7.

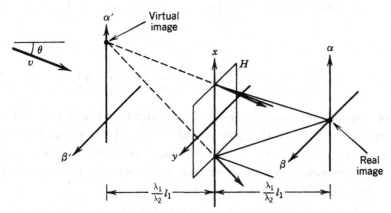

FIGURE 10.9 Reconstruction of the image in an off-axis hologram, with an oblique conjugate plane wave illumination.

If this hologram is illuminated by the oblique conjugate plane wave of λ_2, as illustrated in Figure 10.9, the holographic image diffractions can be evaluated as

$$g(\alpha, \beta; \lambda_2) = \left[\exp\left(-i\frac{2\pi}{\lambda_2} x \sin \theta\right) T(x, y; \lambda_1)\right] * h_{l2}(x, y; \lambda_2), \qquad (10.23)$$

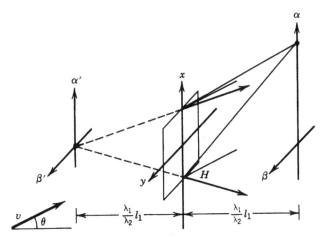

FIGURE 10.10 Hologram image reconstruction, with an oblique plane wave illumination.

where the asterisk denotes the convolution operation and $h_{l2}(x, y; \lambda_1)$ is the spatial impulse response.

By substituting the amplitude transmittance function of Eq. 10.22 in Eq. 10.23, we obtain

$$g_0(\alpha, \beta; \lambda_2) = C_1 \exp\left(-i\frac{2\pi}{\lambda_2}\alpha \sin\theta\right), \tag{10.24}$$

$$g_v(\alpha', \beta'; \lambda_2) = C_2\delta\left(\alpha' - \frac{\lambda_1 + \lambda_2}{\lambda_2} l_1 \sin\theta, \beta'\right), \tag{10.25}$$

and

$$g_r(\alpha, \beta; \lambda_2) = C_3\delta\left(\alpha - \frac{\lambda_2 - \lambda_1}{\lambda_2} l_1 \sin\theta, \beta\right). \tag{10.26}$$

The corresponding reconstructions of the hologram virtual and real images are shown in Figure 10.9. Thus, we see that the real image can be separated from the other diffractions. Similarly, reconstructions of the hologram images with the same oblique reference plane wave, as shown in Figure 10.10, can be evaluated. The results are given by

$$g_0(\alpha, \beta; \lambda_2) = C_1 \exp\left(i\frac{2\pi}{\lambda_2}\alpha \sin\theta\right), \tag{10.27}$$

$$g_v(\alpha', \beta'; \lambda_2) = C_2\delta(\alpha' - \frac{\lambda_2 - \lambda_1}{\lambda_2} l_1 \sin\theta, \beta'), \tag{10.28}$$

and

$$g_r(\alpha, \beta; \lambda_2) = C_3\delta\left(\alpha - \frac{\lambda_1 + \lambda_2}{\lambda_2} l_1 \sin\theta, \beta\right). \tag{10.29}$$

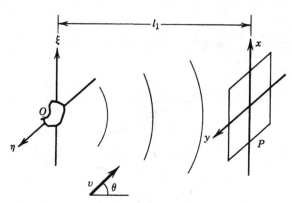

FIGURE 10.11 Holographic construction of an extended object. O, object; v, oblique plane wave; P, photographic plate.

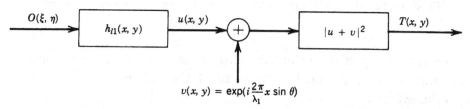

FIGURE 10.12 An analog system diagram of the holographic construction shown in Figure 10.11.

Thus, we see that the virtual image can be viewed through the holographic aperture without interference from the other diffractions.

Let us consider a hologram construction of an extended object, as shown in Figure 10.11. The complex light field distributed on the recording medium can be written as

$$u(x, y) = \iint_S O(\xi, \eta) h_{l1}(x - \xi, y - \eta; \lambda_1) \, d\xi \, d\eta,$$

where $O(\xi, \eta)$ is the object function and S denotes the surface integration over (ξ, η). An analog system diagram representing the construction process is shown in Figure 10.12. The amplitude transmittance function of the recorded hologram can be calculated as follows,

$$T(x, y; \lambda_1) = |u(x, y)|^2 + 1 + 2|u(x, y)\cos[\phi(x, y) - k_1 x \sin \theta], \qquad (10.30)$$

where we have ignored the proportionality constant for simplicity; here $k_1 = 2\pi/\lambda_1$ is called the wavenumber; $u(x, y) = |u(x, y)|e^{i\phi(x,y)}$ represents the object beam; and $\phi(x, y)$ is the phase distribution.

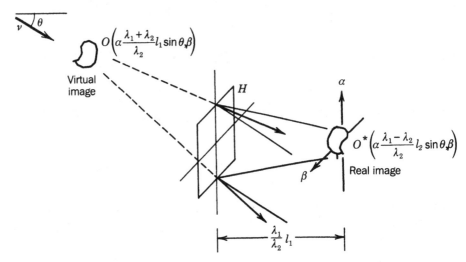

FIGURE 10.13 Off-axis hologram imaging. v, Conjugate plane wave; H, hologram.

Needless to say, if the recorded hologram is illuminated by a conjugate plane wave of λ_2, as depicted in Figure 10.13, the real-image diffractions can be evaluated as

$$g_r(x, y; \lambda_2) = |u(x, y)| \exp\{-i[\phi(x, y) - (k_1 - k_2)x \sin \theta]\} * h_{l2}(x, y; \lambda_2),$$

which can be shown to be equal to

$$g_r(x, y; \lambda_2) = O^*(\alpha, \beta; \lambda_2) \qquad \text{at} \quad l_2 = \frac{\lambda_1 l_1}{\lambda_2}, \tag{10.31}$$

where the asterisk denotes the complex conjugate. Similarly, the process of virtual-image reconstruction is

$$g_v(x, y; \lambda_2) = O\left(\alpha - \frac{\lambda_1 + \lambda_2}{\lambda_2} l_1 \sin \theta, \beta\right) \qquad \text{at} \quad l_2 = -\frac{\lambda_1 l_1}{\lambda_2}. \tag{10.32}$$

This constructed virtual image is located at the front of the hologram. The image reconstructions are illustrated in Figure 10.13. Again we see that the real-image diffraction can be separated from the other diffractions. In addition, we notice that the real image is *pseudoscopic*, that is, the image is in reversed relief. The virtual image is *orthoscopic*, that is, in proper relief.

The *diffraction efficiency* (DE) of a hologram can be defined as

$$\text{DE} = \frac{\text{Output (hologram image intensity)}}{\text{Input (incident intensity)}}. \tag{10.33}$$

In practice, this quantity is frequently used for determining the quality of a hologram.

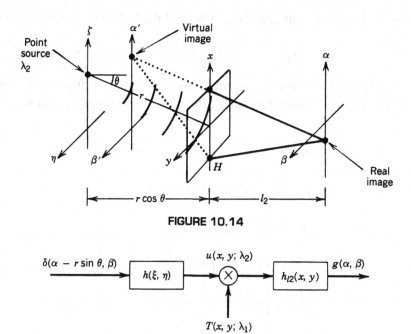

FIGURE 10.14

FIGURE 10.15

EXAMPLE 10.3

If the off-axis object point hologram of Eq. 10.22 is illuminated by a conjugate monochromatic point source, as shown in Figure 10.14.

(a) Draw an analog system diagram to represent the holographic reconstruction process.
(b) Evaluate the real- and virtual-image reconstructions.

Answers
(a) The location of the monochromatic point source can be represented by a delta function, $\delta(\alpha - r\sin\theta, \beta)$. The analog system diagram is shown in Figure 10.15.
(b) The illuminating wavefront $u(x, y; \lambda_2)$ can be shown as follows:

$$u(x, y; \lambda_2) = \delta(\alpha - r\sin\theta, \beta) * h_l(\xi, \eta; \lambda_2)$$
$$= c \exp\left\{\frac{ik_2}{2l}[(x - r\sin)^2 + y^2]\right\},$$

where

$$h_l(\xi, \eta; \lambda_2) = \exp\left[\frac{ik_2}{2l}(\xi^2 + \eta^2)\right], \qquad k_2 = \frac{2\pi}{\lambda_2}, \qquad l = r\cos\theta.$$

Thus, the complex light distribution behind the hologram can be evaluated:

$$g(\alpha, \beta; \lambda_2) = u(x, y; \lambda_2) T(x, y; \lambda_1) * h_{l2}(x, y; \lambda_2).$$

For simplicity, we use a one-dimensional notation for the following analysis. The real-image reconstruction can be calculated as follows,

$$
\begin{aligned}
g_r(\alpha; \lambda_2) &= u(x; \lambda_2) \exp\left[-ik_1\left(\frac{x^2}{2l_1} - x\sin\theta\right)\right] * h_{l2}(x; \lambda_2) \\
&= \int \exp\left[i\frac{k_2}{2l}(x - r\sin\theta)^2\right] \exp\left[-ik_1\left(\frac{x^2}{2l_1} - x\sin\theta\right)\right] \\
&\quad \times \exp\left[i\frac{k_2}{2l_2}(\alpha - x)^2\right] dx \\
&= C\exp\left(i\frac{k_2}{2l_2}\alpha^2\right) \int \exp\left[i\left(\frac{k_2}{2l} - \frac{k}{2l_1} + \frac{k_2}{2l_2}\right)x^2\right] \\
&\quad \times \exp\left\{i\left[\frac{k_2}{l}r\sin\theta - k_1\sin\theta + \frac{k_2}{l_2}\alpha\right]x\right\} dx,
\end{aligned}
$$

where C is the proportionality constant.

If we let the quadratic phase factor equal zero, the longitudinal distance of the hologram image can be determined as

$$l_2 = \frac{l_1 l \lambda_1}{\lambda_2 l - l_1 \lambda_1}, \tag{10.34}$$

where $l = r\cos\theta$. Thus, $g_r(\alpha; \lambda_2)$ can be reduced to the following form,

$$
\begin{aligned}
g_r(\alpha; \lambda_2) &= C\exp\left(i\frac{k_2}{2l_2}\alpha^2\right) \int \exp\left\{i\frac{k_2}{2l_2}\left[\frac{l_2}{l}r\sin\theta - \frac{k_1 l_2}{k_2}\sin\theta + \alpha\right]x\right\} dx \\
&= C_1\delta\left[\alpha + \left(\tan\theta - \frac{\lambda_2}{\lambda_1}\sin\theta\right)l_2\right], \tag{10.35}
\end{aligned}
$$

where $l = r\cos\theta$. Similarly, the virtual-image reconstruction can be computed by

$$
\begin{aligned}
g_v(\alpha; \lambda_2) &= \int \exp\left[i\frac{k_2}{2l}(x - r\sin\theta)^2\right] \exp\left[ik_1\left(\frac{x^2}{2l_1} - x\sin\theta\right)\right] \\
&\quad \times \exp\left[i\frac{k_2}{2l_2}(\alpha - x)^2\right] dx,
\end{aligned}
$$

which can be shown as

$$g_v(\alpha; \lambda_2) = C\delta\left[\alpha + \left(\tan\theta + \frac{\lambda_2}{\lambda_1}\sin\theta\right)l_2\right], \tag{10.36}$$

where

$$l_2 = -\frac{l_1 l \lambda_1}{\lambda_2 l + l_1 \lambda_1}, \qquad (10.37)$$

and $l = r \cos \theta$.

The corresponding reconstructions of real and virtual images are depicted in Figure 10.14.

EXAMPLE 10.4

We assumed that the hologram in Example 10.3 was constructed with $\lambda_1 = 500$ nm and $l_1 = 5$ cm. If, instead of a monochromatic source, the hologram is illuminated by a white-light point source at an oblique angle $\theta = 45°$ and with a radial distance $r = 0.5$ m away from H, as shown in Figure 10.14:

(a) Calculate the smearing length of the reconstructed hologram images.
(b) Sketch the smeared hologram images and identify the colors at the extreme ends. Assume that the white-light source has a continuous spectral line that varies from 350 to 700 nm.

Answers

(a) Using Eq. 10.34, we can compute the longitudinal locations of the violet and red real images as

$$l_v = \frac{(0.05)(0.5)\cos 45°(550)}{(350)(0.5)\cos 45° - (0.05)(500)}$$

$$= \frac{9.72}{123.73 - 25} = 0.0985 \text{ m}$$

and

$$l_r = \frac{9.72}{247.45 - 25} = 0.044 \text{ m}.$$

The longitudinal smearing length of the real image is therefore

$$\Delta l = l_v - l_r = 0.0985 - 0.044 = 0.0545 \text{ m}.$$

Equation 10.37 helps us to compute the violet and the red virtual images:

$$l_v' = -\frac{(0.05)(0.5)\cos 45°(550)}{(350)(0.5) + (0.05)(500)} = -0.065 \text{ m}$$

and

$$l_v' = -\frac{9.72}{247.45 + 25} = -0.036 \text{ m}.$$

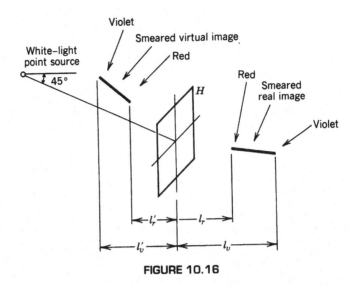

FIGURE 10.16

Notice that the minus sign indicates that the image is reconstructed at the front of the hologram. Thus, the longitudinal smearing length of the virtual image is

$$\Delta l = 0.065 - 0.036 = 0.029\,\text{m}.$$

From this we see that, with divergent illumination, the smearing length for the real image appears longer than that for the virtual image.

(b) A sketch of the reconstructed hologram images is given in Figure 10.16. From this sketch we show that the red images appear closer to the holographic plate and the violet images appear farther away.

EXAMPLE 10.5

Assuming that the modulation index of the object point hologram of Eq. 10.22 is 100%, determine the diffraction efficiency of the hologram.

Since the modulation index is 100%, Eq. 10.22 can be written as

$$T(x,y) = \frac{1}{2}\left\{1 + \frac{1}{2}\cos\left[\frac{2\pi}{\lambda_1}\left(\frac{x^2+y^2}{2l_1} - x\sin\theta\right)\right]\right\}.$$

From Eqs 10.25 and 10.26, the virtual- and real-image reconstructions are

$$g_v = (I_i)^{1/2}\left(\frac{1}{4}\right)\delta\left(\alpha' - \frac{\lambda_1+\lambda_2}{\lambda_2}l_1\sin\theta, \beta'\right).$$

and

$$g_r = (I_i)^{1/2}\left(\frac{1}{4}\right)\delta\left(\alpha - \frac{\lambda_2-\lambda_1}{\lambda_2}l_1\sin\theta, \beta\right),$$

where I_i is the input intensity.

Thus, the diffraction efficiency of the hologram is

$$DE = \frac{(\frac{1}{4})^2 I_i}{I_i} = \frac{1}{16} = 6.25\%.$$

10.3 Holographic Magnifications

Since hologram images are generally three-dimensional in nature, in this section, we shall discuss the phenomenon of holographic magnifications.

To compute the *lateral magnification*, we again use the elementary object point holographic concept, as depicted in Figure 10.17. A simpler one-dimensional analog system diagram of Figure 10.17 is given in Figure 10.18.

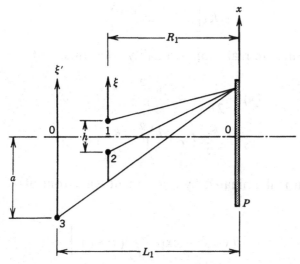

FIGURE 10.17 Recording geometry for determining the lateral magnification. P, photographic plate.

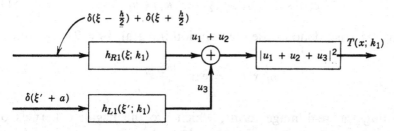

FIGURE 10.18 An analog system diagram of the recording process shown in Figure 10.17.

The complex light fields distributed over the holographic aperture can be written as

$$u_1(x; k_1) = \exp\left[\frac{ik_1}{2R_1}\left(x - \frac{h}{2}\right)^2\right], \tag{10.38}$$

$$u_2(x; k_1) = \exp\left[\frac{ik_1}{2R_1}\left(x + \frac{h}{2}\right)^2\right], \tag{10.39}$$

and

$$u_3(x; k_1) = \exp\left[\frac{ik_1}{2L_1}(x + a)^2\right], \tag{10.40}$$

where $k_1 = 2\pi/\lambda_1$, and λ_1 designates the construction wavelength. Thus, the amplitude transmittance of the recorded hologram is

$$T(x; k_1) = K_0 + K_1 \cos\frac{k_1}{R_1}hx + K_2\left[e^{i(*)} + e^{-i(*)}\right]$$
$$+ K_3\left[e^{i(**)} + e^{-i(**)}\right], \tag{10.41}$$

where the K terms are real proportionality constants, and

$$(*) = \frac{k_1}{2R_1}\left(x - \frac{h}{2}\right)^2 - \frac{k}{2L_1}(x + a)^2, \tag{10.42}$$

$$(**) = \frac{k_1}{2R_1}\left(x + \frac{h}{2}\right)^2 - \frac{k_1}{2L_1}(x + a)^2. \tag{10.43}$$

If the hologram is illuminated by a divergent wavefront of λ_2, as shown in Figure 10.19,

$$u_4(x; k_2) = \exp\left[\frac{ik}{2L_2}(x - b)^2\right], \tag{10.44}$$

then the complex light field behind the hologram can be calculated as

$$g(\alpha; k_2) = u_4(x; k_2)T(x; k_1) * h_l(x; k_2), \tag{10.45}$$

where the asterisk denotes the convolution integral, $k_2 = 2\pi/\lambda_2$, and

$$h_l(x; k_2) = \exp\left(\frac{ik_2}{2l}x^2\right). \tag{10.46}$$

Substituting the real-image terms, which are the negative kernels of Eq. 10.41, we can evaluate the real-image diffractions:

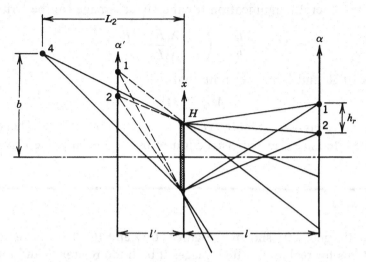

FIGURE 10.19 Reconstruction geometry for determining the lateral magnification. A monochromatic point source is located at 4. From it emanates the illuminating divergent wavefront of λ_2. h_r, Separation of the real images 1 and 2.

$$
g_r(x; k_2) = \int \exp\left\{ -i\frac{k_2}{2}\left(\frac{\lambda_2}{\lambda_1 R_1} - \frac{\lambda_2}{\lambda_1 L_1} - \frac{1}{L_2} - \frac{1}{l} \right)x^2 \right.
$$
$$
\left. + \frac{2}{l}\left[\alpha + 1\left(\frac{b}{L_2} - \frac{\lambda_2 h}{2\lambda_1 R_1} - \frac{\lambda_2 a}{\lambda_1 L_1} \right) \right]x \right\}\,\mathrm{d}x
$$
$$
+ \int \exp\left\{ -i\frac{k_2}{2}\left(\frac{\lambda_2}{\lambda_1 R_1} - \frac{\lambda_2}{\lambda_1 L_1} - \frac{1}{L_2} - \frac{1}{l} \right)x^2 \right.
$$
$$
\left. + \frac{2}{l}\left[\alpha + 1\left(\frac{b}{L_2} + \frac{\lambda_2 h}{2\lambda_1 R_1} - \frac{\lambda_2 a}{\lambda_1 L_1} \right) \right]x \right\}\,\mathrm{d}x. \tag{10.47}
$$

Eliminating the quadratic phase factors in Eq. 10.47 allows the real images to be reconstructed at

$$
l = \frac{\lambda_1 R_1 L_1 L_2}{\lambda_2 L_1 L_2 - \lambda_2 R_1 L_2 - \lambda_1 R_1 L_1}, \tag{10.48}
$$

and to take the form (see Figure 10.19)

$$
g_r(x; k_2) = \delta\left[\alpha + l\left(\frac{b}{L_2} - \frac{\lambda_2 h}{2\lambda_1 R_1} - \frac{\lambda_2 a}{\lambda_1 L_1} \right) \right] + \delta\left[\alpha + l\left(\frac{b}{L_2} + \frac{\lambda_2 h}{2\lambda_1 R_1} - \frac{\lambda_2 a}{\lambda_1 L_1} \right) \right]. \tag{10.49}
$$

Thus, the lateral magnification of the real image can be written as

$$
M_{lat}^r = \frac{h_r}{h} = \left(1 - \frac{\lambda_1 R_1}{\lambda_2 R_2} - \frac{R_1}{L_1} \right)^{-1}. \tag{10.50}
$$

Similarly, the lateral magnification for the virtual image can be written as

$$M_{lat}^v = \frac{h_v}{h} = \left(1 + \frac{\lambda_1 R_1}{\lambda_2 L_2} - \frac{R_1}{L_1}\right)^{-1}. \tag{10.51}$$

From Eqs 10.50 and 10.51, we note that

$$M_{lat}^r \geq M_{lat}^v$$

for *divergent* reference and *divergent* reconstruction beams. The equality holds if the reference and reconstruction beams are both plane waves.

EXAMPLE 10.6

Referring to the geometry shown in Figures 10.17 and 10.19, evaluate the lateral magnifications for real and virtual images if both the reference and the reconstruction beams are oblique plane waves.

Letting L_1 and L_2 approach infinity in Eqs 10.50 and 10.51, we have

$$M_{lat}^r = \lim_{\substack{L_1 \to \infty \\ L_2 \to \infty}} \left(1 - \frac{\lambda_1 R_1}{\lambda_2 L_2} - \frac{R_1}{L_1}\right)^{-1} = 1$$

and

$$M_{lat}^v = \lim_{\substack{L_1 \to \infty \\ L_2 \to \infty}} \left(1 + \frac{\lambda_1 R_1}{\lambda_2 L_2} - \frac{R_1}{L_1}\right)^{-1} = 1.$$

Thus, we see that $M_{lat}^r = M_{lat}^v$.

EXAMPLE 10.7

We assume that a hologram is constructed by a plane reference beam of λ_1. If the hologram images are reconstructed with a *convergent* monochromatic wavefront of λ_2, as illustrated in Figure 10.20, determine the lateral magnifications.

Since the illuminating wavefront converges to a point at a distance behind the hologram, this distance L_2 is regarded as a *negative* quantity in Eqs 10.50 and 10.51:

$$M_{lat}^r = \left(1 + \frac{\lambda_1 R_1}{\lambda_2 L_2'}\right)^{-1}$$

and

$$M_{lat}^v = \left(1 - \frac{\lambda_1 R_1}{\lambda_2 L_2'}\right)^{-1}$$

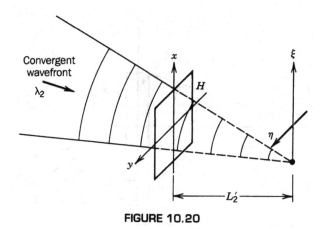

FIGURE 10.20

Thus, we have

$$M_{lat}^r < M_{lat}^v$$

for the plane reference beam and the convergent construction beam.

10.4 Reflection Holography

By a simple rearrangement of the holographic reconstruction process, it is possible to obtain hologram images using a simple white-light illumination. This reconstruction process is entirely dependent on reflection from the recorded hologram, rather than on transmission through the hologram. Since this technique utilizes a thick emulsion on the photographic plate, reflection holography is also known as thick-emulsion holography. The reflection holography that we discuss is similar in concept to Lippmann's color photography.[1] A reflection hologram is constructed by directing the object and reference beams in opposite directions into the photographic plate. The light waves traveling in opposite directions form standing waves. The interference fringes generated by these standing waves are recorded in the thick emulsion. Similar to Lippmann's photograph, the hologram image can be read out by illumination from a simple white-light source. If

---

[1] Lippmann's color photography depends on the interference fringes in standing electromagnetic waves generated when light is reflected by a mercury coating at the back of a special fine-grained photographic emulsion on the camera's photographic plate. Gabriel Lippmann was awarded the 1908 Nobel prize for physics for discovering this principle, first communicated in 1891.

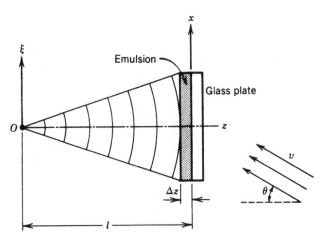

FIGURE 10.21 Construction of an object point reflection hologram. O, Object point; Δz, thickness of photographic emulsion.

polychromatic coherent light is used in the construction process, a color hologram image will be read out by the white-light illumination. Thus, reflection holography is also known as white-light color holography; the concept was first suggested by Yuri Denisyuk in 1962.

In the construction process of object point reflection holography, the object beam and the reference beam are combined from opposite directions within the recording medium, as shown in Figure 10.21. The complex light fields of the object and the reference beams as they come together and interfere within the recording plate can be expressed as

$$u(x; k_1) = \exp\left\{ ik_1 \left[z + \frac{x^2}{2(l+z)} \right] \right\} \qquad (10.52)$$

and

$$\sqrt{}(x; k_1) = \exp[-ik_1(z - x\sin\theta)], \qquad (10.53)$$

for $-\Delta z \leq z \leq 0$, where $k_1 = 2\pi/\lambda_1$.

If we assume that the *developed* photographic density is proportional to the intensity of the construction process, the density distribution of the encoded hologram is

$$D(x; k_1) = K_1 + K_2 \cos\left\{ k_1 \left[2z + \frac{x^2}{2(l+z)} - x\sin\theta \right] \right\}, \qquad (10.54)$$

where the K terms are proportionality constants. From this equation we can see that there are thin holograms arranged in parallel layers within the emulsion, and that these thin holograms are spaced about half the recording wavelength apart, or $\lambda_1/2$. If we assume further that reflectance of these

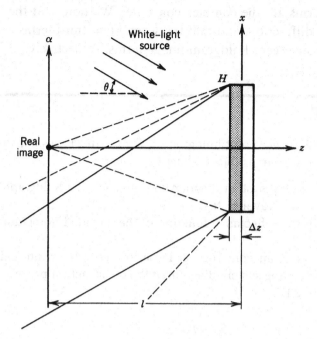

FIGURE 10.22 Reconstruction of the image of a reflection hologram.

thin holograms is proportional to the photographic density, the reflectance function of the encoded hologram is

$$r(x; k_1) = K_1' + K_2' \cos\left\{ k_1 \left[2z + \frac{x^2}{2(l+z)} - x \sin\theta \right] \right\},\qquad (10.55)$$

for $-\Delta z \le z \le 0$, where the K terms are proportionality constants.

If the recorded hologram is illuminated by a white-light source, as shown in Figure 10.22, the wavelength satisfying Bragg's law will be reflected. The complex light field of the selected wavelength can be computed by

$$g(x; k_1) = \exp[ik_1(z - x_1 \sin\theta)]r(x; k_1) * h_l(x; k_1),\qquad (10.56)$$

with the asterisk denoting the convolution integral, and

$$h_l(x; k_1) = C \exp\left(\frac{ik_1}{2l} x^2 \right).$$

Substituting Eq. 10.55 into Eq. 10.56, we have

$$
\begin{aligned}
g(x; k_1) = \; & C_1 \exp(-ik\alpha \sin\theta) \\
& + C_2 \exp\left\{ \frac{k_1}{4(l+z)} [\alpha - 2(l+z)\sin\theta]^2 \right\} \\
& + C_3 \delta(\alpha, \beta) \qquad \text{for} \quad -\Delta z \le z \le 0,\qquad (10.57)
\end{aligned}
$$

where the C terms are the complex constants. We note that the first term is the zero-order diffraction, and that the second and third terms are the divergent and the convergent hologram-image terms, respectively.

EXAMPLE 10.8

Suppose that the object point hologram in Figure 10.21 was constructed using two coherent light sources of wavelengths λ_1 and λ_2.

(a) Draw an analog system diagram to represent this holographic construction with two wavelengths.
(b) What will the reflectance function of the encoded hologram be?

Answers
(a) Since the hologram construction is performed by two mutually coherent sources, the analog system diagram of the construction process is as shown in Figure 10.23.

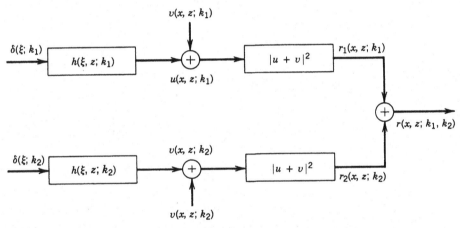

FIGURE 10.23

(b) Using Eq. 10.55, we can write the resulting reflectance function of the thick-emulsion hologram as

$$r(x, z; k_1, k_2) = r_1(x, z; k_1) + r_2(x, z; k_2)$$

$$= K_1 + K_2 \cos\left\{k_1\left[2z + \frac{x^2}{2(l+z)} - x\sin\theta\right]\right\}$$

$$+ K_3 \cos\left\{k_2\left[2z + \frac{x^2}{2(l+z)} - x\sin\theta\right]\right\},$$

where the K terms are proportionality constants.

EXAMPLE 10.9

Suppose that the reflection hologram in Eq. 10.55 is constructed with a coherent light source of wavelength $\lambda_1 = 500\,\text{nm}$.

(a) Calculate the separation of the subholograms within the emulsion.
(b) Assuming that the photographic emulsion is about $10\,\mu\text{m}$ thick, how many subholograms will be constructed within the holographic plate?

Answers

(a) Referring to Eq. 10.55, we can see that the subholograms are constructed when

$$\cos\left(\frac{4\pi}{\lambda_1}z\right) = 1,$$

which can be shown as

$$\frac{4\pi}{\lambda_1}z = 2n\pi, \qquad n = 0, 1, 2, \ldots$$

Thus, the separation of the subholograms is

$$z = \frac{\lambda_2}{2} = \frac{500}{2} = 250\,\text{nm}.$$

(b) Since the emulsion is about $10\,\mu\text{m}$ thick, the number of subholograms would be

$$N = \frac{10}{0.25} = 40.$$

10.5 Rainbow Holography

We now discuss a technique for producing hologram images using a simple, inexpensive white-light source. This type of hologram is capable of producing brighter and more colorful hologram images than those previously discussed. Since these types of hologram image are observed through the transmitted light field, and because they produce rainbow color images, they are called *rainbow holograms*.

As indicated in Section 10.2, a real hologram image can be reconstructed with a conjugate coherent illumination. In fact, hologram images can be reconstructed using a very small holographic aperture, with only a minor degree of resolution loss. In other words, it is possible to reconstruct the

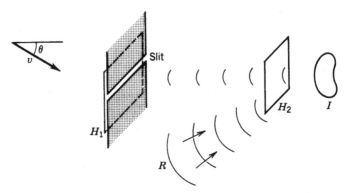

FIGURE 10.24 A two-step rainbow holographic construction. v, Conjugate coherent illumination; H_1, primary hologram; R, convergent reference beam; H_2, holographic plate; I, real image.

entire hologram image when the aperture is reduced to a narrow slit, as shown in Figure 10.24. For convenience of discussion, we call hologram H_1 the primary hologram.

In rainbow holographic recording, we insert a fresh holographic plate H_2. To minimize color blurring in the rainbow hologram image, we recommend that this holographic plate be placed near the hologram image plane. Then, if the holographic plate is properly recorded in the linear region of the T–E curve, the resultant hologram H_2 will be a rainbow hologram.

In order to see the rainbow effect, we first reconstruct the hologram image from the rainbow hologram H_2. If we look through the real slit image, we expect to see a virtual hologram image behind holographic plate H_2. If holographic plate H_2 is inserted behind the hologram image I during the rainbow holographic construction, a real hologram image will be seen through the slit image because the reconstruction process shown in Figure 10.25 takes place. Since the real slit image is convergently reconstructed, we see a brighter image.

As we recall from our discussion of holographic magnification in Section 10.3, the location of the real slit image varies as a function of the reconstruction wavelength of the light source. In other words, when the reconstruction wavelength is longer, the slit width is wider and the slit image appears to be located higher and closer to the hologram aperture H_2. The same effect applies to the hologram image seen through the slit image. When the reconstruction wavelength is longer, the hologram image appears to be larger and closer to the holographic plate H_2.

Now, if hologram H_2 is illuminated by a conjugate divergent white-light source, as shown in Figure 10.26, the hologram slit images produced by the different wavelengths of the white-light source will separate into rainbow colors in the space of the real slit image. The hologram image of the object

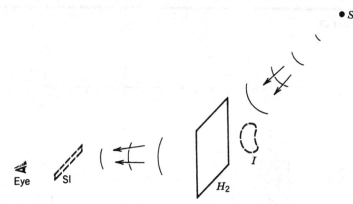

FIGURE 10.25 Holographic reconstruction with a coherent source. SI, Slit image; H_2, holographic plate; I, virtual hologram image; S monochromatic point source.

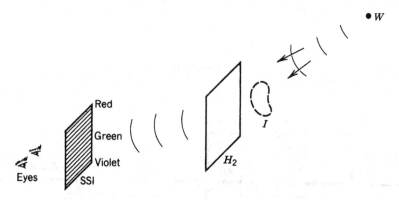

FIGURE 10.26 Rainbow holographic reconstruction with a white-light source. SSI, Smeared slit image; H_2, hologram; I, virtual hologram image; W, divergent white-light source.

behind the hologram will take on the same rainbow effect. If we view this image transversely through the smeared slit image, we will see it in a succession of rainbow colors. In other words, if we view the image through the red-colored slit image, we will see a red-colored hologram image, and if we peer through the green-colored slit image, we will see a smaller green hologram image.

We are not able, however, to see the "over" or "under" of the object image by moving our eyes transversely up and down against the smeared slit image, as we would in a conventional hologram. Thus, one of the consequences of using this type of rainbow holography is that the vertical parallax is lost, although the full horizontal parallax is retained. This means that we still have a right-to-left view for binocular stereopsis and motion parallax, and that the sensation of a three-dimensional scene is preserved.

EXAMPLE 10.10

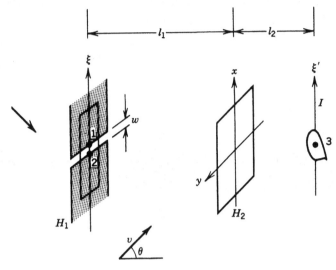

FIGURE 10.27

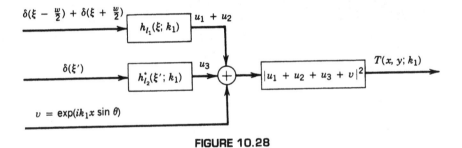

FIGURE 10.28

Given a rainbow holographic construction, as shown in Figure 10.27:

(a) Draw an analog system diagram to represent the holographic construction.

(b) If the real hologram image, derived from the primary hologram, represents an object point, calculate the amplitude transmittance function of the rainbow hologram.

Answers

(a) A one-dimensional analog system diagram of the rainbow holographic construction is given in Figure 10.28.

(b) Using the analog system diagram, we have

$$u_1(x; k_1) = \exp\left[\frac{ik_1}{2l_1}\left(x - \frac{w}{2}\right)^2\right],$$

$$u_2(x; k_1) = \exp\left[\frac{ik_1}{2l_1}\left(x + \frac{w}{2}\right)^2\right],$$

$$u_3(x; k_1) = \exp\left(-i\frac{k}{2l_2}x^2\right),$$

and

$$v(x; k_1) = \exp(ik_1 x \sin\theta).$$

Thus, the amplitude transmittal of the rainbow hologram, as an object point representation, is

$$T(x, k_1) = |u_1 + u_2 + u_3 + v|^2$$

$$= K_1 + K_2 \cos\left(\frac{k_1}{l_1}xw\right) + K_3 \cos\left\{-\frac{k_1}{2}\left[\frac{x^2}{l_2} + \frac{(x - w/2)^2}{l_1}\right]\right\}$$

$$+ K_4 \cos\left\{-\frac{k_1}{2}\left[\frac{x^2}{l_2} + \frac{(x + w/2)^2}{l_1}\right]\right\}$$

$$+ K_5 \cos\left\{k_1\left[\frac{(x - w/2)^2}{2l_1} - x\sin\theta\right]\right\}$$

$$+ K_6 \cos\left\{k_1\left[\frac{(x + w/2)^2}{2l_1} - x\sin\theta\right]\right\}$$

$$+ K_7 \cos\left[-k_1\left(\frac{x^2}{2l_2} + x\sin\theta\right)\right].$$

EXAMPLE 10.11

Let us assume that the parameters of the rainbow holographic construction in Figure 10.27 are $l_1 = 30$ cm, $l_2 = 2$ cm, $w = 2$ mm, $\lambda_1 = 600$ nm, and $\theta = 30°$. The rainbow hologram is illuminated by a broad-band plane wave of uniform spectral distribution, from $\lambda = 600$ to 400 nm, as shown in Figure 10.29.

(a) Evaluate the length of the smeared slit image.
(b) If the rainbow hologram image is viewed through the smeared slit image by an unaided eye with a 2.5-mm pupil, calculate the wavelength spread over the pupil.

Answers
(a) The fifth and sixth terms of the amplitude transmittance function in Example 10.10 represent the construction of the real slit image, that is, object

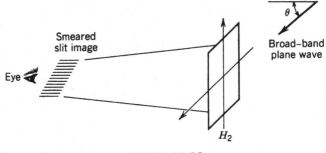

FIGURE 10.29

points 1 and 2 in Figure 10.27. The location of the real slit image can be computed, as

$$g(\alpha; k) = \exp(-ikx \sin \theta) \exp\left\{ -ik_1 \left[\frac{(x \pm w/2)^2}{2l_1} - x \sin \theta \right] \right\} * \exp\left(i\frac{k}{2l} x^2 \right),$$

which can be written as

$$g(\alpha; k) = C \exp\left(\frac{ik}{2l} \alpha^2 \right) \int \exp\left[\frac{i}{2}\left(\frac{k}{l} - \frac{k_1}{l_1} \right) x^2 \right]$$
$$\times \exp\left\{ \frac{-ik}{l} \left[\alpha \pm \frac{k_1 l w}{2kl_1} - \frac{k_1 - k}{k} l \sin \theta \right] x \right\} dx.$$

When the quadratic phase factor is eliminated, the longitudinal distance of the slit image is $l = (\lambda_1/\lambda)l_1$. The transverse locations of the slit edge images can be found by

$$\alpha = \pm \frac{k_1 l w}{2kl_1} + \frac{k_1 - k}{k} l \sin \theta.$$

By substituting $l = (\lambda_1/\lambda)l_1$, we have

$$\alpha = \pm \frac{w}{2} + \frac{\lambda - \lambda_1}{\lambda} l_1 \sin \theta.$$

By substituting $\lambda = \lambda_1 = 600$ nm, we have $l = l_1$ and $\alpha_{12} = \pm w/2 = \pm 1$ mm. The slit image is formed at the same location as the slit. However, for $\lambda = 400$ nm, we have

$$l = \frac{600}{400} \times 30 = 40 \, \text{cm},$$
$$\alpha_1 = 1 - \frac{600 - 400}{400}(300)(0.5) = -74 \, \text{mm},$$

and

$$\alpha_2 = -1 - \frac{6-4}{4}(300)(0.5) = -76\,\text{mm}.$$

Thus, the length of the smeared slit image is

$$L = [(1+76)^2 + (400-300)^2]^{1/2} = 126.2\,\text{mm}.$$

(b) From part (a) we can see that the slit image spreads uniformly about 77 mm, as projected along the vertical axis. Thus, the wavelength spread over the pupil is

$$\frac{(200)(2.5)}{77} = 6.5\,\text{nm}.$$

10.6 One-step Rainbow Holograms

In the previous section we discussed the general concept of the rainbow holographic process. This process requires two recording steps: first, use the conventional off-axis holographic technique to make a primary hologram from a real object; and, secondly, record a rainbow hologram from the real hologram image of the primary hologram. Placing a narrow-slit aperture behind the primary hologram in the second step of the holographic process means that the reconstruction light source need not be coherent. However, a two-step holographic recording process is cumbersome and requires a separate optical setup for each step, a major undertaking for laboratories with limited resources for optical components.

In this section we illustrate a one-step technique for producing rainbow holograms. This technique offers certain flexibilities in the construction of rainbow holograms, and the optical arrangement is simpler than that for the conventional two-step process.

As noted, making a rainbow hologram requires recording a real hologram image of the object through a narrow slit. If the rainbow hologram is illuminated by a monochromatic light source, a real hologram image of the object is produced, but the vertical parallax of the image is limited by the narrow slit. If the rainbow hologram is illuminated by a white-light source, however, the hologram image of the slit will disperse in rainbow colors. Therefore, the basic goal of rainbow holography is to form the image of the slit aperture between the hologram image of the object and the observer.

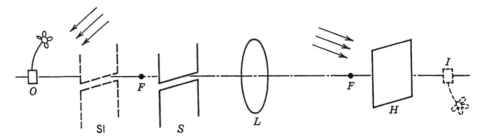

FIGURE 10.30 One-step rainbow holographic construction for pseudoscopic imaging. SI, Slit image; *F*, focal point of the lens; *S*, slit; *L*, imaging lens; *H*, holographic plate.

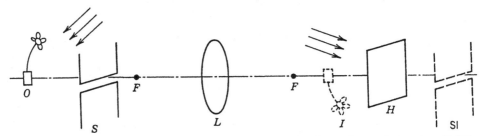

FIGURE 10.31 One-step rainbow holographic construction for orthoscopic imaging. *S*, Slit; *F*, focal point of the lens; *L*, imaging lens; *H*, holographic plate; SI, slit image.

Figures 10.30 and 10.31 show how a lens or lens system can in a one-step process simultaneously image both the object and the slit. In this manner a rainbow hologram can be made without a primary hologram.

EXAMPLE 10.12

Consider the one-step rainbow holographic construction shown in Figure 10.32. If the narrow slit is located at $f/4$ away from the imaging lens:

(a) Draw an analog system diagram to evaluate the slit image produced by the imaging lens.
(b) Determine the location of the slit image.
(c) Draw an equivalent optical setup to replace the one shown in Figure 10.32.
(d) Draw an analog system diagram of the rainbow holographic construction.

Answers
(a) The analog system diagram is shown in Figure 10.33.

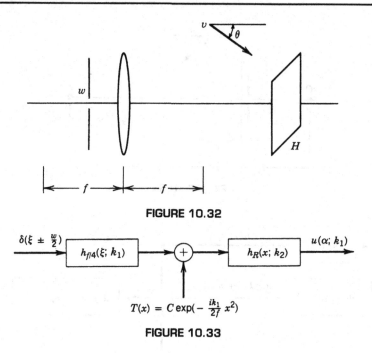

FIGURE 10.32

$$\xrightarrow{\delta(\xi \pm \tfrac{w}{2})} \boxed{h_{f/4}(\xi; k_1)} \longrightarrow \oplus \longrightarrow \boxed{h_R(x; k_2)} \xrightarrow{u(\alpha; k_1)}$$

$$T(x) = C \exp(-\tfrac{ik_1}{2f} x^2)$$

FIGURE 10.33

(b) The location of the slit image can be calculated as

$$u(\alpha; k_1) = \left\{ \left[\delta\left(\xi \pm \frac{w}{2}\right) * h_{f/4}(\xi; k_1) \right] T(x; k_1) \right\} * h_R(x; k_1)$$

$$= C \exp\left(i \frac{k_1}{2R} \alpha^2 \right) \int \exp\left[i \frac{k_1}{2} \left(\frac{3}{f} + \frac{1}{R} \right) x^2 \right]$$

$$\exp\left\{ -i \frac{k_1}{R} \left[\left(\alpha \pm \frac{2R}{f} w \right) x \right] \right\} dx\, dy.$$

Thus, we see that the slit image will be located at

$$R = -\frac{f}{3}$$

and

$$\alpha_{1,2} = \pm \tfrac{2}{3} w.$$

(c) An equivalent optical setup without the lens is shown in Figure 10.34.

(d) An analog system diagram of the rainbow holographic construction is given in Figure 10.35.

EXAMPLE 10.13

Show that the rainbow hologram image produced by the one-step process in Example 10.12 is pseudoscopic.

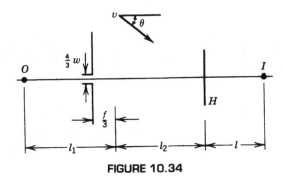

FIGURE 10.34

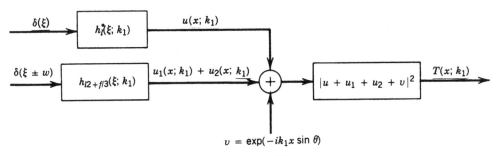

FIGURE 10.35

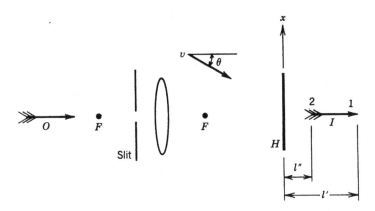

FIGURE 10.36

For the rainbow hologram image to be pseudoscopic, we assume that the rainbow holographic construction takes place with a longitudinal object, as shown in Figure 10.36. An analog system diagram of the construction process is given in Figure 10.37. Thus, the holographic amplitude transmittance function is

$$T(x; k_1) = K_1 + K_2 \cos\left[\frac{k_1}{2}\left(\frac{1}{l''} - \frac{1}{l'}\right)x^2\right] + K_3 \cos\left[k_1\left(\frac{x^2}{2l''} - x\sin\theta\right)\right]$$
$$+ K_4 \cos\left[k_1\left(\frac{x^2}{2l'} - x\sin\theta\right)\right].$$

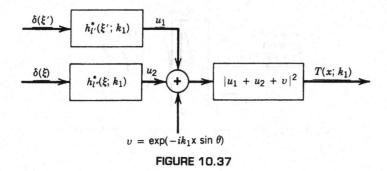

$v = \exp(-ik_1 x \sin \theta)$

FIGURE 10.37

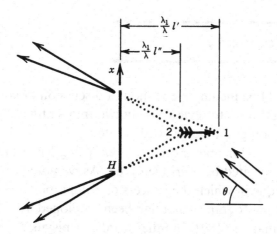

FIGURE 10.38

If the hologram is illuminated by a conjugate plane wave of λ, as shown in Figure 10.38, the divergent or virtual-image diffractions can be computed as

$$g_2(\alpha; k) = \int \exp(ikx \sin \theta) \exp\left[ik_1\left(\frac{x^2}{2l''} - x \sin \theta\right)\right] \exp\left[i\frac{k}{2l}(\alpha - x)^2\right] dx$$

$$= \exp\left(i\frac{k}{2l}\alpha^2\right) \int \exp\left[i\left(\frac{k_1}{2l''} + \frac{k}{2l}\right)x^2\right] \exp\left(i\frac{k}{l}\alpha x\right)$$

$$\exp[i(k - k_1)x \sin \theta \, dx,$$

$$g_2(\alpha; k)\Big|_{l=-(\lambda_1/\lambda)l''} = C_2\delta\left(\alpha - \frac{\lambda - \lambda_1}{\lambda}l'' \sin \theta\right),$$

and

$$g_1(\alpha; k)\Big|_{l=-(\lambda_1/\lambda)l'} = C_1\delta\left(\alpha - \frac{\lambda - \lambda_1}{\lambda}l' \sin \theta\right).$$

Since hologram image point 2 appears closer to the holographic plate than image point 1, we conclude that the rainbow hologram image is pseudoscopic.

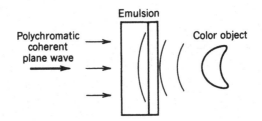

FIGURE 10.39 A reflection holographic construction.

10.7 Color Holography

This chapter would be incomplete without a discussion of color holography. The aim of this section is to review briefly two frequently used techniques for generating color hologram images with white light.

The best-known color holographic process using white light is the reflection holography invented by Yuri Denisyuk. As we noted earlier, in 1962 he reported a technique in which the process of holography was combined with the form of color photography that had been developed by the French physicist Gabriel Lippmann in 1891. In other words, Denisyuk's work is one of the cornerstones of white-light holography, combining as it does the work of Lippmann and of Dennis Gabor by using coherent light for holographic construction and white light for hologram image reconstruction.

In this method a coherent polychromatic wave field, with the primary colors of light, passes through a recording plate, falls on a diffused color object, and then is reflected back to the recording plate, as shown in Figure 10.39. As in the Lippmann color photography process, interferometric fringes are formed throughout the depth of the emulsion covering the plate. The color hologram, which has the characteristics of a photograph produced by the Lippmann process, can be viewed with a white-light source of limited spatial extent, for example, an ordinary high-intensity desk lamp or a slide projector, as illustrated in Figure 10.40. Although such color hologram images have been widely demonstrated, reflection color holography does have several drawbacks which prevent widespread practical applications. Two of these drawbacks are that: (1) an elaborate film-processing technique is required to prevent the emulsion from shrinking; and (2) the efficiency of the hologram image diffraction is rather low. Nevertheless, the reflection hologram image can be viewed by direct white-light illumination, and it is useful for decorative display purposes.

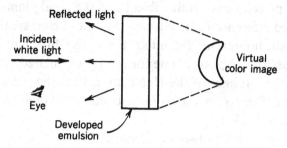

FIGURE 10.40 Reconstruction of a reflection hologram image using white-light illumination.

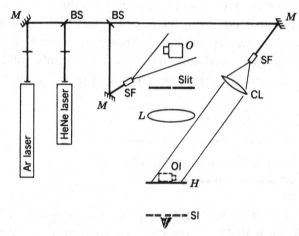

FIGURE 10.41 A one-step technique for constructing a color rainbow hologram. *M*, Mirror; BS, beam splitter; SF, spatial filter; *O*, object; *L*, imaging lens; CL, collimated lens; OI, object image; SI, slit image.

In 1969, another type of color holography was reported by Stephen Benton. The hologram produced has been called a white-light transmission hologram, but it is best known as a rainbow hologram (Section 10.5).

Benton's two-step technique for constructing a true-color rainbow hologram is rather cumbersome. First, three primary holograms have to be constructed using three primary color coherent sources. Then the projected real images of these three primary holograms are sequentially multiplexed onto a fourth hologram, again with three-color coherent readout. These three primary holograms must be aligned very carefully to make certain that their reconstructions fit exactly one on top of another. Thus, this technique is complicated and not easily implemented.

We shall now describe a one-step technique for generating a color rainbow hologram. The optical setup is illustrated in Figure 10.41. A HeNe laser is used to provide the red light (6328 Å), and an argon laser provides the green and blue lights (5145 and 4765 Å). The illuminated object is imaged through an imagining lens to a plane just in front of the hologram. A narrow slit of about 1.5 mm is placed between the object and the focal plane of the imaging lens. A collimated reference beam ensures that the carrier spatial frequency is the same across the hologram. The intensities of the three lights are measured independently, and the exposure time for each is calculated. The hologram is first exposed to the red light of the HeNe laser, then the green light (5145 Å) of the argon laser. The argon laser is then tuned to the blue line (4765 Å), and a third exposure is made.

A Kodak 649F plate is used because it has a relatively flat spectral response. As the plate is developed, a rainbow hologram is formed. When the hologram is viewed with a white-light point source, a very bright color image can be reconstructed. As in the two-step technique, a true-color hologram image can be observed when the hologram is viewed in the correct plane. If the viewer moves off this plane, different shades of color can still be seen, but the color will be different from that of the original object.

10.8 Photorefractive Holograms

Two of the most widely used white-light holograms must be the reflection hologram (see Section 10.4) and the rainbow hologram (see Section 10.5). In the reflection hologram, a thickness emulsion of about 20 μm has a wavelength selectivity of about $\Delta\lambda/\lambda = 1/40$, which is high enough to produce color hologram images without significant color blur. However, the physical requirements for constructing reflection holograms are rather stringent, which prevent their widespread application. In addition, the construction of a rainbow hologram requires a narrow slit, and thus the parallax information in the hologram image is partly lost. We shall now demonstrate that color holograms can be constructed with a photorefractive crystal using a "white-light" laser. Since a photorefractive crystal is much thicker than a conventional photographic emulsion, it provides higher wavelength selectivity by which the color blur can be minimized. The obvious advantages of using photorefractive material are that the construction of holograms is in the real-time mode and shrinkage of the emulsion is prevented.

Applying the coupled wave theory to the thick crystal hologram, as illustrated in Figure 10.42, wavelength selectivities (under the weak coupling assumption) for the transmission-type and reflection-type holograms can be shown to be [6],

$$\left(\frac{\Delta\lambda}{\lambda}\right)_t = \frac{(\eta^2 - \sin^2\alpha)^{1/2}}{\sin^2\alpha}\frac{\lambda}{d} \tag{10.58}$$

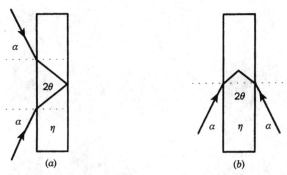

FIGURE 10.42 Photorefractive hologram constructions: (a) transmission type; (b) reflection type. 2α, External construction angle; 2θ, internal construction angle.

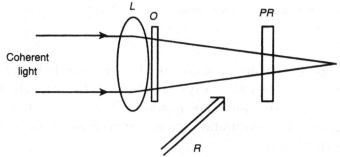

FIGURE 10.43 Construction of a transmission-type photorefractive hologram. L, Transform lens; O, object transparency; R, reference beam; PR, photorefractive crystal.

and

$$\left(\frac{\Delta\lambda}{\lambda}\right)_r = \frac{1}{(\eta^2 - \cos^2\alpha)^{1/2}}\frac{\lambda}{d}, \qquad (10.59)$$

where d is the thickness of the crystal, α is the incident angle, and η is the refractive index of the hologram.

For a transmission-type hologram, we have used a 1-mm thick LiNbO$_3$ PR crystal with a writing angle of $\alpha = 15°$. To reduce the noise disturbance caused by defects in the crystal, a piece of ground glass is placed in front of a color object transparency, as shown in Figure 10.43. The recording light source is a Lexel Kr–Ar "white-light" laser, which has nine spectral lines distributed within the visible region. The strongest spectral lines of this laser are 488, 514, and 647 nm. Since the spectral lines oscillate in different transverse modes, spectral filters are used in both the object and the reference beams to insure color uniformity. The object-to-reference beam ratio for this hologram construction is about 1 : 8, for which a high quality photorefractive hologram is constructed. Figure 10.44 shows a reconstructed color hologram image using the "white-light" laser, in which both the vertical and the horizontal parallax are preserved. If the hologram image is viewed at different

FIGURE 10.44 A reconstructed color photorefractive hologram image (here shown in black-and-white).

angles, it does not change in color as in a rainbow holographic image. Excellent color fidelity can always be obtained by properly controlling the exposure and spectral-line polarizations. Nevertheless, if the exposure is incorrect, the color saturation would be dominated by the strongest spectral line of the laser.

There is, however, some degree of color cross-talk, which is primarily due to a holographic grating formed at one wavelength being read out by other wavelengths. Since the wavelength selectivities given by Eqs 10.58 and 10.59 are mainly concerned with the center lobe frequency detuning, the weak side lobes would cause serious color cross-talk. The phenomenon of color cross-talk can be easily seen as spatial frequency beating, which deteriorates the hologram image quality.

We have shown that holographic imaging can be accomplished using a thick photorefractive crystal. The major advantages are that the holographic construction is in the real-time mode and shrinkage of the emulsion is avoided. Because of the high angular and wavelength selectivities of the photorefractive material, high-capacity holographic data can be multiplexed in a thick crystal.

EXAMPLE 10.14

Let us assume that the thickness and the refractive index of an LiNbO$_3$ crystal are $d = 1$ mm and $\eta = 2.3$, respectively. If the construction wavelength and the writing angle of the photorefractive hologram are $\lambda = 440$ nm and $\alpha = 90°$, respectively:

(a) Calculate the wavelength selectivities for transmission-type and reflection-type photorefractive holograms.

(b) Repeat (a) for $\alpha = 20°$ and $40°$.

(c) Determine the optimum (selectivity) construction angles for the photo-reactive hologram constructions.

(d) In view of (c), how can these optimum angles be implemented in practice?

Answers

(a) By substituting the data in Eqs 10.58 and 10.59, the wavelength selectivity for a transmission-type hologram is

$$\left(\frac{\Delta\lambda}{\lambda}\right)_t = \frac{[2.3^2 - \sin^2(90°)]^{1/2}}{\sin^2(90°)} \frac{440 \times 10^{-9}}{1 \times 10^{-3}} = 911 \times 10^{-6}$$

and for a reflection-type hologram it is

$$\left(\frac{\Delta\lambda}{\lambda}\right)_r = \frac{1}{[2.3^2 - \cos^2(90)]^{1/2}} \frac{440 \times 10^{-9}}{1 \times 10^{-3}} = 191 \times 10^{-6}.$$

Thus the wavelength selectivity for the reflection-type hologram is about 4.77 times higher than for the transmission type.

(b) For $\alpha = 20°$, we have

$$\left(\frac{\Delta\lambda}{\lambda}\right)_t = \frac{[2.3^2 - \sin^2(20°)]^{1/2}}{\sin^2(20°)} \frac{440 \times 10^{-9}}{1 \times 10^{-3}} = 8555 \times 10^{-6}$$

and

$$\left(\frac{\Delta\lambda}{\lambda}\right)_r = \frac{1}{[2.3^2 - \cos^2(20)]^{1/2}} \frac{440 \times 10^{-9}}{1 \times 10^{-3}} = 209.6 \times 10^{-6}.$$

The wavelength selectivity for the reflection-type hologram is about 41 times higher than for the transmission type.

For $\alpha = 40°$, we have

$$\left(\frac{\Delta\lambda}{\lambda}\right)_t = 2352 \times 10^{-6}$$

and

$$\left(\frac{\Delta\lambda}{\lambda}\right)_r = 2029 \times 10^{-6}.$$

(c) Referring to the results of part (b), the optimum construction angles occur at $2\alpha = 180°$ for both the transmission and the reflection holograms.

(d) It is trivial to see that the optimum angle $2\alpha = 180°$ cannot actually be implemented for transmission-type holograms.

REFERENCES

1. F. T. S. YU, *Optical Information Processing*, Wiley-Interscience, New York, 1983, Chapters 10 and 11.
2. J. W. GOODMAN, *Introduction to Fourier Optics*, McGraw-Hill, New York, 1968, Chapter 8.
3. D. GABOR, A new microscope principle, *Nature*, 161 (1948) 777.
4. Y. N. DENISYUK, Photographic reconstruction of the optical properties of an object in its own scattered radiation field, *Soviet Physics Doklady*, 7 (1962) 543.
5. S. A. BENTON, Hologram reconstructions with extended light sources, *Journal of the Optical Society of America*, 59 (1969) 1545.
6. F. T. S. YU and S. JUTAMULIA, *Optical Signal Processing, Computing and Neural Networks*, Wiley-Interscience, New York, 1992, Chapter 7.

PROBLEMS

10.1 Let the on-axis object point hologram of Eq. 10.4 be normally illuminated by a *divergent* monochromatic point source, which is located at a distance $2l_1$ from the hologram.

(a) Draw an analog system diagram of the holographic reconstruction process.

(b) Calculate the locations of the virtual and real hologram images, if the reconstruction wavelength is $0.8\lambda_1$.

10.2 Assume a *convergent* monochromatic wavefront, which converges at a distance $\frac{1}{2}l_1$ behind an on-axis object point hologram, is normally incident on the hologram.

(a) Draw a schematic diagram to represent the holographic reconstruction.

(b) Draw an analog system diagram of the reconstruction process.

(c) Evaluate the virtual and real images, assuming that the illuminating wave is the same as the construction wavelength.

10.3 Use the on-axis object point hologram of Eq. 10.4

(a) Plot the amplitude transmittance function of the hologram along the x axis.

(b) If the hologram was constructed with wavelength $\lambda_1 = 500\,\text{nm}$, at a distance of $l_1 = 20\,\text{cm}$, and with a $10 \times 10\,\text{cm}^2$ square recording aperture, compute the spatial frequency near the edge of the hologram.

10.4 Consider the on-axis hologram in Example 10.1. If the hologram is illuminated obliquely from a 45° angle by a collimated white light, calculate the smearing lengths of the hologram images. The spectral content of the white-light source is assumed to be uniformly distributed over the range 350–700 nm.

10.5 For the on-axis object point hologram in Example 10.1, we assume that the separation between the object point and the holographic aperture is

0.2 m (i.e., $l_1 = 0.2$ m). The hologram is illuminated by normally incident white light.

(a) Calculate the hologram image reconstruction.

(b) What is the smearing length of the reconstructed images?

(c) By comparing this result with the result obtained in Example 10.2, draw an explicit conclusion.

10.6 An on-axis object point hologram is constructed with a reference beam for which the object-to-reference beam (intensity) ratio is $1:3$. If we assume that the hologram is recorded in the linear region of the T–E curve, what will the diffraction efficiency of the hologram be?

10.7 Consider the on-focus holographic construction shown in Figure 10.45.

(a) Draw an analog system diagram of the holographic construction.

(b) Show that the hologram image can be viewed by a simple white-light illumination.

(c) What color would you expect the hologram image to be? Explain briefly.

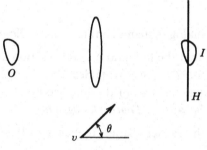

FIGURE 10.45

10.8 An off-axis object point hologram is constructed with a convergent reference beam as depicted in Figure 10.46.

(a) Draw an analog system diagram of the holographic construction.

(b) Evaluate the amplitude transmittance function of the recorded hologram.

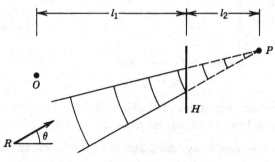

FIGURE 10.46

10.9 Assume that the off-axis hologram shown in Figure 10.46 is reconstructed with a conjugate divergent point source, located at point P, of the same wavelength.

(a) Draw an analog system diagram of the holographic reconstruction.

(b) Evaluate the real and virtual hologram images.

10.10 Assume that the distance parameters of the off-axis hologram in Problem 10.8 are $l_1 = 30\,\text{cm}$, $l_2 = 20\,\text{cm}$, and that the oblique angle θ of the convergent reference beam is $30°$. If the hologram is illuminated by a conjugate white-light point source located at point P, as shown in Figure 10.46:

(a) Calculate the vertical smearing length of the real image.

(b) Identify the red and violet color images. Notice that the white-light point source has a uniform spectral density over the range 350–700 nm.

10.11 In Section 2.3 we saw that a positive lens is capable of performing a two-dimensional Fourier transformation. If a holographic construction takes place because of the Fourier transform property of a lens, as shown in Figure 10.47:

(a) Draw an analog system diagram of the holographic recording.

(b) Assuming that the holographic construction is recorded in the linear region of the T–E curve (see Section 3.2.2), calculate the amplitude transmittance function of the encoded hologram. (*Note*: This type of hologram is called a *Fourier hologram*.)

(c) Calculate the carrier spatial frequency of the hologram.

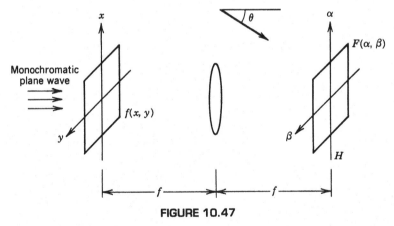

FIGURE 10.47

10.12 If the Fourier hologram in Problem 10.11 is inserted at the front focal length of a Fourier transform lens, as depicted in Figure 10.48,

(a) Draw an analog system diagram of the hologram image reconstruction.

(b) Evaluate the output of the complex light distribution.

(c) Sketch the locations of the hologram images.

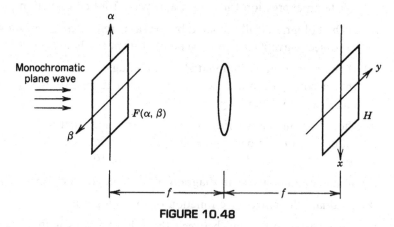

FIGURE 10.48

10.13 Assume that the wavelength of the off-axis holographic construction in Figure 8.11 is $\lambda_1 = 500\,nm$, that the separation between the extended object and the holographic aperture is $l_1 = 30\,cm$, and that the oblique angle θ is 45°.

(a) Calculate the carrier spatial frequency of the hologram.

(b) Assuming that the Fresnel diffraction from the object is spatial-frequency-limited (i.e., finite in spatial-frequency bandwidth), sketch the spectral distributions (i.e., the spatial-frequency contents) of the hologram along the x axis.

(c) If the spatial-frequency bandwidth of the extended object beam is about 50 lines/mm, compute the minimum oblique angle θ of the reference beam needed to separate the spectral contents of the hologram images.

10.14 The off-axis hologram in Problem 10.13 is illuminated by a conjugate divergent point source, located at $l_2 = 35\,cm$, and the reconstruction wavelength is $\lambda_2 = 600\,nm$.

(a) Calculate the lateral magnifications of the real and virtual hologram images.

(b) If the reconstruction wavelength is $\lambda_2 = 400\,nm$, compute the corresponding lateral magnifications of hologram images.

(c) Calculate the longitudinal locations of the hologram images in parts (a) and (b), and state the effects of the hologram images produced by the reconstruction wavelength.

10.15 Repeat Problem 10.14 for hologram images that are reconstructed using a conjugate convergent wavefront that converges to a point at a distance of 35 cm (i.e., $l_2 = -35\,cm$) behind the hologram. Compare the results with those obtained in Problem 10.14.

10.16 Hologram image reconstruction is also known as wavefront reconstruction.

(a) Write an expression for the off-axis wavefront construction process.

(b) If the hologram is illuminated by the same reference beam, show that an object wavefront can be generated from the hologram.

(c) If the hologram is illuminated by a conjugate reference beam, show that a conjugate object wave field, which represents the real-image reconstruction, can be generated.

10.17 The hologram image blurring (i.e., light dispersion) that is due to white-light illumination can be minimized by making a reflection hologram with near-field object recording, as shown in Figure 10.49.

(a) Draw an analog system diagram of the holographic construction.

(b) Evaluate the reflectance function of the hologram.

(c) Determine the spacing between the subholograms within the emulsion.

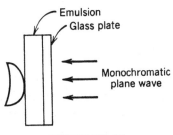

FIGURE 10.49

10.18 If the reflection hologram in Problem 10.17 is illuminated by a normally incident white light, show that color blurring (i.e., light dispersion) of the object will be minimized when the object point is closer to the recording emulsion.

10.19 Show that a true-color reflection hologram can be constructed with the optical setup shown in Figure 10.49, if the holographic construction is carried out with red, green, and blue coherent light.

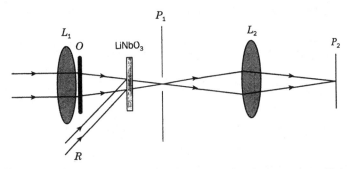

FIGURE 10.50 Geometry for multiplexing a photorefractive hologram. P_1, Pinhole array; L_1, L_2, transform lenses; O, object transparency; P_2, output imaging plane.

10.20 For the two-step rainbow holographic construction depicted in Figure 10.24, show that the hologram image is *orthoscopic* if the rainbow hologram is illuminated from behind by a conjugate plane wave.

10.21 If the oblique angle of the reference beam for the rainbow hologram in Example 10.11 is $50°$ (i.e., $\theta = 50°$), show that the color blurring of the hologram image observed by the unaided eye would be reduced.

10.22 Determine the location of the real-slit image if the rainbow hologram in Problem 10.21 is illuminated by a conjugate monochromatic plane wave of wavelength $\lambda_2 = 700\,\text{nm}$. Notice that $l_1 = 30\,\text{cm}$, $\lambda_1 = 600\,\text{nm}$, $w = 2\,\text{mm}$, and $\theta = 50°$.

10.23 If the slit width of the rainbow holographic construction in Example 10.11 is reduced to 1 mm instead of 2 mm, show that the color blurring of the rainbow hologram image under observation would be reduced.

10.24 If the narrow slit of the one-step rainbow holographic construction shown in Figure 10.30 is located at $f/2$ (i.e., a half of the focal length):

(a) Calculate the location and slit width of the slit image.

(b) Draw an equivalent schematic diagram representing the holographic construction.

(c) Draw an object point analog system diagram for the schematic diagram in part (b).

10.25 If the rainbow hologram described in Problem 10.24 is viewed with a white-light source:

(a) Calculate the location of the white-light source for observing the rainbow hologram images. Notice that, in practice, the holographic emulsion is not negligibly thin.

(b) Draw a schematic diagram representing the reconstruction of the rainbow hologram images. Sketch the location of the smeared slit image.

10.26 To encode an orthoscopic rainbow hologram imaging, during the holographic construction, we place a narrow slit at $l = 1.5f$ (i.e., 1.5 times the focal length) in front of the imaging lens.

(a) Calculate the location and the slit width of the slit image produced by the imaging lens.

(b) Compare the preceding results with those of part (a) in Problem 10.24.

(c) Sketch an equivalent schematic diagram representing the one-step rainbow holographic construction.

10.27 (a) Draw a schematic diagram for the rainbow hologram imaging in Problem 10.26.

(b) Show that the rainbow hologram imaging is orthoscopic.

10.28 By referring to the photorefractive hologram construction shown in Figure 10.43:

   (a) Calculate the wavelength selectivities as a function of the writing angle α. Assume that the thickness and the refractive index of the photorefractive crystal are $d = 1\,\text{mm}$ and $\eta = 2.28$, respectively, and that the construction wavelength is $\lambda = 630\,\text{nm}$.

   (b) Calculate and plot the normalized wavelength selectivity (i.e., $(d/\lambda)(\Delta\lambda/\lambda)$) as a function of α, on a logarithmic scale.

10.29 **(a)** Repeat Problem 10.28, for a reflection-type hologram.

   (b) By comparing the result from (a) with Problem 10.28, draw a brief conclusion.

10.30 Referring to the geometry of multiplexing photorefractive holographic storage shown in Figure 10.50:

   (a) Show that a high density photorefractive holographic memory can be constructed using wavelength multiplexing.

   (b) Assume that the size of the pinhole used at P_1 is given by $m\lambda f/a$, where $m \times m$ is the number of object pixels, a is the size of the object, and f is the focal length of L_1. If one uses an $N \times N$ pinhole array, show that N^2 objects can be multiplexed in a photorefractive crystal.

SIGNAL PROCESSING

The recent advances in real-time spatial-light modulators and electro-optic devices have brought optical signal processing to a new stage of development. Much attention has been focused on high-speed optical signal processing and on computing at a high rate of data processing. In this chapter we discuss the basic principles of optical signal processing under coherent, incoherent, and partially coherent illuminations.

11.1 An Optical System Under Coherent and Incoherent Illumination

Let us consider the hypothetical optical system depicted in Figure 11.1. The light emitted by the source Σ is monochromatic. The complex light distribution at the input plane comes from the incremental light source $d\Sigma$. To determine the light field at the output plane, we let this complex light distribution at the input plane be $u(x,y)$. If the complex amplitude transmittance of the input plane is $f(x,y)$, the complex light field immediately behind the signal plane would be $u(x,y)f(x,y)$.

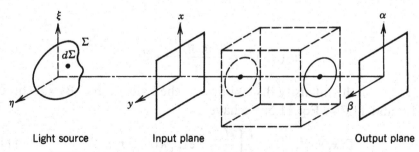

Light source   Input plane   Output plane

FIGURE 11.1 A hypothetical optical signal-processing system.

We assume that the optical system under consideration is a linear and spatially invariant with a spatial impulse response of $h(x, y)$. The complex light field at the output plane of the system, which comes from $d\Sigma$, can then be determined by the following convolution equation,

$$g(\alpha, \beta) = [u(x, y)f(x, y)] * h(x, y), \tag{11.1}$$

where the asterisk denotes the convolution operation.

The corresponding intensity distribution at the output plane, which is due to $d\Sigma$, is

$$dI(\alpha, \beta) = g(\alpha, \beta)g^*(\alpha, \beta)\, d\Sigma, \tag{11.2}$$

where the asterisk represents the complex conjugate. Thus, the overall intensity distribution at the output plane is

$$I(\alpha, \beta) = \iint_{\Sigma} |g(\alpha, \beta)|^2\, d\Sigma, \tag{11.3}$$

which can be written in the following convolution form,

$$I(\alpha, \beta) = \int\int\int\int_{-\infty}^{\infty}\int \Gamma(x, y; x', y')h(\alpha - x, \beta - y)h^*(\alpha - x', \beta - y')$$
$$\times f(x, y)f^*(x', y')\, dx\, dy\, dx'\, dy', \tag{11.4}$$

where

$$\Gamma(x, y; x', y') = \iint_{\Sigma} u(x, y)u^*(x', y')\, d\Sigma \tag{11.5}$$

is the *spatial coherence factor*, which is also known as the *mutual intensity function*.

Let us now choose two arbitrary points Q_1 and Q_2 at the input plane. If r_1 and r_2 are the respective distances from Q_1 and Q_2 to $d\Sigma$, the complex light disturbances at Q_1 and Q_2 that come from $d\Sigma$ can be written as

$$u_1(x, y) = \frac{[I(\xi, \eta)]^{1/2}}{r_1} e^{ikr_1} \tag{11.6}$$

and

$$u_2(x', y') = \frac{[I(\xi, \eta)]^{1/2}}{r_2} e^{ikr_2} \tag{11.7}$$

where $I(\xi, \eta)$ is the intensity distribution of the light source. By substituting Eqs 11.6 and 11.7 in Eq. 11.5, we have

$$\Gamma(x, y; x', y') = \iint_{\Sigma} \frac{I(\xi, \eta)}{r_1 r_2} \exp[ik(r_1 - r_2)]\, d\Sigma. \tag{11.8}$$

When the light rays are paraxial, $r_1 - r_2$ can be approximated by

$$r_1 - r_2 \simeq \frac{1}{r}[\xi(x - x') + \eta(y - y')], \tag{11.9}$$

where r is the distance between the source plane and the signal plane. Then Eq. 11.8 can be written as

$$\Gamma(x, y; x', y') = \frac{1}{r^2} \int\int I(\xi, \eta) \exp\left\{ i\frac{k}{r}[\xi(x - x') + \eta(y - y')] \right\} d\xi\, d\eta, \tag{11.10}$$

which represents the inverse Fourier transform for intensity distribution at the source plane. Equation 11.10 is also known as the Van Cittert–Zernike theorem. The normalized form of this theorem is given in Eq. 8.47.

Now let us consider two extreme situations. In one we let the light source become infinitely large, and we assume that it is uniform, that is, $I(\xi, \eta) \simeq K$. Thus, Eq. 11.10 becomes

$$\Gamma(x, y; x', y') = K_1 \delta(x - x', y - y'), \tag{11.11}$$

where K_1 is a proportionality constant. This equation describes a completely *incoherent* optical system.

On the other hand, if we let the light source be vanishingly small, then $I(\xi, \eta) \simeq K\delta(\xi, \eta)$ and Eq. 11.10 becomes

$$\Gamma(x, y; x', y') = K_2, \tag{11.12}$$

where K_2 is an arbitrary constant. This equation describes a completely *coherent* optical system. In other words, a monochromatic point source describes a strictly coherent regime, whereas an extended source describes a strictly incoherent system. Furthermore, an extended monochromatic source is also known as a *spatially incoherent* source.

For the completely incoherent optical system described in Eq. 11.11, $\Gamma(x, y; x', y') = K_1 \delta(x - x', y - y')$, the intensity distribution at the output plane, given in Eq. 11.4, becomes

$$I(\alpha, \beta) = \int\int\int\int_{-\infty}^{\infty} \delta(x' - x, y' - y)h(\alpha - x, \beta - y)$$
$$\times h^*(\alpha - x', \beta - y')f(x, y)f^*(x', y')\, dx\, dy\, dx'\, dy', \tag{11.13}$$

which can be reduced to

$$I(\alpha, \beta) = \int\int_{-\infty}^{\infty} |h(\alpha - x, \beta - y)|^2 |f(x, y)|^2\, dx\, dy. \tag{11.14}$$

It is therefore apparent that for incoherent illumination, the intensity distribution at the output plane is the convolution of the input signal's intensity in relation to the intensity of the spatial impulse response. In other words, an incoherent optical system is linear in *intensity*, or

$$I(\alpha, \beta) = |h(x, y)|^2 * |f(x, y)|^2, \tag{11.15}$$

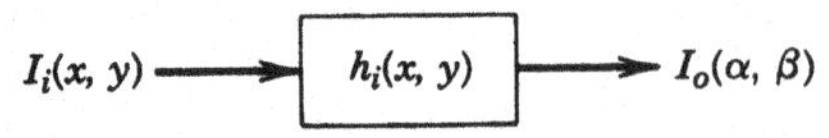

$$I_i(x, y) \longrightarrow \boxed{h_i(x, y)} \longrightarrow I_o(\alpha, \beta)$$

FIGURE 11.2 An analog system diagram of an incoherent system. $h(x, y)$, Intensity of the spatial impulse response.

$$f(x, y) \longrightarrow \boxed{h(x, y)} \longrightarrow g(\alpha, \beta)$$

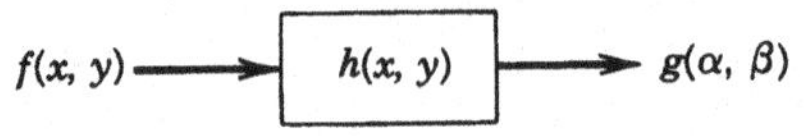

FIGURE 11.3 An analog system diagram of a coherent system. $h(x, y)$, Spatial impulse response.

where the asterisk denotes the convolution operation. An analog system diagram of an incoherent system is shown in Figure 11.2. The output intensity response can be determined as

$$I_o(\alpha, \beta) = \int\!\!\!\int\limits_{-\infty}^{\infty} I_i(x, y) h_i(\alpha - x, y - \beta) \, dx \, dy, \tag{11.16}$$

where $I_i(x, y)$ is the input intensity excitation and $h_i(x, y) = |h(x, y)|^2$ is the intensity of the spatial impulse response.

However, for the strictly coherent illumination described in Eq. 11.12, $\Gamma(x, y; x', y') = K_2$, Eq. 11.4 becomes

$$I(\alpha, \beta) = g(\alpha, \beta) g^*(\alpha, \beta) = \int\!\!\!\int\limits_{-\infty}^{\infty} h(\alpha - x, \beta - y) f(x, y) \, dx \, dy$$

$$\times \int\!\!\!\int\limits_{-\infty}^{\infty} h^*(\alpha - x', \beta - y') f^*(x', y') \, dx' \, dy'. \tag{11.17}$$

It is therefore apparent that the coherent optical system is linear in *complex amplitude*, or

$$g(\alpha, \beta) = \int\!\!\!\int\limits_{-\infty}^{\infty} h(\alpha - x, \beta - y) f(x, y) \, dx \, dy. \tag{11.18}$$

An analog system diagram of Eq. 11.18 is given in Figure 11.3.

EXAMPLE 11.1

Consider the spatial impulse response of an input–output optical system that is given by

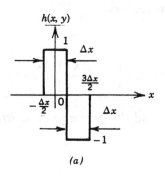

(a)

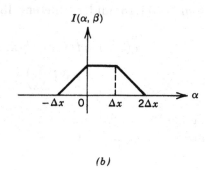

(b)

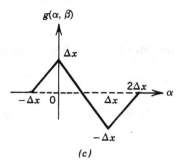

(c)

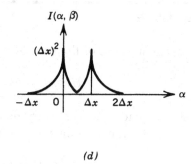

(d)

FIGURE 11.4

$$h(x,y) = \text{rect}\left(\frac{x}{\Delta x}\right) - \text{rect}\left(\frac{x - \Delta x}{\Delta x}\right)$$

and shown in Figure 11.4a.

(a) If the optical system is illuminated by a spatially limited incoherent wavefront such as

$$f(x,y) = \text{rect}\left(\frac{x}{\Delta x}\right),$$

calculate the corresponding irradiance at the output plane.

(b) If the spatially limited illumination of the optical system is a coherent wavefront, compute the corresponding complex light field at the output plane.

Answers

(a) From Eq. 11.15, under incoherent illumination we have

$$I(\alpha,\beta) = |f(x,y)|^2 * |h(x,y)|^2$$
$$= \text{rect}\left(\frac{x}{\Delta x}\right) * \text{rect}\left(\frac{x - \Delta x/2}{2\,\Delta x}\right).$$

The graphical sketch of this result is given in Figure 11.4b.

(b) From Eq. 11.18, under coherent illumination we have

$$g(\alpha, \beta) = f(x, y) * h(x, y)$$

$$= \text{rect}\left(\frac{x}{\Delta x}\right) * \left[\text{rect}\left(\frac{x}{\Delta x}\right) - \text{rect}\left(\frac{x - \Delta x}{\Delta x}\right)\right].$$

A graphical sketch is given in Figure 11.4c, and the corresponding irradiance is shown in Figure 11.4d.

EXAMPLE 11.2

The transfer function of the optical system shown in Figure 11.1 is given by

$$H(p, q) = K_1 \exp[i(\alpha_0 p + \beta_0 q)],$$

where K_1, α_0, and β_0 are arbitrary positive constants. We assume that the complex amplitude transmittance at the input plane is

$$f(x, y) = K_2 e^{i\phi(x,y)},$$

where K_2 is an arbitrary positive constant. Calculate the output responses under incoherent and coherent illuminations, respectively.

To compute the output responses, we first evaluate the spatial impulse response of the optical system by

$$h(x, y) = \mathscr{F}^{-1}[H(p, q)] = K_1 \delta(x + \alpha_0, y + \beta_0).$$

We refer to Eq. 11.15 to calculate the output response under incoherent illumination,

$$I(\alpha, \beta) = |f(x, y)|^2 * |h(x, y)|^2$$

$$= K_2^2 * K_1^2 \delta^2(x + \alpha_0, y + \beta_0)$$

$$= K_1^2 K_2^2.$$

We use Eq. 11.18 to calculate the output response under coherent illumination,

$$g(\alpha, \beta) = f(x, y) * h(x, y)$$

$$= K_2 e^{i\phi(x,y)} * K_1 \delta(x + \alpha_0, y + \beta_0)$$

$$= K_1 K_2 \exp[i\phi(x + \alpha_0, y + \beta_0)].$$

Notice that the phase distribution is preserved under coherent illumination.

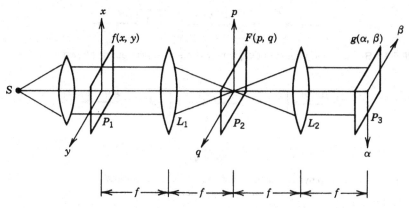

FIGURE 11.5 A coherent optical signal processor.

11.2 Coherent Optical Signal Processing

Referring to the phase transform properties of lenses discussed in Section 2.3, we see that transform lenses can be put together to construct a coherent optical signal processor, as depicted in Figure 11.5, where P_1, P_2, and P_3 represent the input, the Fourier, and the output planes, respectively, and a monochromatic point source S is located at the front focal length of a collimating lens. If an object transparency of amplitude transmittance $f(x,y)$ is inserted at the input plane P_1, the complex light field distributed at P_2 would be the Fourier transform of $f(x,y)$, or

$$F(p,q) = \mathscr{F}[f(x,y)], \tag{11.19}$$

where $p = (2\pi/f\lambda)x$ and $q = (2\pi/f\lambda)y$ are the angular spatial-frequency coordinates. A positive lens will perform a direct Fourier transformation, in which the transform kernel, $e^{-i(px+qy)}$, is negative. But a positive transform kernel, $e^{i(px+qy)}$, is needed for an inverse Fourier transformation. It is therefore apparent that a positive transform kernel can be introduced, with the help of a second positive lens, by simply inverting the output coordinate system (α, β), as shown in the output plane P_3 in Figure 11.5. Thus, the complex light distributions at the output plane P_3 can be shown to be

$$f(\alpha, \beta) = \mathscr{F}^{-1}[F(p,q)]. \tag{11.20}$$

An analog system diagram of Figure 11.5 is given in Figure 11.6.

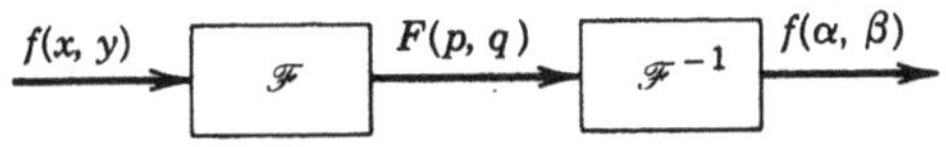

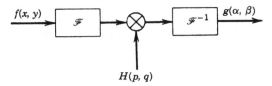

FIGURE 11.6 An analog system diagram of the signal processor shown in Figure 11.5.

$$f(x, y) \longrightarrow \boxed{\mathscr{F}} \longrightarrow \bigotimes \longrightarrow \boxed{\mathscr{F}^{-1}} \longrightarrow g(\alpha, \beta)$$

$$H(p, q)$$

FIGURE 11.7 An analog system diagram of the signal processor shown in Figure 11.5, with a spatial filter insertion.

Let us now assume that a spatial filter of complex amplitude transmittance $H(p, q)$ is inserted in the Fourier plane P_2. This complex light field P_2 immediately behind the spatial filter is then

$$E(p, q) = KF(p, q)H(p, q), \tag{11.21}$$

where K is a proportionality constant.

Since the second lens L_2 performs an inverse Fourier transformation of the complex light field $E(p, q)$ to the output plane P_3, the complex-amplitude light distribution at P_3 is given by

$$g(\alpha, \beta) = K \iint_S F(p, q)H(p, q)e^{i(p\alpha + q\beta)} \, dp \, dq, \tag{11.22}$$

where the surface integration is taken over the spatial-frequency domain P_2.

Alternatively, according to the Fourier multiplication property, Eq. 11.22 can be written as

$$g(\alpha, \beta) = K \iint_S f(x, y)h(\alpha - x, \beta - y) \, dx \, dy = Kf(x, y) * h(x, y), \tag{11.23}$$

where the integral is taken at the input spatial domain, and $h(x, y)$ is the spatial impulse response of the filter:

$$h(x, y) = \mathscr{F}^{-1}[H(p, q)]. \tag{11.24}$$

An analog system diagram of the optical signal processor is shown in Figure 11.7.

It is important to stress that the spatial filter $H(p, q)$ can consist of apertures or slits of any shape. Depending on the arrangement of the apertures, it can act as a low-pass, high-pass, or band-pass spatial filter. Clearly, any opaque portion in the filter represents a rejection of the spatial-frequency band. In addition, inclusion of a phase plate with the filter would produce a phase delay. Since we are able to construct amplitude filters and phase filters separately, in principle we are able to construct any complex spatial filter. However, Anthony Vander Lugt developed an interferometric technique for constructing a complex spatial filter, as we show in the next section. Vander Lugt constructed the filter using a Fourier hologram.

EXAMPLE 11.3

Assume that the amplitude transmittance of an object function is given by

$$f(x,y)[1 + \cos(p_0 x)],$$

where p_0 is an arbitrary angular carrier spatial frequency, and $f(x,y)$ is assumed to be spatial-frequency limited. If this object transparency is inserted at the input plane of the coherent optical signal processor in Figure 11.5:

(a) Determine the corresponding spectral distribution at the Fourier plane P_2.
(b) Design a stop band filter for which the light distribution at the output plane P_3 will be $f(x,y)$.

Answers
(a) Since lens L_1 will perform a direct Fourier transform, the complex light distribution at P_2 can be shown to be

$$\mathscr{F}\{f(x,y)[1 + \cos(p_0 x)]\} = F(p,q) + \tfrac{1}{2}F(p - p_0, q) + \tfrac{1}{2}F(p + p_0, q).$$

This distribution is sketched in Figure 11.8a.

(b) A stop band filter for the spectral distribution shown in Figure 11.8a is sketched in Figure 11.8b. This is essentially a low-pass filter which allows

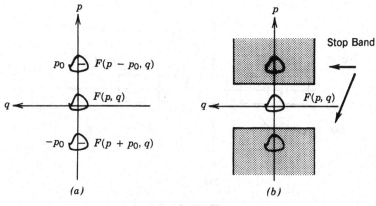

(a)   *(b)*

FIGURE 11.8

$F(p,q)$ to pass through. Since L_2 will perform an inverse Fourier transformation, the complex light field at P_3 is

$$\mathscr{F}^{-1}[F(p,q)] = f(x,y).$$

EXAMPLE 11.4

Consider the coherent optical signal processor shown in Figure 11.9*a*. The spatial filter is a one-dimensional sinusoidal grating,

$$H(p) = \tfrac{1}{2}[1 + \sin(\alpha_0 p)],$$

where α_0 is an arbitrary constant that is equal to the separation of the input object functions $f_1(x,y)$ and $f_2(x,y)$. Compute the complex light field at the output plane P_3.

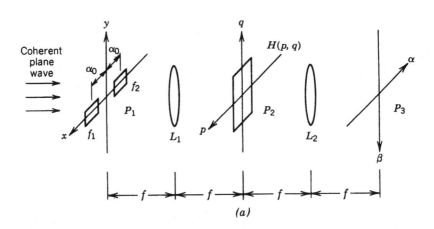

(a)

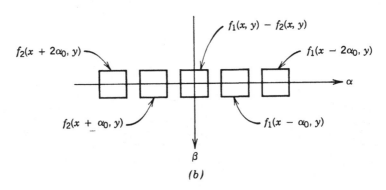

(b)

FIGURE 11.9

By applying the Fourier translation property, we can show that the complex light field at P_2 is

$$F_1(p, q)e^{-i\alpha_0 p} + F_2(p, q)e^{i\alpha_0 p}.$$

The complex light distribution immediately behind the filter is

$$
\begin{aligned}
E(p, q) &= [F_1(p, q)e^{-i\alpha_0 p} + F_2(p, q)e^{i\alpha_0 p}]H(p) \\
&= \tfrac{1}{2}F_1(p, q)e^{-i\alpha_0 p} + \tfrac{1}{2}F_2(p, q)e^{i\alpha_0 p} \\
&\quad + \frac{1}{4i}F_1(p, q) - \frac{1}{4i}F_1(p, q)e^{-i2\alpha_0 p} - \frac{1}{4i}F_2(p, q) \\
&\quad + \frac{1}{4i}F_2(p, q)e^{i2\alpha_0 p}.
\end{aligned}
$$

The light distribution at the output plane is therefore

$$
\begin{aligned}
g(\alpha, \beta) &= \mathscr{F}^{-1}[E(p, q)] \\
&= \tfrac{1}{2}f_1(x - \alpha_0, y) + \tfrac{1}{2}f_2(x + \alpha_0, y) \\
&\quad + \frac{1}{4i}[f_1(x, y) - f_2(x, y)] \\
&\quad - \frac{1}{4i}f_1(x - 2\alpha_0, y) + \frac{1}{4i}f_2(x + 2\alpha_0, y).
\end{aligned}
$$

A sketch of $g(\alpha, \beta)$ is given in Figure 11.9b. Note that the coherent optical signal processor is capable of performing image subtraction, that is, $f_1(x, y) - f_2(x, y)$, which is diffracted at the origin of the output plane.

11.3 Synthesis of a Complex Spatial Filter

In general, a spatial filter can be described by a complex amplitude transmittance distribution:

$$H(p, q) = |H(p, q)|e^{i\phi(p, q)}. \tag{11.25}$$

In practice, optical spatial filters are generally of the passive type. The physically realizable conditions of optical spatial filters are

$$|H(p, q)| \leq 1 \tag{11.26}$$

and

$$0 \leq \phi(p, q) < 2\pi. \tag{11.27}$$

We note that such a transmittance function can be represented by a set of points within or on a unit circle in the complex plane, as shown in Figure 11.10. The amplitude transmission of the filter changes with the optical

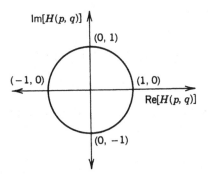

FIGURE 11.10 A complex amplitude transmittance.

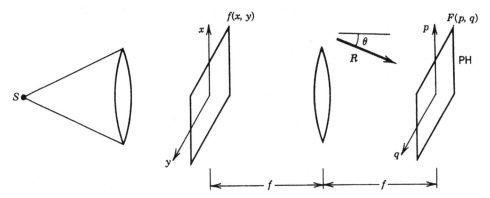

FIGURE 11.11 Construction of a complex spatial filter. $f(x, y)$, Input object transparency; R, reference plane wave; PH, photographic plate, the recording medium.

density, and the phase delay varies with the thickness. Thus, a complex spatial filter may be constructed by combining an amplitude filter and a phase delay filter.

Let us now discuss the technique developed by Vander Lugt for constructing a complex spatial filter using an interferometric method, as shown in Figure 11.11. The complex light field over the spatial-frequency plane is

$$E(p, q) = F(p, q) + e^{-i\alpha_0 p}, \qquad (11.28)$$

where $\alpha_0 = f \sin \theta$, f is the focal length of the transform lens, and $F(p, q) = |F(p, q)| e^{i\phi(p, q)}$.

The corresponding intensity distribution over the recording medium is

$$I(p, q) = 1 + |F(p, q)|^2 + 2|F(p, q)| \cos[\alpha_0 p + \phi(p, q)]. \qquad (11.29)$$

We assume that if the recording is linear in amplitude transmittance, the corresponding amplitude transmittance function of the spatial filter is

$$H(p, q) = K\{1 + |F(p, q)|^2 + 2|F(p, q)| \cos[\alpha_0 p + \phi(p, q)]\}, \qquad (11.30)$$

which is, in fact, a *real positive function*.

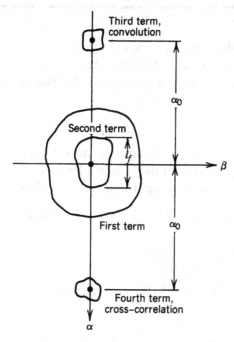

FIGURE 11.12 Sketch of output diffractions.

If this complex spatial filter is inserted in the Fourier plane of a coherent optical signal processor, as shown in Figure 11.5, the complex light immediately behind the spatial filter would be

$$
\begin{aligned}
E(p,q) &= F(p,q)H(p,q) \\
&= K[F(p,q) + F(p,q)|F(p,q)|^2 + F(p,q)F(p,q)e^{ip\alpha_0} \\
&\quad + F(p,q)F^*(p,q)e^{-ip\alpha_0}],
\end{aligned}
\tag{11.31}
$$

where the asterisk denotes the complex conjugate. The complex light field at the output plane can be determined by

$$
\begin{aligned}
g(\alpha,\beta) &= \iint E(p,q)e^{-i(\alpha p+\beta q)}\,dp\,dq \\
&= K[f(x,y) + f(x,y)*f(x,y)*f^*(-x,-y) \\
&\quad + f(x,y)*f(x+\alpha_0,y) + f(x,y)*f^*(-x+\alpha_0,-y)],
\end{aligned}
\tag{11.32}
$$

where the first and second terms represent the zero-order diffraction, which appears at the origin of the output plane, and the third and fourth terms are the convolution and cross-correlation terms, which are diffracted in the neighborhood of $\alpha = -\alpha_0$ and $\alpha = \alpha_0$, respectively, as sketched in Figure 11.12.

EXAMPLE 11.5

Assume that a noisy object transparency, such as $f(x,y) + n(x,y)$, is inserted in the input plane of the coherent optical signal processor shown in Figure 11.5. if a matched spatial filter $H(p,q)$ is inserted in the Fourier plane of the optical processor,

$$H(p,q) = K_1 + 2K_2|F(p,q)|\cos[\alpha_0 p + \phi(p,q)],$$

where the K terms are proportionality constants and $F(p,q) = \mathscr{F}[f(x,y)]$, calculate the light distribution at the output plane.

The light distribution at the output plane can be obtained as

$$g(\alpha, \beta) = [f(x,y) + n(x,y)] * h(x,y),$$

where

$$\begin{aligned} h(x,y) &= \mathscr{F}^{-1}[H(p,q)] \\ &= K_1\delta(x,y) + K_2 f(x + \alpha_0, y) + K_2 f^*(-x + \alpha_0, -y). \end{aligned}$$

Thus, we have

$$\begin{aligned} g(\alpha, \beta) = {} &K_1[f(x,y) + n(x,y)] * \delta(x,y) \\ &+ K_2[f(x,y) + n(x,y)] * f(x + \alpha_0, y) \\ &+ K_2[f(x,y) + n(x,y)] * f^*(-x + \alpha_0, -y). \end{aligned}$$

If we assume that the additive noise is white and Gaussian-distributed (notice that $n(x,y) * f(x,y) = 0$), the light distribution at the output plane reduces to

$$\begin{aligned} g(\alpha, \beta) = {} &K_1[f(x,y) + n(x,y)] + K_2 f(x,y) * f(x + \alpha_0, y) \\ &+ K_2 f(x,y) \circledast f^*(x - \alpha_0, y), \end{aligned}$$

where $\circledast$ denotes the correlation operation.

EXAMPLE 11.6

If the input object function is translated to a new location, that is, $f(x - x_0, y - y_0)$, show that the output correlation peak is also translated to the same location.

Using the last term of Eq. 11.32, we show that

$$\iint f(x - x_0, y - y_0) f^*(x + \alpha - \alpha_0, y + \beta)\, dx\, dy = R_{11}(\alpha - \alpha_0 - x_0, \beta - y_0).$$

Thus, we see that the autocorrelation function R_{11}, the correlation peak, moves to the location to which the input object function moved. A sketch of the output diffraction is shown in Figure 11.13.

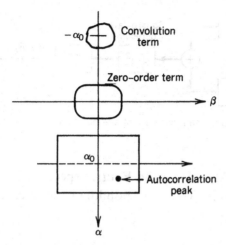

FIGURE 11.13

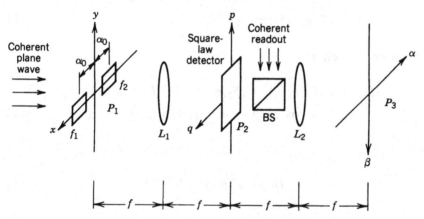

FIGURE 11.14 An optical joint Fourier transform processor. BS, Beam splitter.

11.4 The Joint Transform Correlator

Complex spatial filtering can also be performed with an optical joint Fourier transform processor, as shown in Figure 11.14. An analog system diagram of the optical architecture appears in Figure 11.15. Since the input objects are illuminated by a coherent plane wave, the complex light distribution arriving at the square-law detector in the Fourier plane P_2 will be

$$E(p, q) = F_1(p, q)e^{-i\alpha_0 p} + F_2(p, q)e^{i\alpha_0 p}, \tag{11.33}$$

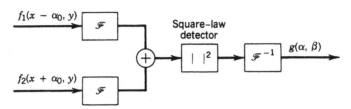

FIGURE 11.15 An analog system diagram of the optical architecture of the joint transform processor shown in Figure 11.14.

where $F_1(p,q)$ and $F_2(p,q)$ are the Fourier spectra of input objects $f_1(x,y)$ and $f_2(x,y)$, respectively. The corresponding irradiance at the output end of the square-law detector is

$$
\begin{aligned}
I(p,q) &= |E(p,q)|^2 \\
&= |F_1(p,q)|^2 + |F_2(p,q)|^2 + F_1(p,q)F_2^*(p,q)e^{-i2\alpha_0 p} \\
&\quad + F_1^*(p,q)F_2(p,q)e^{i2\alpha_0 p},
\end{aligned} \tag{11.34}
$$

which can be written as

$$
\begin{aligned}
I(p,q) &= |F_1(p,q)|^2 + |F_2(p,q)|^2 + 2|F_1(p,q)||F_2(p,q)| \\
&\quad \times \cos[2\alpha_0 p - \phi_1(p,q) + \phi_2(p,q)],
\end{aligned} \tag{11.35}
$$

where

$$
F_1(p,q) = |F_1(p,q)|e^{i\phi_1(p,q)}. \tag{11.36}
$$

and

$$
F_2(p,q) = |F_2(p,q)|e^{i\phi_2(p,q)}. \tag{11.37}
$$

If the irradiance of Eq. 11.35 is read out by a coherent plane wave, the complex light distribution at the output plane P_3 will be

$$
\begin{aligned}
g(\alpha,\beta) &= f_1(x,y) \circledast f_1^*(x,y) + f_2(x,y) \circledast f_2^*(x,y) \\
&\quad + f_1(x,y) \circledast f_2^*(x - 2\alpha_0, y) + f_1^*(x,y) \circledast f_2(x + 2\alpha_0, y),
\end{aligned} \tag{11.38}
$$

where $\circledast$ denotes the correlation operation. The first two terms represent overlapping correlation functions, $f_1(x,y)$ and $f_2(x,y)$, which are diffracted at the origin of the output plane. The last two terms are the two cross-correlation terms, which are diffracted around $\alpha = 2\alpha_0$ and $\alpha = -2\alpha_0$, respectively. Notice that a square-law converter, such as a photographic plate, a liquid crystal light valve, or a charge-coupled camera, can be used.

EXAMPLE 11.7

In the optical joint Fourier transform processor of Figure 11.14, we assume that $f_1(x - \alpha_0, y)$ is embedded in additive white, Gaussian noise, that is, $f_1(x - \alpha_0, y) + n(x - \alpha_0, y)$, and $f_2(x + \alpha_0, y)$ is replaced by $f_1(x + \alpha_0, y)$.

(a) Draw an analog system diagram of the optical joint Fourier transform correlator.

(b) Evaluate the complex light distribution at the output plane.

Answers
(a) An analog system diagram of this problem appears in Figure 11.16.

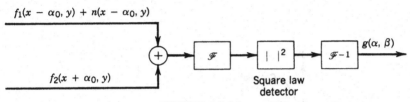

FIGURE 11.16

(b) The complex light field arriving at the input end of the square-law detector is

$$E(p, q) = F_1 e^{-i\alpha_0 p} + N e^{-i\alpha_0 p} + F_1 e^{i\alpha_0 p}.$$

The corresponding irradiance can be shown as

$$I(p, q) = 2|F_1|^2 + |N|^2 + F_1 N^* + N F_1^* + (F_1 F_1^* + N F_1^*)e^{-2\alpha_0 p}$$
$$+ (F_1 F_1^* + F_1 N^*)e^{i2\alpha_0 p}.$$

Since the noise is assumed to be additive and Gaussian distributed, we note that

$$\iint f_1(x, y) n(\alpha + x, \beta + y) \, dx \, dy = 0.$$

Thus, the complex light field at the output plane would be

$$g(\alpha, \beta) = 2 f_1(x, y) \circledast f_1^*(x, y) + n(x, y) \circledast n^*(x, y)$$
$$+ f_1(x, y) \circledast f_1^*(x - 2\alpha_0, y)$$
$$+ f_1(x, y) \circledast f_1^*(x + 2\alpha_0, y).$$

The first two terms will be diffracted around the origin of the output plane, and the two autocorrelation peaks, the third and fourth terms, will be diffracted at $\alpha = 2\alpha_0$ and $\alpha = -2\alpha_0$, respectively.

EXAMPLE 11.8

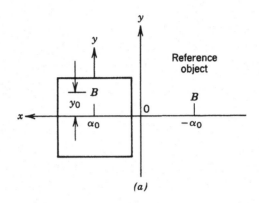

(a)

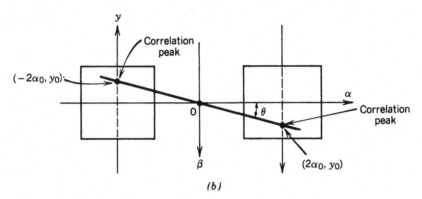

(b)

FIGURE 11.17

Using the optical joint Fourier transform processor of Figure 11.14, show that the spatial frequency and orientation of the interference fringes within a joint power spectrum determine the spatial content of the object. Sketch the locations of the autocorrelation peaks in the output plane.

Let the reference function be $f_2(x, y) = f_1(x, y)$, which is centered at $x = -\alpha_0$. If the input object function is $f_1(x - \alpha_0, y - y_0)$, as shown in Figure 11.17a, the complex light distribution at the input end of the square-law detector is

$$E(p, q) = F_1 \exp[-i(\alpha_0 p + y_0 q)] + F_1 e^{i(\alpha_0 p)}.$$

The corresponding power spectral distribution at the output plane can be shown as

$$I(p, q) = 2|F_1|^2 + F_1 F_1^* \exp[-i(2\alpha_0 p + y_0 q)] + F_1 F_1^* \exp[i(2\alpha_0 p + y_0 p)],$$

which can also be written as

$$I(p, q) = 2|F_1|^2[1 + \cos(2\alpha_0 p + y_0 q)].$$

From this result we see that there is a linear-phase distribution that can be expressed in terms of the fringe direction. Thus, by orienting the spatial-frequency coordinate of p to p', we have

$$I(p', q') = 2|F_1(p', q')|^2 [1 + \cos(\sqrt{(2\alpha_0)^2 + (y_0)^2} \, p')],$$

where the angular orientation of the fringes is

$$\theta = \tan^{-1}\left(\frac{y_0}{2\alpha_0}\right).$$

The positions of the autocorrelation peaks in the output plane, which are located at $(r, -\theta)$ and $(r, \pi - \theta)$, respectively, can therefore be determined by

$$\gamma = \sqrt{(2\alpha_0)^2 + (y_0)^2}$$

and

$$\theta = \tan^{-1}\left(\frac{y_0}{2\alpha_0}\right).$$

By taking the inverse Fourier transform of $I(p, q)$, we have the following complex light distribution at the output plane,

$$g(\alpha, \beta) = 2f_1(x, y) \circledast f_1^*(x, y) + f_1(x, y) \circledast f_1^*(x - 2\alpha_0, y - y_0)$$
$$+ f_1^*(x, y) \circledast f_1(x + 2\alpha_0, y + y_0),$$

which is sketched in Figure 11.17b. Thus, we see that the spatial frequency and the orientation of the fringes determine the spatial content of the object.

11.5 White-light Optical Signal Processing

Although coherent optical signal processors can perform a variety of complex signal operations, coherent processing systems are usually plagued by coherent artifact noise. This difficulty has prompted optical engineers to look for an alternative, for example, using a partially coherent source for optical signal processing. The basic advantages of partially coherent processing are that: (1) it can suppress the coherent artifact noise; (2) partially coherent sources are inexpensive; (3) the processing environment is very relaxed; (4) partially coherent processors are relatively easy and economical to operate; and (5) they are suitable for color image processing.

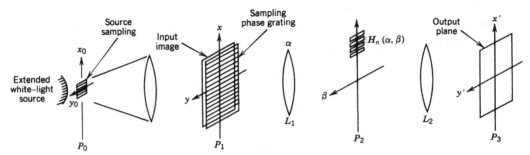

FIGURE 11.18   A white-light optical signal processor.

We now discuss an achromatic, partially coherent processing technique that can be carried out by a white-light source, as shown in Figure 11.18. This partially coherent processing system is similar to a coherent processing system, except that it uses an extended white-light source, a source-encoding mask, a signal-sampling grating, multiple spectral-band filters, and achromatic transform lenses. For example, if we place an input object transparency $s(x, y)$ in contact with an image-sampling phase grating, for every wavelength λ the complex wave field at the Fourier plane P_2 would be, assuming a white-light point source,

$$E(p, q; \lambda) = \int \int s(x, y)e^{ip_0 x}e^{-i(px+qy)} \, dx \, dy$$
$$= S(p - p_0, q). \tag{11.39}$$

Here the integral is over the spatial domain of the input plane P_1, (p, q) denotes the angular spatial-frequency coordinate system, p_0 is the angular spatial frequency of the sampling phase grating, and $S(p, q)$ is the Fourier spectrum of $s(x, y)$. The preceding equation can be written in (α, β) spatial variables,

$$E(\alpha, \beta; \lambda) = S\left(\alpha - \frac{\lambda f}{2\pi}p_0, \beta\right), \tag{11.40}$$

where $p = (2\pi/\lambda f)\alpha$, $q = (\pi/\lambda f)\beta$, and f is the focal length of the achromatic transform lens. Thus, we see that the Fourier spectra disperse into rainbow colors along the α axis, and that each Fourier spectrum for a given wavelength λ is centered at $\alpha = (\lambda f/2\pi)p_0$.

In complex signal filtering, a set of complex spatial filters with a narrow spectral bandwidth is provided. We assume that the input object is spatial-frequency limited; and that the spatial bandwidth of each spectral-band filter $H(p_n, q_n)$ is given by

$$H(p_n, q_n) = \begin{cases} H(p_n, q_n), & \alpha_1 < \alpha < \alpha_2, \\ 0, & \text{otherwise,} \end{cases} \tag{11.41}$$

where $p_n = (2\pi/\lambda_n f)\alpha$, $q_n = (2\pi/\lambda_n f)\beta$; λ_n is the main wavelength of the filter; $\alpha_1 = (\lambda_n f/2\pi)(p_0 + \Delta p)$ and $\alpha_2 = (\lambda_n f/2\pi)(p_0 - \Delta p)$ are the upper and lower spatial limits of $H(p_n, q_n)$, respectively; and Δp is the spatial bandwidth of the input object $s(x, y)$.

Since the limiting wavelengths of each $H(p_n, q_n)$ can be shown to be

$$\lambda_h = \lambda_n \frac{p_0 + \Delta p}{p_0 - \Delta p} \quad \text{and} \quad \lambda_l = \lambda_n \frac{p_0 - \Delta p}{p_0 + \Delta p}, \tag{11.42}$$

its spectral bandwidth can be approximated by

$$\Delta\lambda_n = \lambda_n \frac{4p_0\,\Delta p}{p_0^2 - (\Delta p)^2} \simeq \frac{4\Delta p}{p_0}\lambda_n. \tag{11.43}$$

If we place this set of spectral-band filters side by side and position them properly over the smeared Fourier spectra, the complex light field at the output plane, which comes from λ_n, would be

$$g(x, y) = s(x, y; \lambda_n) * h(x, y; \lambda_n), \tag{11.44}$$

where the asterisk represents the convolution operation, and $h(x, y; \lambda_n)$ is the spatial impulse response of $H_n(p_n, q_n)$. Since the spectral lines of the white-light source are *mutually incoherent*, the intensity distribution at the output plane is

$$I(x, y) = \sum_{n=1}^{N} \Delta\lambda_n |s(x, y; \lambda_n) * h(x, y; \lambda_n)|^2. \tag{11.45}$$

Thus, the partially coherent processor is capable of processing the signal in a complex wave field. Since the output intensity is the sum of the mutually incoherent narrow-band spectral irradiances, the coherent artifact noise can be eliminated. It is also apparent that the white-light source emits all the visible wavelengths, and that it is suitable for color image processing.

EXAMPLE 11.9

Draw an analog system diagram of the processing operation carried out by the partially coherent optical signal processor of Figure 11.18. Since we assume a white-light point source, the object transparency inserted at the input plane is illuminated by a *spatially coherent* white-light plane wave. The complex wave field immediately behind the object transparency is

$$\sum_n s(x, y; \lambda_n).$$

An analog system diagram of the white-light process is shown in Figure 11.19.

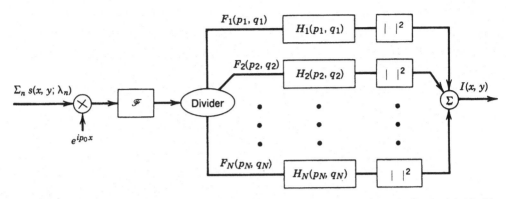

FIGURE 11.19 An analog system diagram of the signal processor shown in Figure 11.18. We assume a white-light point source.

EXAMPLE 11.10

The amplitude transmittance function of a multiplexed transparency, with a positive and a negative image, is given by

$$t(x, y) = t_1(x, y)(1 + \cos p_0 x) + t_2(x, y)(1 + \cos q_0 y),$$

where t_1 and t_2 are the positive and negative image functions, and p_0 and q_0 are the spatial sampling frequencies along the x and y axes, respectively.

This encoded transparency is inserted at the input plane of the white-light processor shown in Figure 11.20a.

(a) Evaluate the smeared spectra at the Fourier plane.
(b) If the focal length of the transform lens is $f = 300$ mm, and the sampling frequencies are $p_0 = 80\pi$ and $q_0 = 60\pi$ rad/mm, compute the smearing length of the Fourier spectra. Assume that the spectral lines of the white-light source vary in length from 350 to 750 nm.
(c) Design a set of transparent color filters by which the density or gray levels of the image can be encoded in pseudocolors at the output plane.
(d) Compute the irradiance of the pseudocolor image.

Answers
(a) The smeared Fourier spectra can be evaluated as follows,

$$\mathscr{F}[t(x, y)] = T_1(p, q) + \tfrac{1}{2}T_1(p - p_0, q) + \tfrac{1}{2}T_1(p + p_0, q)$$
$$+ T_2(p, q) + \tfrac{1}{2}T_2(p, q - q_0) + \tfrac{1}{2}T_2(p, q + q_0),$$

where T_1 and T_2 are the Fourier spectra of t_1 and t_2, respectively. Using Eq. 11.40, we get

$$T_1(p \mp p_0, q) = T_1\left(\alpha \mp \frac{\lambda f}{2\pi} p_0, \beta\right)$$

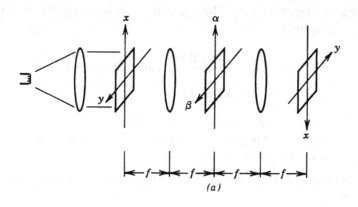

(a)

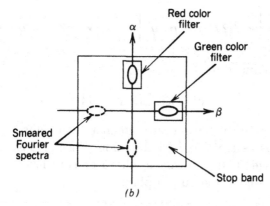

(b)

FIGURE 11.20

and

$$T_2(p, q \mp q_0) = T_2\left(\alpha, \beta \mp \frac{\lambda f}{2\pi} q_0\right).$$

Thus, we see that the positive and negative image spectra smear into rainbow colors along the α and the β axes, respectively.

(b) The smearing length along the α axis can be calculated as follows:

$$\alpha_1 - \alpha_2 = \frac{f p_0}{2\pi}(\lambda_1 - \lambda_2).$$

Substituting $f = 300\,\text{mm}$, $p_0 = 80\pi$, $\lambda_1 = 750\,\text{nm}$, and $\lambda_2 = 350\,\text{nm}$, we get

$$\alpha_1 - \alpha_2 = \frac{(300)(80\pi)}{2\pi}(750 - 350)(10^{-6})$$
$$= 4.8\,\text{mm}.$$

as shown in Figure 11.20b. The smearing length along the β axis can be similarly computed:

$$\beta_1 - \beta_2 = \frac{(300)(60\pi)}{2\pi}(750 - 350) \times 10^{-6}$$

$$= 3.6\,\text{mm}.$$

(c) A set of transparent color filters for encoding the density of the image is shown in Figure 11.20b. Notice that the positive image is encoded in red and the negative image in green.

(d) The irradiance of the pseudocolor image at the output plane can be evaluated as

$$I(x,y) = \left|\mathscr{F}^{-1}[2T_1(p - p_0, q)]\right|^2_{\lambda=\lambda_r} + \left|\mathscr{F}^{-1}[2T_2(p, q - q_0)]\right|^2_{\lambda=\lambda_g},$$

which can be reduced to

$$I(x,y) = K[T_{1r}^2(x,y) + T_{2g}^2(x,y)],$$

where λ_r and λ_g represent the red and green wavelengths, respectively. The output irradiance is essentially a combination of a red positive image and a green negative image. Thus, an image whose density levels are pseudocolor encoded can be viewed at the output plane.

11.6 Hybrid Optical Signal Processing

Two-dimensional optical processing has been used in a wide variety of applications in which the large-capacity and parallel-processing capabilities of optics can be exploited. Although optical processing relies on general holographic filtering techniques, which have been discussed in the preceding sections, the concept of programmability in signal processing will be introduced. This is primarily due to the advances in sophisticated spatial light modulators that allow us to construct various types of near-real-time hybrid optical processors. The systems can be compactly packaged, and so, in principle, can be used for on-board processing stations.

It is apparent that a purely optical processor has some drawbacks, which make certain tasks difficult or impossible to implement. The first is that optical systems are difficult to program as general-purpose electronic

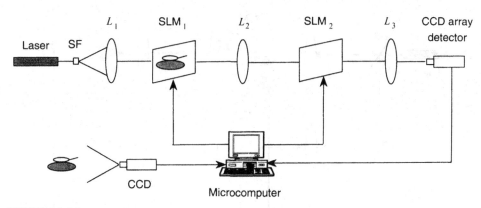

FIGURE 11.21 A hybrid optical correlator. L_1, Collimated lens; L_2, L_3, Fourier transform lenses.

computers. Although an optical system can be designed to perform specific tasks, it cannot be used where more flexibility is required. A second problem is that a system based on Fourier optics is naturally analog, in which mode great accuracy is difficult to achieve. A third problem is that optical systems by themselves cannot be used to make decisions, as some electronic counterparts can. Even the simplest type of decision-making is based on the comparison of the output with a stored value, such that the operation cannot be performed without the intervention of electronics.

The first approach to the (microcomputer based) hybrid optical processor is to use conventional optical configuration, in which the input object and the spatial filter are generated by programmable spatial light modulators (SLMs), as shown in Figure 11.21. For instance, a programmable complex conjugate Fourier transform of reference patterns is generated with the SLM_2 using a microcomputer, through which cross-correlation between the input object and the reference pattern can be detected by a charge-coupled device (CCD) array detector. The detected signal can be fed back to the computer for display and for decision-making. Thus, we see that a programmable hybrid optical processor can be realized, if an SLM with sufficient space–bandwidth product (SBP) and resolution is available for the display of the computer-generated complex spatial filter.

The second approach to the hyrbid-optical processor is to use a joint Fourier transform configuration, in which both the input object and the special impulse response of the filter function can be simultaneously displayed at the input SLM_1, as shown in Figure 11.22. For instance, programmable spatial reference function can be generated side by side with the input object, such that the joint transform power spectrum

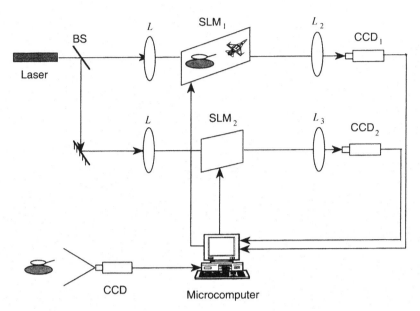

FIGURE 11.22 A hybrid optical joint-transform correlator. BS, Beam Splitter; L, collimated lens; L_2, L_3, Fourier transform lenses.

(JTPS) can be detected by CCD_1. By displaying the JTPs on the SLM_2, via the computer, cross-correlation between the input object and the reference function can be obtained at the back focal plane of the Fourier transform lens L_3.

Although the operations of the preceding hybrid optical configurations are basically the same, there is one major distinction between them. The spatial filter synthesis (e.g., Fourier hologram) is independent of the input signal, whereas the joint power spectra displayed on the SLM_2 (i.e., joint transform filter) is *dependent* on the input signal. Thus, nonlinear filtering may not be generally applied to joint transform architecture, since it may produce false alarms and lower output signal-to-noise ratio.

Let us now illustrate a technique for performing a real-time joint-transform correlation, as shown in Figure 11.23, in which a programmable spatial light modulator (e.g., magneto-optic light modulator (MOSLM) is used to display both the real-time input object and a set of reference images. The rate of operation is dependent upon the duty cycle of the liquid crystal light valve (LCLV).

The aforementioned architecture can be implemented using a single liquid crystal television (LCTV) to replace the MOSLM and the LCLV shown in Figure 11.24. There are, however, three major objections to using commercially available LCTVs for practical implementations: (1) their low contrast ratio; (2) their phase nonuniformity; and (3) their low resolution and low SBP.

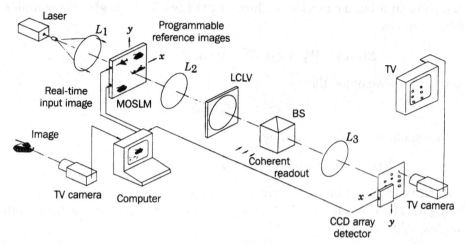

FIGURE 11.23   A real-time programmable joint-transform correlator. BS, Beam splitter; L_1, collimated lens; L_2, L_3, Fourier transform lenses.

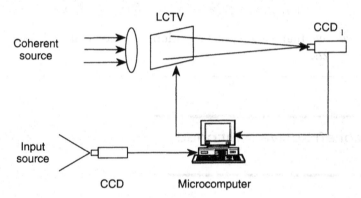

FIGURE 11.24   A single spatial light modulator joint-transform correlator.

EXAMPLE 11.11

By referring to the single SLM joint transform correlator shown in Figure 11.24, if the input object and reference function displayed on the LCTV panel are given by

$$f_1(x - \alpha, y) + f_2(x + \alpha, y),$$

in which we assume $f_1 = f_2$ and α is an arbitrary constant, then evaluate the operation of the hybrid optical processor.

Since the LCTV panel is not located at the front focal length of the transform lens, the Fourier transform at the back focal length introduces a quadratic phase factor, as given by

$$F(p, q)e^{i\beta(p^2 + q^2)},$$

where β is an arbitrary constant. Thus, at the back focal length, the complex field is given by

$$E(p,q) = [F_1(p,q)e^{-i\alpha P} + F_2(p,q)e^{i\alpha P}]e^{i\beta(p^2+q^2)}.$$

Since we have assumed that

$$F_1(p,q) = F_2(p,q),$$

the corresponding JTPS is given by

$$I(p,q) = E(p,q)E^*(p,q) = 2|F_1(p,q)|^2[1 + \cos(2\alpha p)],$$

in which we see that the quadratic phase factor has been eliminated. By displaying the JTPS back to the LCTV panel in the next half-duty cycle, the output complex light can be shown to be

$$
\begin{aligned}
g(x,y) &= \mathcal{F}^{-1}[I(p,q)] \\
&= 2[f_1(x,y) \circledast f_1(x,y)] \\
&\quad + f_1(x,y) \circledast f_1^*(x - 2\alpha, y) + f_1^*(x,y) \circledast f_1(x + 2\alpha, y),
\end{aligned}
$$

in which we see that two autocorrelation functions are diffracted at $x = +2\alpha$ and $x = -2\alpha$.

11.7 Photorefractive Matched Filters

Recently, multiplexing of filters in photorefractive materials has attracted a great deal of attention. By measuring the maximum phase shift achieved in a deep volume hologram, and calculating an exposure schedule for recording multiple equally diffracting holograms, over 1000 holograms have been shown recorded in a 4 mm thick BaTiO$_3$ crystal [5].

In Section 10.8, we have shown that the reflection-type photorefractive hologram has a higher wavelength selectivity. In this section we shall illustrate that a large-capacity wavelength-multiplexed spatial filter can be synthesized with a photorefractive crystal, as shown in Figure 11.25. Without loss of generality, we shall use a one-dimensional analysis, in which $q_1(x_1)$ represents the reference image located at the front focal plane of the transform lens L. For a given point x_1, a wave represented by the wavevector $k_1(x_1, \lambda_1)$ is incident at the photorefractive crystal. The reference beam is a plane wave represented by the wavevector $k_0(\lambda_1)$, which is incident at the *opposite* side of the crystal. It is apparent that a reflection-type matched filter can be recorded in the photorefractive crystal. Let $q_2(x_2)$ be the readout

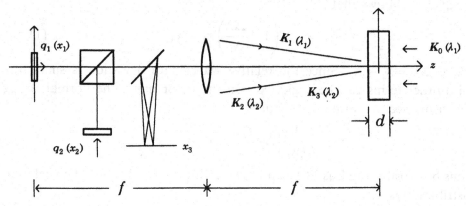

FIGURE 11.25 A photorefractive-based optical correlator. x_1, x_2, x_3, Reference, input, and output planes; K_0, K_1, K_2, K_3 reference, writing, reading, and readout wavevectors.

beam representing the input object in which each point of x_2 produces a readout wavevector $k_2(x_2, \lambda_2)$. The corresponding reconstructed wavevector is denoted by $k_3(x_3, \lambda_2)$, where x_3 represents the output coordinate plane. The wavevector representations in the crystal can be written as

$$k_0(\lambda_1) = -(\eta/\lambda_1)z,$$

$$k_1(\lambda_1) = \frac{x_1}{\lambda_1 f}u + \frac{\eta}{\lambda_1}\left(1 - \frac{x_1^2}{2\eta^2 f^2}\right)z,$$

$$k_2(\lambda_2) = \frac{x_2}{\lambda_2 f}u + \frac{\eta}{\lambda_2}\left(1 - \frac{x_2^2}{2\eta^2 f^2}\right)z,$$

$$k_3(\lambda_2) = \frac{x_3}{\lambda_2 f}u - \frac{\eta}{\lambda_2}\left(1 - \frac{x_3^2}{2\eta^2 f^2}\right)z, \qquad (11.46)$$

where z and u are the normal and the transverse unit vectors. It can be shown that the output correlation distribution is

$$R(x_3) = A\left|\iiiint q_1^*(x_1)q_2(x_2)\exp[i2\pi\,\Delta k(uu + zz)]\,dx_1\,dx_2\,du\,dz\right|^2, \qquad (11.47)$$

where Δk is the Bragg *dephasing vector*, as given by

$$\Delta k = k_0 - k_1 + k_2 - k_3. \qquad (11.48)$$

For simplicity, we assume that the size of the crystal is transversally extended such that the Bragg dephasing wavevector in the u direction would be zero. Thus we have

$$\frac{x_1}{\lambda_1} - \frac{x_2 + x_3}{\lambda_2} = 0. \qquad (11.49)$$

in which it follows that

$$x_2 = \left(1 - \frac{\Delta\lambda}{\lambda_1}\right)x_1 - x_3, \qquad (11.50)$$

where $\Delta\lambda = \lambda_1 - \lambda_2$. Since the relative wavelength deviation is small, the third-order terms can be neglected. On the other hand, the normal Bragg dephasing vector can be written as

$$\Delta k_z = \frac{1}{\eta\lambda_1}\left[2\eta^2\frac{\Delta\lambda}{\lambda_1} - x_3(x_1 - x_3)\right]z. \qquad (11.51)$$

Thus by substituting Eqs 11.50 and 11.51 in Eq. 11.47, the output correlation distribution is

$$R(x_3) = A\left|\int q_1^*(x_1)q_2\left[\left(1 - \frac{\Delta\lambda}{\lambda}\right)x_1 - x_3\right]\right.$$
$$\left. \times \operatorname{sinc}\left\{\frac{\pi d}{\eta\lambda_1}\left[2\eta^2\frac{\Delta\lambda}{\lambda_1} - x_3(x_1 - x_3)\right]\right\}dx_1\right|^2. \qquad (11.52)$$

where the asterisk denotes the complex conjugate. Needless to say, by referring to the sinc factor in the preceding result, the wavelength selectivity and the shift invariant properties of the photorefractive filter can be evaluated. If the input object is centered with respect to the reference function (i.e., $x_1 = x_2$), then the filter would respond to the wavelength variations that satisfy the following inequality:

$$\left|\frac{\Delta\lambda}{\lambda_1}\right| < \frac{\lambda_1}{\eta d}, \qquad (11.53)$$

where d is the thickness of the photorefractive crystal. Note that the preceding result is essentially the same as the wavelength selectivity of Eq. 10.59, when the construction beams are counterpropagated within the crystal.

On the other hand, if the readout wavelength is the same as the writing beams, the filter would respond to object shifting which satisfies the following inequality:

$$\left|\frac{x_3(x_1 - x_3)}{f^2}\right|_{max} < \frac{\lambda_1}{\eta d}, \qquad (11.54)$$

where f represents the focal length of the transform lens.

Thus by substituting Eq. 11.50 into Eq. 11.54, and letting $x_2 = x_1 - \Delta x$, we have

$$|\Delta x(x_1 - \Delta x)|_{max} < \lambda_1 f^2/\eta d. \qquad (11.55)$$

This inequality shows that the allowable shift Δx is dependent on the half-width of the input object $(x_1 - \Delta x)_{max}$, the thickness of the crystal, and the focal length of the lens. It follows that the wavelength selectivity and the shift invariance property of the photorefractive matched filter are dependent on the thickness of the crystal. None the less, by changing the effective focal length of the correlator, a desired shift invariance for a given thickness of the crystal can be obtained.

EXAMPLE 11.12

By referring to the wavelength selectivities of the transmission-type and reflection-type holograms in Eqs 10.58 and 10.59 in Section 10.8:

(a) Plot the selectivities (i.e., $\lambda/\Delta\lambda$) as a function of the construction angle (2α), for $d = 1, 7,$ and 10 mm. Assume that the refractive index of the photorefractive crystal is $\eta = 2.28$.

(b) What would be the *optimum* (wavelength selectivity) construction angles for the transmission-type and the reflection-type filters?

(c) If the photorefractive crystal is used as a large high-capacity matched filter, which type of matched filter would you recommend?

Answers

(a) With reference to Eqs 10.58 and 10.59, the wavelength selectivities for the transmission and the reflection-type photorefractive holograms are plotted in Figure 11.26, in which we see that the reflective-type photorefractive matched filter offers a higher selectivity and it increases as the thickness of crystal increases.

(b) Although both the construction angles for the transmission- and the reflection-type filters occur at $2\alpha = 180°$, the reflection-type filter offers a more uniform and higher selectivity over the transmission type. Notice that it is impossible to implement a 180° construction angle for a transmission-type filter.

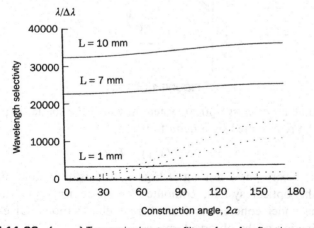

FIGURE 11.26 (- - - -) Transmission-type filter; (———) reflection-type filter.

(c) In view of parts (a) and (b), we have to agree that the reflection-type matched filter would be the best choice.

REFERENCES

1. J. W. GOODMAN, *Introduction to Fourier Optics*, McGraw-Hill, New York, 1968.
2. F. T. S. YU, *Optical Information Processing*, Wiley-Interscience, New York, 1983.
3. F. T. S. YU, *White-Light Optical Signal Processing*, Wiley-Interscience, New York, 1985.
4. F. T. S. YU and S. JUTAMULIA, *Optical Signal Processing, Computing and Neural Networks*, Wiley-Interscience, New York, 1992, Chapter 7.
5. J. H. HONG, P. YEH, D. PSALTIS and D. BRADY, Diffraction efficiency of strong volume holograms, *Optics Letters*, 6 (1990) 344.

PROBLEMS

11.1 Given an input–output linear and spatially invariant optical system under strictly incoherent illumination:

(a) Derive the incoherent transfer function in terms of the spatial impulse response of the system.

(b) State some basic properties of the incoherent transfer function.

11.2 Derive the cut-off frequency of an optical system under coherent and incoherent illuminations.

11.3 Given an optical imaging system that is capable of combining two Fraunhofer diffractions resulting from input objects $f_1(x,y)$ and $f_2(x,y)$, as shown in Figure 11.27, compute the irradiance at the output plane under coherent and incoherent illuminations.

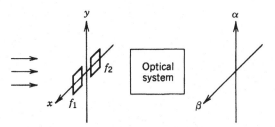

FIGURE 11.27

11.4 Assume an *all-pass* optical system in which the spatial impulse response can be represented by a delta function,

$$h(x,y) = \delta(x,y).$$

If the binary-phase grating of Figure 11.28 is inserted at the input plane of the optical system, compute the image irradiance at the output plane under coherent and incoherent illuminations. Give a conclusive distinction of the observed images.

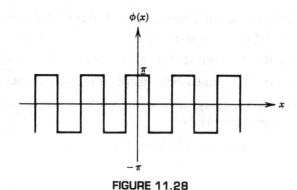

$\phi(x)$

π

x

$-\pi$

FIGURE 11.28

11.5 We assume that the input object function for the coherent optical signal processor in Figure 11.5 is given as

$$f(x,y) = \tfrac{1}{2}[1 + \cos(60\pi x)],$$

and that the focal length of the processor $f = 300\,\mathrm{mm}$.

(a) Evaluate the complex light field at the Fourier plane when the processor is illuminated by a monochromatic plane wave of $\lambda = 400\,\mathrm{nm}$.

(b) Repeat part **(a)** for $\lambda = 600\,\mathrm{nm}$. What scale changes does the shift in wavelength cause?

11.6 If the zero-order spectral distribution for the coherent optical signal processor described in Problem 11.5 is blocked by a high-pass spatial filter:

(a) Calculate the intensity distribution $I(\alpha,\beta)$ at the output plane.

(b) Compute the basic spatial frequency of the intensity distribution calculated in part **(a)**.

11.7 We assume that the coherent optical signal processor shown in Figure 11.5 has a monochromatic point source S of $\lambda = 600\,\mathrm{nm}$, and that the focal length of its transform lens is $f = 1000\,\mathrm{mm}$. If the input object $f(x,y)$ is an open square aperture of size $s = 5\,\mathrm{mm}$:

(a) Calculate the size of its Fourier spectrum (i.e., the first lobe) at the spatial-frequency plane P_2.

(b) Repeat part **a** for focal length $f = 100\,\mathrm{mm}$.

11.8 Assume that an object function embedded in an additive white, Gaussian noise, $f(x,y) + n(x,y)$, is inserted in the input plane of the coherent optical signal processor of Figure 11.5. A matched spatial filter, $H(p,q) = KF^*(p,q)$, is inserted in the Fourier plane.

(a) Compute the complex light distribution at the output plane. Assume that $f(x,y) * n(x,y) = 0$.

(b) If $f(x,y)$ is a complicated object function, such as a tank, sketch the shape of irradiance at the output plane.

11.9 If the object function referred to in Problem 11.8 is moved to a new location, that is, $f(x - x_0, y - y_0)$,

(a) Show that the corresponding spectrum is shift invariant.

(b) Evaluate the complex light distribution at the output plane.

11.10 (a) Draw an analog system diagram of the coherent optical signal processor in Figure 11.29.

(b) Evaluate the complex light distribution at its output plane.

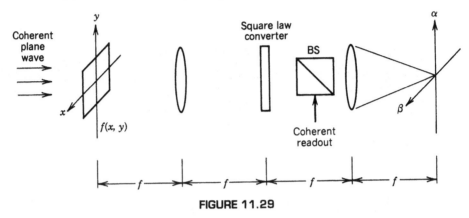

FIGURE 11.29

11.11 A spatial filter, $H(p, q) = i(p + q)$, is inserted at the Fourier plane of the coherent optical signal processor in Figure 11.5.

(a) Calculate the complex light distribution at the output plane. *Hint*:

$$\mathscr{F}\left[\frac{\partial^2 f(x, y)}{\partial x \, \partial y}\right] \simeq [i(p + q)][F(p, q)].$$

(b) If the input object is an open rectangular aperture, what would the image irradiance at the output plane be?

11.12 Suppose that the input object function of the coherent optical signal processor in Figure 11.5 is a rectangular grating of spatial frequency p_0, as shown in Figure 11.30.

(a) Evaluate and sketch the spectral content of the object at the Fourier plane.

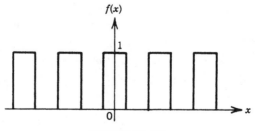

FIGURE 11.30

(b) If we insert a π-phase filter,

$$H(r) = \begin{cases} e^{i\pi}, & r \leq r_0, \\ 0, & \text{otherwise,} \end{cases}$$

at the origin of the spatial-frequency plane, sketch the light distribution at the output plane and state your conclusion.

11.13 For construction of the matching filter in Figure 11.11, assume that the focal length of the transform lens is $f = 300\,\text{mm}$, that the wavelength is $\lambda = 6500\,\text{Å}$, and that the oblique angle of the reference beam is $\theta = 30°$.

(a) Compute the carrier spatial frequency of the matching filter.

(b) What resolution must the recording medium have (e.g., film)?

11.14 Given an image transparency which has been distorted by linear motion and whose image points are each described by a small line segment. The corresponding transmittance of a smeared object point is represented by a rectangular function,

$$f(x) = \text{rect}\left(\frac{x}{\Delta x}\right),$$

where Δx is the smeared length. If we wish to restore the image with the coherent optical signal processor of Figure 11.5, show that the inverse filter $H(p)$ is, strictly speaking, physically unrealizable.

11.15 The matching filter of Eq. 11.30 is placed at the Fourier plane of the coherent optical signal processor of Figure 11.5, and no object transparency is inserted in the input plane.

(a) Compute the complex light distribution at the output plane.

(b) Sketch your observation at the output plane.

11.16 The separation of the input objects for the optical joint Fourier transform processor of Figure 11.14, is $2\alpha_0 = 2\,\text{cm}$, the wavelength of the coherent source is $\lambda = 6000\,\text{Å}$, and the focal length of the transform lenses is $f = 500\,\text{mm}$.

(a) Calculate the spatial frequency of the fringes at the Fourier plane.

(b) What resolution must the square-law converter have?

11.17 The reference object f_2 of a joint transform correlator is located at the origin, $f_2(x, y)$, and the input object is located at $x = 10\,\text{mm}$, $y = 10\,\text{mm}$, that is, $f_1(x - 10, y - 10)$.

(a) Calculate the intensity distribution at the input end of the square-law converter, if $\lambda = 6000\,\text{Å}$ and $f = 500\,\text{mm}$.

(b) Using the result of part **a**, determine the locations in the output plane of the cross-correlation functions with readout illuminations of $\lambda = 6000\,\text{Å}$ and $\lambda = 4000\,\text{Å}$, respectively.

11.18 Given an input transparency with an object size of 5 mm that is to be processed by the white-light processor of Figure 11.18. We assume that the focal length of the transform lenses is $f = 500$ mm.

(a) Calculate the size of the source required.

(b) If the spatial frequency of the input object is assumed to be two lines per millimeter, estimate the required size of the spatial filter H_n and the required spatial frequency of the phase grating.

11.19 We wish to perform spatial-frequency pseudocolor encoding of an input object's transparency with a white light.

(a) Sketch a white-light processor that can perform this task.

(b) We wish to encode low spatial frequency in red and high spatial frequency in blue. Sketch a spatial filter that can perform this operation.

11.20 It is known that a color image can be encoded in a black-and-white transparency in such a way that the encoded transmittance is represented by the following equation,

$$t(x, y) = K + t_r(x, y) \cos(p_0 x) + tg(x, y) \cos(p_0 y) + t_b(x, y) \cos(2p_0 y),$$

where K is an arbitrary constant and t_r, t_g, and t_b represent the red, green, and blue color images. If the encoded transparency is inserted in the white-light processor of Figure 11.20a:

(a) Evaluate the corresponding smeared Fourier spectra at the spatial-frequency plane.

(b) Design a set of transparent color filters for retrieving color images.

(c) Compute the irradiance of the color image at the output plane.

11.21 Consider a hybrid optical JTC as shown in Figure 11.21, in which the SLMs are LCTVs. What would the duty cycle of the JTC be? Assume that the frame ratios of the LCTVs and CCDs are 60 frames/s.

11.22 (a) Repeat Problem 11.21 for the single SLM architecture shown in Figure 11.24.

(b) Comment on the advantages and disadvantages of using the architectures shown in Figures 11.21 and 11.24.

11.23 By referring to the hybrid optical correlation in Figure 11.22, and assuming that the frame ratios of the SLMs and CCDs are 60 frames/s,

(a) Calculate the operating speed of the system.

(b) Comment on your result as compared with those obtained in Problems 11.21 and 11.22.

(c) In terms of filter synthesis, what are the major differences between the Vander Lugt and JTC filters. For examples, spatial carrier frequencies, and signal dependent filterings.

11.24 By referring to the photorefractive-based correlator shown in Figure 11.25, assume that $\lambda = 488\,\text{nm}$, $d = 1\,\text{mm}$, and $\eta = 2.25$.

(a) Calculate the wavelength response of the matched filter.

(b) If the photorefractive crystal is to be multiplexed by a tunable diode laser, what would be the requirement for the spectral lines.

11.25 The focal length of the transform lens in Problem 11.24 is given as $f = 1\,\text{m}$.

(a) Calculate the shift requirement of the input object, so that the photorefractive filter can respond.

(b) If the input object is spatially limited to about $1.5\,\text{mm}$ in size, what would be the shift-invariant constraint of this correlator?

12 FIBER OPTICS

Over the past two decades, considerable efforts have been made to use optical fibers for imaging, sensing, and, in particular, for communications. With the advances in opto-electronics and semiconductor lasers, data transmission rates of over 1 Gbit/s with losses less than 1 dB/km have been realized with single mode fibers. Intercontinental fiber communication systems have been installed, and computer networks have been widely implemented using optical fibers. The merits of using optical fibers for data transmission include the huge bandwidth, the electrical insulation between transmitter and receiver, immunity from electromagnetic interference and tapping, low operation cost in terms of available bandwidth, light weight and small size, good mechanical properties, and low signal alternation and dispersion. In this chapter we shall briefly discuss some basic aspects of fiber optics, such as the fabrication and optical properties of fibers, the propagation of light waves through optical fibers, and the applications of fibers in sensing and telecommunications.

12.1 Fiber Construction

The structure of a typical optical fiber is illustrated schematically in Figure 12.1. The center of the fiber is a cylindrical *core* with refractive index η_1, surrounded by a layer of material, called the *cladding*, with a lower refractive index η_2. The light waves are restricted by the cladding within the core and propagate along the fiber. The outer jacket layers not only protect the fiber from moisture and abrasion, but also provide the strength that is required practically.

The core and the cladding are made of either silicate material (e.g., glass) or plastic. Three major combinations of these two types of material are used to make optical fibers: plastic core with plastic cladding, glass core with plastic cladding, and glass core and glass cladding. A plastic core is generally

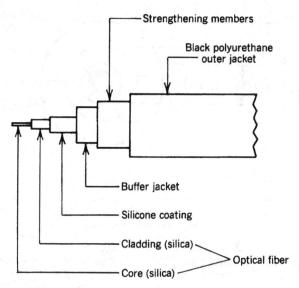

FIGURE 12.1 Cross-section of an optical fiber.

made of polystyrene or polymethyl methacrylate, while a plastic cladding is typically made of silicone or Teflon. For glass cores and claddings, the silica must be extremely pure; however, very small amounts of dopants such as boron, germanium, or phosphorous may be added to change the refractive indices. In some claddings, boron oxide is often added to silica to form borosilicate glass.

In comparison with glass fibers, plastic fibers are flexible, inexpensive, and easy to install and connect. Furthermore, they can withstand greater stress and weigh 50% less than glass fibers. However, they do not transmit light as efficiently. Due to their considerably high losses, they are used only for short runs (such as networks within buildings) and some fiber sensors. Since glass core fibers are much more widely used than plastic ones, subsequent references to fibers in this chapter are to glass fibers. Glass fibers, although slightly heavier than plastic fibers, are much lighter than copper wires. For example, a 40 km long glass fiber core weighs only 1 kg, whereas a 40 km long copper wire with a 0.32 mm outer diameter weighs about 30 kg.

In a fiber communication network, information is transmitted by means of the propagation of electromagnetic waves along the core of the optical fiber. To minimize the transmission loss, the electromagnetic wave must be restricted within the fiber core, and not be allowed to leak into the cladding. As we shall see, the refractive index of the core η_1 must be greater than that of the cladding η_2 if such a requirement is to be met.

The diameters of the core and cladding determine many of the optical and physical characteristics of the fiber (see Section 12.3). For example, the diameter of a fiber should be large enough to allow splicing and the attachment

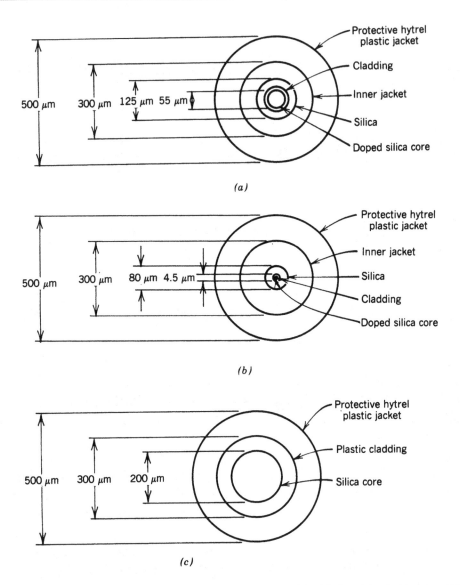

FIGURE 12.2 Nominal dimensions of typical optical fibers; (a) wide-band graded-index multi-mode fiber; (b) step-index single-mode fiber; (c) large-core plastic-clad optical fiber.

of connectors. However, if the diameter is too large, the fiber will be too stiff to bend and will take up too much material and space. In practice, the diameter of fiber cores ranges from 5 to 500 μm, and outer diameters of fiber claddings vary from 100 to 700 μm. To keep the light wave within the fiber core, the cladding must have a minimum thickness of one or two wavelengths of the light transmitted. The protective jackets may add as much as 100 μm to the fiber's total diameter. Typical fiber dimensions are shown in Figure 12.2.

Although fibers with the minimum core diameter (about 5 μm) have a much broader bandwidth than those with larger diameters, their small size may cause severe handling and connection problems. In fact, for any given fiber, tight tolerances are necessary, because even a slight variation in the dimensions can cause significant changes in optical characteristics. The tolerance for the core diameter is typically about ±2 μm.

Although ordinary glass is brittle and is easily broken or cracked, optical glass fibers usually have high tensile strength and are able to withstand hard pulling or stretching. The toughest fibers are as strong as stainless steel wires of the same diameter, and have the same tensile strength as copper wires with twice the diameter. For example, 1 km lengths of these fibers have withstood pulling forces of more than 500 000 lb/in.[2] before breaking. A 10 m long fiber can be stretched by 50 cm and still spring back to its original shape, and a fiber with 400 μm diameter can be bent into a circle with a radius as small as 2 cm.

In order to produce fibers with such tenacity, manufacturers have to keep the glass core and cladding free from microscopic cracks on the surface or flaws in the interior. When a fiber is under stress, it can break at any one of these flaws. Flaws can develop during or after manufacturing. Even a tiny particle of dust or a soft piece of Teflon can lead to a fatal scratch on the surface of the fiber core or cladding. To prevent such abrasion, manufacturers coat the fiber with a protective plastic (or organic polymer) jacket immediately after the fiber has been fabricated. This jacket also protects the fiber surface from moisture and cushions the fiber when it is pressed again irregular surfaces. The cushioning reduces the effect of small random bends (i.e., so-called *microbends*), which would otherwise cause transmission losses. Moreover, the jacket compensates for some of the contractions and expansions caused by temperature variations.

Although a single fiber is used in some applications, a number of fibers are usually placed together to form fiber cables. These cables are optimally designed such that there is little or no stress on the fibers themselves.

12.2 Fiber Waveguides

The basic principle of light transmission through an optical fiber can be simply explained by geometric optics as total internal reflection at the boundary between the core and cladding. To understand this, we first consider a glass plate with a cone of light launched from one end, as depicted in Figure 12.3.

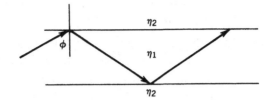

FIGURE 12.3 Total internal reflection.

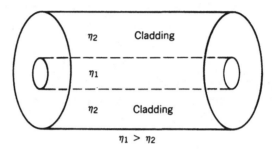

FIGURE 12.4 Typical optical fiber with a cladding.

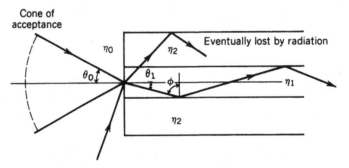

FIGURE 12.5 The acceptance angle of the fiber. ϕ, Incident angle at the core–cladding boundary.

Assume that the refractive index of the glass is η. As discussed in Chapter 1, all light rays with an incident angle ϕ greater than the critical angle

$$\phi_c = \sin^{-1}\left(\frac{1}{\eta}\right), \tag{12.1}$$

will be totally reflected back into the glass and will be guided along the glass plate. For this reason, such a glass plate is also called a *planner waveguide*.

Optical fibers can be considered as cylindrical waveguides. There are two boundaries to be considered in an optical fiber: one at the core–cladding interface, and the other at the cladding–air (or cladding–jacket) interface, as shown in Figure 12.4. The core always has a higher refractive index than the cladding. Figure 12.5 illustrates how light rays are coupled into and are kept within the optical fiber. If the incident angle of a light ray at the end of the fiber core is less than the *angle of acceptance*, as given by (see Example 12.1)

$$\theta_a = \sin^{-1}\left(\frac{\sqrt{\eta_1^2 - \eta_2^2}}{\eta_0}\right), \qquad (12.2)$$

it will be totally reflected at the core–cladding boundary and will be guided along the fiber core. Light rays at greater incident angles may be guided by the cladding, which is also shown in Figure 12.5. However, since the loss in the cladding material is usually much higher than that of the core, the light rays guided along the cladding layer will diminish due to absorption after a very short distance. Therefore, information can be transmitted only along the core of an optical fiber.

EXAMPLE 12.1

Given an optical fiber as shown in Figure 12.5, evaluate the numerical aperture and angle of acceptance using Snell's law of refraction.

If the incident angle at the end of the fiber core is denoted by θ_0, the refraction angle θ_1 inside the fiber can be obtained using the Snell's law:

$$\eta_0 \sin \theta_0 = \eta_1 \sin \theta_1,$$

where η_0 and η_1 are the refractive indices of the air and the core, respectively. It should be noted that the incident angle at the core–cladding interface is

$$\phi = \frac{\pi}{2} - \theta_1,$$

where ϕ should be greater than, or at least equal to, the critical angle to insure that the light ray is totally reflected back into the fiber core. By substituting the preceding equation into the Snell's law of refraction, and using $\sin^2 \phi + \cos^2 \phi = 1$, we have

$$\eta_0 \sin \theta_0 = \eta_1 \sqrt{1 - \sin^2 \phi}.$$

Let ϕ equal the critical angle ϕ_c, then the numerical aperture of the fiber can be obtained as

$$\eta_0 \sin \theta_a = \sqrt{\eta_1^2 - \eta_2^2}.$$

Thus the acceptance angle of the fiber can be written as

$$\theta_a = \sin^{-1}\left(\frac{\sqrt{\eta_1^2 - \eta_2^2}}{\eta_0}\right),$$

which is the same as Eq. (12.2).

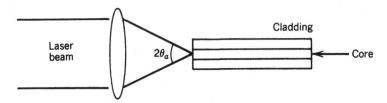

FIGURE 12.6 Generation of the correct angular spread of light for transmission.

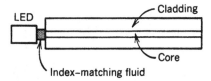

FIGURE 12.7 Light transmission through an optical fiber with a light-emitting diode (LED).

In order to efficiently couple light energy into an optical fiber, it is necessary to generate a cone of light that matches the acceptance angle of the fiber. This can be achieved by using a condenser lens with a suitable numerical aperture, as shown in Figure 12.6. When small-sized semiconductor lasers or light emitting diodes are used, they can also be put in close proximity to the fiber, as depicted in Figure 12.7. In this direct coupling scheme, the light source and the fiber are generally linked by index-matching liquid to reduce the losses. The coupling efficiency depends on the degree of matching between the numerical apertures of the light source and the fiber.

After the light has become properly coupled into the fiber, two physical mechanisms take place during the transmission: *attenuation* and *dispersion*. Attenuation is the reduction in light intensity as light propagates through the fiber, which is mainly due to the chemical impurities in the fiber core. If the absorption resonance of these impurities is within the range of optical frequencies being transmitted, there will be considerable attenuation through absorption. A tremendous research and development effort has been made to reduce the amounts of these impurities. At present, optical fibers with an attenuation of less than 0.5 dB/km have been made, and an attenuation of 3 dB/km has become standard for fiber communication networks.

There are, however, two other sources of attenuation, both of which are related to the effect of bending, as illustrated in Figures 12.8 and 12.9. These diagrams depict the escape of light rays from the fiber core, which are just within the critical angle before bending or microbending occurs. By "bending" we mean the actual bending of the whole fiber, but a microbend is in fact a slight kink at the boundary between the core and the cladding. Notice that a considerable amount of light will leak when the bending (or microbend) is sharp. In this context, a sharp bend is one having a radius of about 50 mm or less. Such bends can lead to a loss of a few decibels per kilometer.

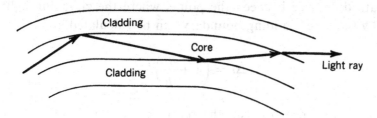

FIGURE 12.8 Attenuation caused by bending.

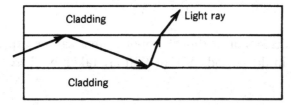

FIGURE 12.9 Attenuation caused by microbending.

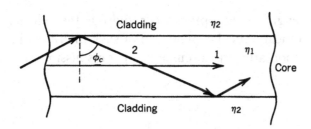

FIGURE 12.10 Dispersion when light transmission takes two different optical paths.

Dispersion of light transmission in the optical fiber has two general causes. The first one is called *material dispersion*, and results in the separation of a complex electromagnetic wave into its various frequency components. It is the result of the dependence of the velocity of light on the wavelength, or in other words, the dependence of the refractive index of the fiber material on the wavelength $\eta(\lambda)$, a circumstance that is similar to the chromatic aberration of a lens.

The second type of dispersion is called *intermodal dispersion*, which occurs even with purely monochromatic light waves, and is caused by different paths of various light rays in the fiber. This can be easily understood from the diagram shown in Figure 12.10, in which two light rays are shown: one travels along the optical axis in the middle, and the other zig-zags up and down at the critical angle. It is apparent that these two light rays travel different distances along the core to arrive at the same destination. The optical path difference between the points where the zigzaging light ray is reflected by the core–cladding boundary can be calculated as

$$\Delta l = \left(1 + \frac{\eta_2}{\eta_1}\right) l, \tag{12.3}$$

where l is the distance between the reflections of the light ray. Thus the relative difference is

$$\frac{\Delta l}{l} = \frac{\eta_1 + \eta_2}{\eta_1}. \tag{12.4}$$

The intermodal dispersion within the optical fiber can be expressed in terms of the difference in the periods of time that are taken by these two light rays, as given by

$$\Delta T = \left(\frac{\eta_1 - \eta_2}{\eta_1}\right) \frac{L\eta_1}{c} = (\eta_1 - \eta_2)\frac{L}{c}, \tag{12.5}$$

where L is the length of the optical fiber and c is the velocity of light in a vacuum. The physical optical explanation of such intermodal dispersion will be better understood after the discussions of transmission modes in the next section.

EXAMPLE 12.2

The refractive indices of the core and the cladding of a silica fiber are 1.5 and 1.45, respectively. Calculate the critical angle at the core–cladding boundary surface and the acceptance angle of the fiber.

The critical angle at the core–cladding interface can be determined as

$$\phi_c = \sin^{-1}\left(\frac{\eta_2}{\eta_1}\right) = \sin^{-1}\left(\frac{1.45}{1.5}\right) = 75.16°.$$

The acceptance angle can be obtained by using Eq. 12.2:

$$\theta_a = \sin^{-1}\left(\frac{\sqrt{\eta_1^2 - \eta_2^2}}{\eta_0}\right) = \sin^{-1}\left(\frac{\sqrt{1.5^2 - 1.45^2}}{1}\right) = 22.59°.$$

In practice, the acceptance angles of optical fibers are typically between 10° and 20°.

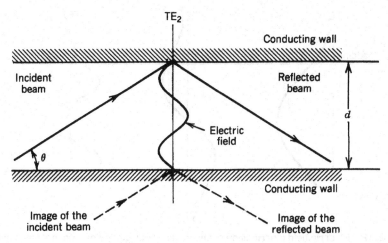

FIGURE 12.11 A conducting waveguide. TE_2, Transverse electric mode of order 2.

12.3 Modal Description

In the preceding section, geometrical optics has been used to explain the propagation of light rays along optical fibers. However, the transmission of light along optical fibers can be better characterized using electromagnetic wave theory. Similar to the laser oscillation modes that we discussed in Section 6.5, the propagation of light waves in an optical fiber is not arbitrary, but has to take some discrete propagation modes which are determined by the boundary conditions. For simplicity, we first assume that the optical fiber behaves as a waveguide with perfect conducting walls. Under such circumstances, the electric field at the waveguide boundary must be zero. An example of the propagation mode that meets such a requirement is illustrated in Figure 12.11.

It should be noted that the electric field *across* the waveguide consists of interference patterns between the incident and reflected beams. To satisfy the boundary conditions (i.e., $E_b = 0$), the diameter of the waveguide must intersect an integral number of half-wavelengths of the incident light. This can be expressed mathematically as

$$\sin\theta = \frac{(n+1)\lambda}{2d\eta}, \qquad n = 0, 1, 2, 3, 4, \ldots, \tag{12.6}$$

where θ is the incident angle, d is the separation of the conducting walls, and η is the index of the propagation mode. Thus, we can see that only certain

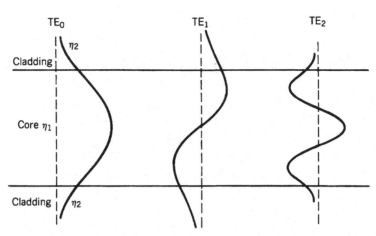

FIGURE 12.12 Electric field distribution of the lower order guided modes in a slab waveguide. Note: $\eta_1 > \eta_2$.

values of the incident angles allow propagation along the waveguide; otherwise the boundary condition would be violated.

An exact solution of the modal description for optical fibers can be obtained by using a set of Maxwell's equations subject to the cylindrical boundary conditions of the fiber. The detailed derivations are beyond the scope of this chapter, and interested readers can find them in a number of texts (e.g., reference [1]). Nevertheless, a similar solution for the conducting waveguide applies to the core–cladding boundary in an optical fiber, although the exact required boundary conditions are slightly different. The consequence is that only a number of specific angular directions of the incident beam allow propagation along the fiber. Each of the allowed beam directions corresponds to a different mode of wave propagation in the optical fiber.

To discuss further the modal description in optical fibers, we now examine qualitatively the appearance of the modal field in a planar dielectric slab waveguide, as shown in Figure 12.12. The planar waveguide is composed of a dielectric slab (corresponding to core in an optical fiber) of refractive index η_1 sandwiched between dielectric cladding layers of refractive index η_2, in which we assume $\eta_1 > \eta_2$. This simple planar waveguide helps us to understand the propagation in an optical fiber, which is difficult to deal with due to its cylindrical shape. A set of low-order transverse electric guided modes obtained from the solutions of Maxwell's equations are shown in Figure 12.12. The order of the mode is equal to the number of transverse field minima (zero field points) across the waveguide, and is also related to the angle of the incident beam as given by Eq. 12.6. It can be seen that the steeper the incident angle, the higher the order of the guided wave mode. Unlike the perfect conducting waveguide, the electric fields of the *guided modes* are not completely confined within the central core in such a dielectric waveguide, but go beyond the core–cladding boundary and decay very rapidly in the cladding layer. For lower order guided modes, the electric fields are more concentrated near the center of the core. However, for higher order modes, the electric field distribution spreads toward the core–cladding boundary and penetrates further into the cladding layer.

In addition to the guided modes, there exist also *leaky modes* in optical fibers. These leaky modes are only partially confined in the core region of the fiber. As they propagate along a fiber their intensity attenuates continuously by radiating out of the fiber core. It can be shown that the upper and lower bounds of the propagation constant β for the existence of such leaky modes is given by:

$$\eta_2 k < \beta < \eta_1 k, \tag{12.7}$$

where $k = 2\pi/\lambda$. Most leaky modes disappear after a few centimeters along the fiber due to significant radiation losses. However, under certain circumstances a few leaky modes may persist in the fiber for a 100 m.

EXAMPLE 12.3

Consider an optical fiber designed to guide monochromic light at a wavelength of 670 nm. If the diameter of the core is 10 μm and $\eta = 1.5$, calculate the allowable incident angles for the guided modes.

By referring to Eq. 12.6, the allowable incident angles can be calculated as

$$\theta = \sin^{-1}\left[\frac{(n+1)\lambda}{2d\eta}\right] = \sin^{-1}\left[\frac{0.67 \times (n+1)}{2 \times 10 \times 15}\right] = \sin^{-1}[0.02233 \times (n+1)],$$

where n is the order of the guided modes. Thus, we have

$$
\begin{aligned}
n &= 0, & \theta &= 1.28° \\
n &= 1, & \theta &= 2.56° \\
n &= 2, & \theta &= 3.84° \\
n &= 3, & \theta &= 5.13° \\
&\text{etc.}
\end{aligned}
$$

EXAMPLE 12.4

Assume that the refractive indices of the core and cladding of the optical fiber described in Example 12.3 are 1.5 and 1.45, respectively.

(a) Calculate the largest allowable incident angle.

(b) Determine the actual acceptance angle of the fiber.

(c) Find the numerical aperture of the fiber.

Answers

(a) The largest allowable incident angle is determined by the critical angle, as well as the highest order of the guided mode. The critical angle at the core–cladding interface is

$$\phi_c = \sin^{-1}\left(\frac{\eta_2}{\eta_1}\right) = \sin^{-1}\left(\frac{1.45}{1.5}\right) = 75.16°.$$

Thus the permitting incident angle (inside the fiber core) is given by

$$\theta < 90° - 75.16° = 14.84°.$$

According to Eq. 12.6, the allowable incident angles of guided waves can be calculated as

$$n = 10, \qquad \theta = 14.22°,$$

and

$$n = 11, \qquad \theta = 15.54°.$$

Hence, the largest allowable incident angle inside the fiber core is 14.22°.

(b) The actual acceptance angle can be obtained using Snell's law of refraction, such that

$$\theta_a = \sin^{-1}\left[\frac{\eta_1}{\eta_0}\sin(\theta)\right] = \sin^{-1}\left(\frac{1.5}{1.0}\sin 14.22°\right) = 21.62°.$$

Notice that the acceptance angle calculated using Eq. 11.2 is

$$\theta_a = \sin^{-1}\left(\sqrt{\eta_1^2 - \eta_2^2}\right) = \sin^{-1}\left(\sqrt{1.5^2 - 1.45^2}\right) = 22.59°,$$

which is slightly larger than the actual acceptance angle determined by the highest guided mode.

(c) The numerical aperture of the fiber can therefore be determined

$$NA = \eta_0 \sin \theta_a = \sin 21.62° = 0.37.$$

12.4 Types of Optical Fiber

In general, optical fibers can be divided into two broad categories, namely *single-mode* and *multimode* fibers.

In single-mode fibers, only a single guided mode is allowed to propagate. This transmitting mode has the lowest order ($n = 0$) and corresponds to the transverse electric field profile that has no minima across the fiber cross-section. All the other orders ($n \geq 1$) have an incident angle at the core–cladding boundary that is smaller than the critical angle and, consequently, quickly leak into the cladding layer and are absorbed. By referring to Eq. 12.6, the condition for single-mode operation is

$$\theta \approx \sin \theta < \frac{\lambda}{d} \qquad (12.8)$$

where d is the diameter of the fiber core. In terms of critical angle at the core–cladding boundary, the single-mode operation condition can be expressed as

$$\theta = \frac{\pi}{2} - \phi_c < \frac{\lambda}{d}, \qquad (12.9)$$

or

$$d < \frac{\lambda}{\frac{\pi}{2} - \phi_c}. \qquad (12.10)$$

This implies that, in order to achieve single-mode operation, the fiber core can only have a very small diameter, ranging between 3 to 10 µm depending on the wavelength of light to be transmitted.

The dispersion in a single-mode fiber is primarily due to absorption by the material of the core. Therefore, single-mode fibers are the most suitable media for long-distance communication and are capable of handling very broad band signals. Since the core diameter of a single-mode fiber is generally very small, launching light into the fiber is rather difficult, and keeping the insertion of light constant over a long period of time is even more difficult. Fortunately, these problems have been solved in the past two decades as the result of the advances in optoelectronic technology. Single-mode fibers are being deployed in various high-capacity communications networks.

In a multimode fiber, a number of guided modes (from two to a few thousand) are transmitted simultaneously. There are two types of multimode fiber: step-index fibers and graded-index fibers.

A step-index fiber is, in fact, the type of fiber that we have been discussing so far. In this type of fiber, the refractive index is constant within the core, but changes abruptly at the core–cladding interface, a shown in Figure 12.13.

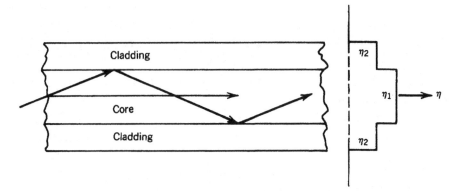

FIGURE 12.13 Step-index fiber.

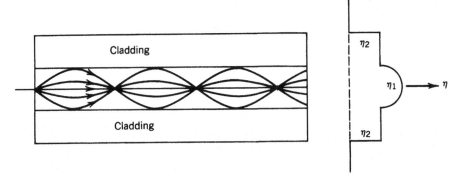

FIGURE 12.14 Graded-index fiber.

Although they have a simple structure, this type of fiber suffers from severe intermodal dispersion. Ray diagrams for two different propagation modes (represented by an axial ray and a reflected ray) are shown in Figure 12.13. One per cent of optical path difference between these two light rays is common, which may lead to a time delay of 40 ns between the two modes after they have propagated 1 km along the fiber. This makes it impossible to transmit information at a data rate of 20 Mbit/s over a 1 km communication link using step-index fibers.

In contrast, a graded-index fiber does not have a constant refractive index within its core, but has a decreasing index $\eta(r)$ as a function of radial distance. As shown in Figure 12.14, the refractive index varies from a maximum value η_1 at the center of the core to a constant value η_2 at the core–cladding boundary. The profile of the refractive index can be generally described by the following expression

$$\eta(r) = \begin{cases} \eta_1 \sqrt{1 - 2\Delta\left(\frac{r}{a}\right)^\alpha}, & r < a \\ \eta_2, & r \geq a \end{cases} \tag{12.11}$$

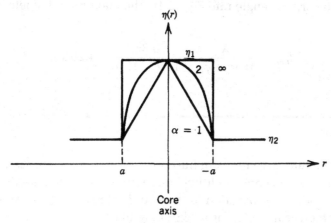

FIGURE 12.15 Possible profiles of the refractive indices of a graded fiber.

where a is the radius of the core, α is the profile parameter, r is the radial distance, and

$$\Delta = \frac{\eta_1^2 - \eta_2^2}{2\eta_1^2} \approx \frac{\eta_1 - \eta_2}{\eta_1} \qquad (12.12)$$

is known as the *relative refractive index difference*. The refractive index profile varies with the profile parameter α, as shown in Figure 12.15. It can be seen that when $\alpha = \infty$, the profile actually represents that of the step-index fiber; $\alpha = 2$ corresponds to a parabolic profile; and $\alpha = 1$ illustrates a triangular profile.

The ray paths in a graded-index fiber core are curved, as illustrated in Figure 12.14. Since the inner rays travel through a higher refractive index and the outer rays travel through a lower index, the difference in the propagation time period for the outer and inner rays to be transmitted through the same fiber length can be zero. In practice, such a time difference ΔT can be reduced to less than 1 ns/km for a core diameter of about 50 µm. In other words, the graded-index fibers have much less intermodal dispersions than the step-index fibers. Since launching the light into a 50 µm core is fairly straightforward, it is possible to use butt-joint coupling with a semiconductor laser or a light emitting diode (see Figure 12.7).

EXAMPLE 12.5

The refractive indices of the core and cladding of an optical fiber are 1.50 and 1.47, respectively. In order to transmit light waves at a wavelength of 850 nm, what is the maximum diameter of the core?

The critical angle at the core–cladding boundary is

$$\phi_c = \sin^{-1}\left(\frac{\eta_2}{\eta_1}\right) = \sin^{-1}\left(\frac{1.47}{1.5}\right) = 1.37 \, \text{rad}.$$

Substituting the critical angle into Eq. 12.10, the maximum diameter of the core is

$$d_{max} = \frac{\lambda}{0.5\pi - \phi_c} = \frac{0.85}{1.57 - 1.37} = 4.25\,\mu m$$

EXAMPLE 12.6

Assume that the refractive index of the core of a step-index fiber is 1.5, and that the critical angle at the core–cladding boundary is 80°. Calculate the difference in the time periods of propagation for two kinds of ray, the axial rays and those reflected at the critical angle, for a 1 km long fiber.

The refractive index of the cladding can be obtained from the critical angle and the index of the core as

$$\eta_2 = \eta_1 \sin\phi_c = 1.5 \times \sin(80°) = 1.4772.$$

By substituting the given parameters into Eq. 12.5, the time difference can be calculated as

$$\Delta T = (\eta_1 - \eta_2)\frac{L}{c} = (1.5 - 1.4772) \times \frac{1}{300000} = 7.6 \times 10^{-8}\,s.$$

Thus, the dispersion is about 76 ns/km.

12.5 Optical Fiber Communications

Great interest in communicating at optical frequencies was created in the 1960s with the advent of lasers, which made available highly coherent optical sources. Since the optical frequencies are of the order of 5×10^{14} Hz, lasers have a theoretical information capacity exceeding that of microwave systems by five orders of magnitude, which is approximately equal to 10 million TV channels. The use of optical fibers as reliable and versatile media to transmit laser light over long distances, however, seemed impossible at the very beginning, due to the extremely large losses. The breakthrough was made by Kao and Hockman in 1966, when they found a way to reduce the loss significantly by controlling the impurities in the fiber materials. Since then, the improvement in optical fibers has been extraordinary. At present, the attenuation of state of the art fibers at 1.55 mm wavelength has been reduced to 0.16 dB/km, which is close to the theoretical limit of 0.14 dB/km. Owing to the successful reduction in optical fiber losses, more than 10 million km of fiber has been installed worldwide in various optical networks. Today, bit rates for long-haul fiber links typically range between 400 and 600 Mbit/s, and in some cases up to 4 Gbit/s, which is far beyond the reach of traditional copper cable communication systems. It is expected that optical fiber communication systems will play a very important role in the development of the information superhighway, which will have a profound impact on many aspects of our society.

12.5.1 FIBER COMMUNICATION SYSTEMS

An optical fiber communication system is illustrated schematically in Figure 12.16. In each channel, a light source which is dimensionally compatible with the fiber core is used as the transmitter. Semiconductor light emitting diodes (LEDs) and laser diodes are used as transmitter source, since their light output can be modulated rapidly simply by varying the bias voltage. The electrical input signal can be either analog or digital in form, although digital signals are utilized in most telecommunications networks. The transmitter circuitry converts electrical input signals into optical signals by varying the current flow through the light source.

Optical outputs from different channels are multiplexed by a multiplexer and then coupled into the optical cable. Connectors and couplers are utilized along the optical fiber link to extend the distance to restore connection after repair and maintenance, or to switch between different channels. These components will be discussed in detail in the next section. At the end of the optical fiber link, the transmitted optical signals are first demultiplexed and then detected by photodetectors in the receivers in each channel. Semiconductor PIN and avalanche photodiodes are the two principal photodetectors used in optical fiber communications. The outputs of the photodetectors are amplified and reshaped before they are delivered to the user (e.g., a telephone or a computer terminal).

After an optical signal has been launched into the fiber, it will be progressively attenuated and distorted as it propagates along the fiber. This is primarily due to scattering, absorption, and dispersion in the fiber. A repeater is needed in the transmission line to amplify and reshape the signal before it becomes too attenuated and distorted. An optical repeater consists of a receiver and a transmitter placed back to back. The receiver section detects the optical signal and converts it to an electric signal, which is amplified, reshaped, and then sent to the electric input of the transmitter section. The transmiter section converts this electric signal back to an optical signal and sends it down the optical fiber link.

Important components in an optical fiber communication system include transmitters, receivers, repeaters, connectors, couplers, multiplexers, and

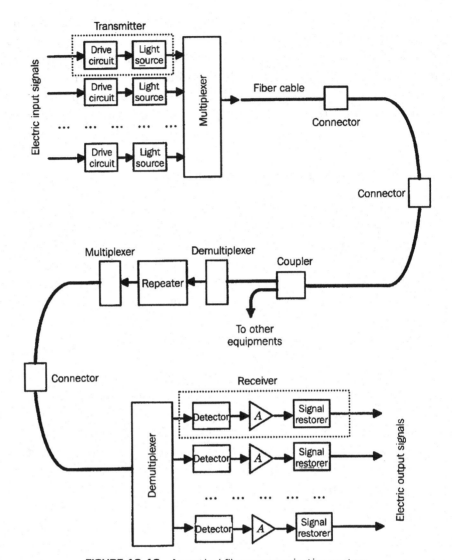

FIGURE 12.16 An optical fiber communication system.

demultiplexers. The key parts in optical transmitters and receivers, such as semiconductor lasers and photodetectors, have been covered in previous chapters (see Chapters 4 and 6) and will not be discussed here. Our discussions in the following sections are focused on optical connectors, couplers, multiplexers, and demultiplexers.

12.5.2 SPLICES AND CONNECTORS

A significant factor in any fiber optic system installation is the interconnection of fibers in a low-loss manner. These interconnections occur at the optical source, the photodetector, the intermediate points within a cable where two fibers are joined, and the intermediate in a link where two cables

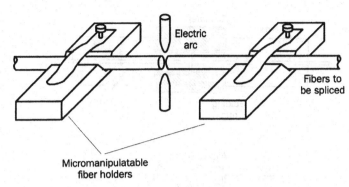

FIGURE 12.17 Fusion splicing of optical fibers.

are connected. The particular technique selected for joining the fibers depends on whether a permanent bond or an easily demountable connection is desired. A permanent bond is generally referred to as a *splice*, whereas a demountable joint is known as a *connector*.

Many different fiber splicing techniques have been developed. As an example, the fusion splice is illustrated schematically in Figure 12.17. The fiber ends to be spliced are first prealigned and butted together, which is done either in a grooved fiber holder or under a microscope with micromanipulator. The butt joint is then heated with an electric arc or a laser pulse so that the fiber ends are momentarily melted and bond together. Fusion splice can produce very low splice losses (typically less than 0.06 dB).

Optical fibers can also be spliced by mechanical techniques. A common method involves the use of an accurately produced rigid alignment tube into which the prepared fiber ends are permanently bonded. Transparent adhesive (e.g., epoxy resin) is used to give mechanical sealing and index matching of the splice. The average insertion loss of mechanical splice is about 0.1–0.15 dB, which is higher than that of fusion splices.

Demountable fiber couplers are more difficult to achieve than optical fiber splices. This is because they must maintain similar tolerance requirements to splices in order to couple the light between fibers efficiently, but they must accomplish it in a removable fashion. The majority of connectors today are of the butt-joint form. These connectors employ a metal, ceramic, or molded plastic ferrule for each fiber, and a precision sleeve into which the ferrule fits. The fiber is glued into a precision hole which has been drilled into the ferrule. The mechanical challenges of ferrule connectors include maintaining both the dimensions of the hole diameter and its position relative to the ferrule outer surface.

Figure 12.18 shows two common butt-joint alignment designs, used in both multimode and single-mode fiber systems which are based on the *straight sleeve* and the *tapered sleeve* (or *biconical*) mechanisms. In the straight sleeve connector, the length of the sleeve and a guide ring on the

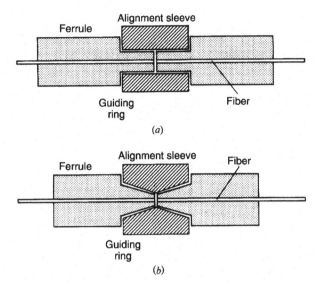

FIGURE 12.18 Two popular butt-joint fiber connectors: (a) straight sleeve connector; (b) tapered sleeve connector.

ferrule determine the end separation of the fibers. The biconical connector uses a tapered sleeve to accept and guide tapered ferrules. Again the sleeve length and the guide rings determine a given fiber end separation.

It is essential using these types of connector that the end faces of the fibers are smooth and perpendicular to the fiber axis. This may be achieved with varying success by either:

(1) Cleaving the fiber before insertion into the ferrule.
(2) Inserting the bonding before cleaving the fiber close to the ferrule end face.
(3) Using either (1) or (2) and polishing the fiber end face until it is flush with the end of the ferrule.

Polishing the fiber end face after insertion and bonding provides the best results, but it tends to be time consuming and inconvenient, especially in the field. The typical *insertion loss* for the fiber connectors shown in Figure 12.18 is within the range 0.5–1 dB.

12.5.3 COUPLERS AND SWITCHES

A communication system uses couplers and switches to direct the light beams that representing the information signals to their appropriate destinations. Couplers always operate on the incoming signals in the same manner; whereas switches are controllable couplers whose coupling topology can be modified by an external command to change.

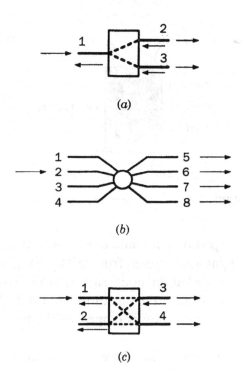

(a)

(b)

(c)

FIGURE 12.19 Examples of couplers: (a) T-coupler; (b) star coupler; (c) directional coupler.

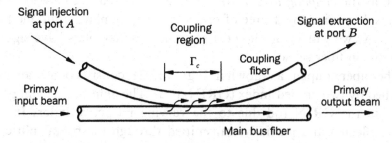

Signal injection at port A

Coupling region

Signal extraction at port B

Γ_c

Coupling fiber

Primary input beam

Primary output beam

Main bus fiber

FIGURE 12.20 Schematic diagram of a directional optical fiber coupler.

Optical couplers are commonly used to provide access to optical data buses. At each data bus terminal, a coupler removes a portion of the light signal from the bus trunk line or injects additional light signal into the trunk. Examples of fiber couplers are shown schematically in Figure 12.19. In the T-coupler, a signal at input port 1 reaches both output points 2 and 3; a signal at either port 2 or port 3 reaches port 1. In the star coupler, the signal at any of the input ports reaches all output ports. In a four-port directional coupler, a signal at any one of the input ports (1 or 2) can reach both output ports, but not the other input port.

A schematic diagram of a directional fiber coupler is shown in Figure 12.20. The coupler has four ports: two connecting the device to the bus,

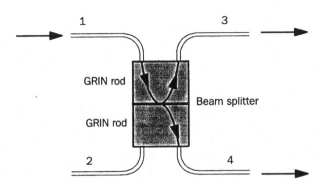

FIGURE 12.21 A four-port directional coupler.

one for receiving the tapped-off data, and one for inserting data onto the bus line. Assume that the light wave travels from left to right in Figure 12.20. When a single coupler is used at the node, signals can be injected into the bus via port A and/or extracted from the bus through port B. The arrows in the figure show that some optical power is tapped off the primary fiber bus into the receiver at port B.

Such a coupler can be simply made from two optical fibers by fusing them together over a short distance. This distance, denoted by Γ_c in Figure 12.20, is known as the *coupling length*. Its length and the spacing between the two fiber cores determine the degree of optical power coupling from one fiber to the other. As we will see in the next section, these couplers have significant applications in fiber sensors.

Another fiber coupler, as shown in Figure 12.21, consists of a beam splitter sandwiched by two graded-index (GRIN) rods. The light beam from port 1 is bent within the GRIN rod due to the varying refractive index. It is then partially reflected and partially transmitted through the beam splitter. The reflected light is coupled into port 2 and the transmitted light is led into port 4. Obviously, if the data bus is connected between ports 2 and 4, light signals can be injected into the bus through port 1 and can be taken out of the bus via port 3.

When operated as a switch, the four-port directional coupler is switched between the parallel state (1–3 and 2–4 connections) and the cross-state (1–4 and 2–3 connections). Figure 12.22 schematically illustrates a 2×2 optical switch implemented by integrated optics. Two waveguides are formed in close proximity on a substrate and the electrodes are laid over the two wave-guides. If the effective distance between the two electrodes is d, an applied voltage creates an electric field $E_1 \approx V/d$ in one waveguide and $E_2 \approx -V/d$ in the other, with the electric field line going downwards in one of the wave-guides and upwards in the other. The result is that the refractive index is incremented in one waveguide and decremented in the other. The variation in

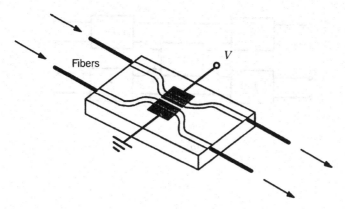

FIGURE 12.22 A 2 × 2 switch using an integrated optic directional coupler.

the refractive index leads to the coupling of light between the two wave-guides. Therefore, by altering the applied voltage, the optical power either remains in the same waveguide or is transferred into the other waveguide. These switches typically operate at a few volts with speeds that can exceed 20 GHz.

An $N \times N$ switch can be built by combining 2 × 2 switches. Figure 12.23 shows a 4 × 4 switch which is composed of five 2 × 2 switches. This config-uration can be built on a single substrate in the geometry shown in Figure 12.23b. Currently, 8 × 8 switches are commercially available, and larger switches are being developed.

12.5.4 TIME- AND WAVELENGTH-DIVISION MULTIPLEXING

To take advantage of the tremendous potential data capacity of an optical fiber, signals from multichannels are transmitted through a single fiber using a multiplexing technique. There are generally two types of multiplexing tech-nique: *time-division multiplexing* and the *wavelength-division multiplexing*.

Although the practical time-division multiplexing and demultiplexing sys-tems may be very sophisticated, their operational principle is simple and straightforward, and can be well understood by looking at the two channel systems illustrated in Figure 12.24. The signals from different channels are first encoded into series of short pulses and are then launched into the fiber cable with a time delay δt. The multiplexer simply consists of a mirror and a beam splitter. It should be mentioned that the light waves launched into the fiber are linearly polarized. At the end of the fiber link, the transmitted signal is demultiplexed by a demultiplexer composed of an electro-optic modulator and a polarization beam splitter. The electro-optic modulator (see Chapter 5)

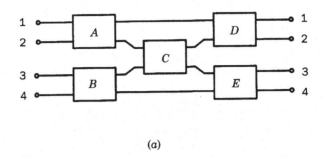

(a)

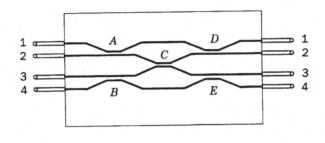

(b)

FIGURE 12.23 A 4 × 4 switch made of five 2 × 2 switches.

rotates the polarization of the light from channel 1 clockwise, and the polarization of the light from channel 2 counterclockwise. The rotated beams are then separated by a polarization beam splitter and are directed to two photodetectors.

Since any optical light source used for fiber communication has a narrow spectral width, time-division multiplexing utilizes only a very small portion of the transmission bandwidth offered by optical fibers. It is apparent that a drastic increase in the information capacity of a fiber link can be achieved by the simultaneous transmission of optical signals over the same fiber from many different light sources having properly spaced peak emission wavelengths. This is the basis of wavelength-division multiplexing. Conceptually, the wavelength-division multiplexing scheme is the same as the frequency-division multiplexing used in the microwave radio and satellite communication systems. A block diagram of the wavelength-division multiplexing system is given in Figure 12.25. It should be noticed that the signal in each wavelength channel may be already multiplexed via a time-division multiplexing scheme.

Angular dispersive elements, such as prisms and gratings, can be used as multiplexers in wavelength division multiplexing systems. As shown in

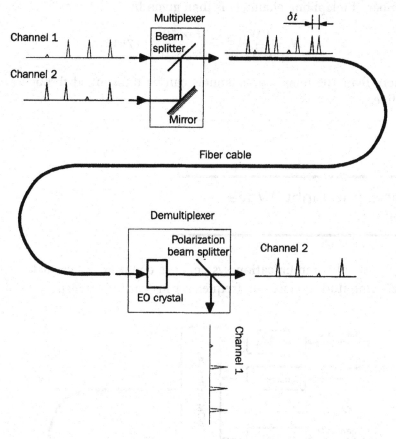

FIGURE 12.24 Time-division multiplexing.

Figure 12.26, light rays from different channels (and hence at different wavelengths) are directed onto the prism or the grating at different incident angles. At the output of the multiplexer, these light rays become co-linear and can be easily launched simultaneously into an optical fiber. A wavelength-division demultiplexer works in exactly the reverse fashion, directing light beams of various wavelengths from a fiber into their respective channels.

EXAMPLE 12.7

A telephone signal is digitized at 8 kHz with an accuracy of 8 bit. If the bandwidth of the light source and detector of a time-division multiplexing system is 50 MHz, how many telephone channels can be transmitted via a single fiber?

The bandwidth of each telephone channel can be calculated as

$$\mathrm{BW}_{phone} = 8000 \times 8 = 6.4 \times 10^4 \, \mathrm{Hz}$$

The number of telephone channels is then given by

$$N = \frac{\text{BW}_{TDM}}{\text{BW}_{phone}} = \frac{5 \times 10^7}{6.4 \times 10^4} \approx 781.$$

Thus more than 700 telephone channels can be transmitted through a single optical fiber.

12.6 Coherent Light Wave Communication

Coherent optical communication systems use complex optical field modulation (i.e., amplitude, phase, or frequency) instead of intensity modulation.

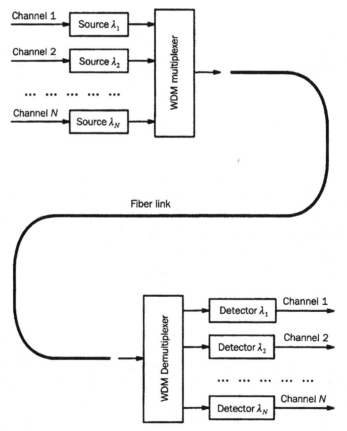

FIGURE 12.25 Schematic diagram of a wavelength-division multiplexing fiber communication system.

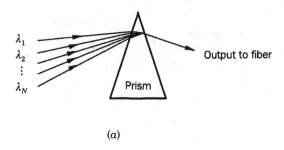

(a)

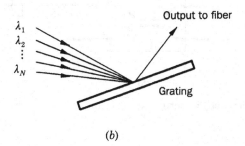

(b)

FIGURE 12.26 Wavelength-division multiplexers: (a) prism multiplexer; (b) grating multiplexer.

They employ highly coherent light sources, single mode fibers, and heterodyne receivers. In comparison to the direct detection communication system discussed in the preceding section, coherent communication systems have the advantage of low output noise, insensitivity to unwanted background light, and a capability of optical phase and frequency measurement. The cost of these advantages is an increase in the system complexity.

Photodetectors detect only the photon flux (or intensity), but not optical phase. However, by mixing an optical signal with a coherent reference optical field of stable phase, called a *local oscillator*, the complex amplitude (both amplitude and phase) of the optical signal can be detected with a photodetector, as illustrated in Figure 12.27. Such a detection scheme is known as optical *heterodyne*, or *coherent optical detection*. The coherent optical receiver is the equivalent of a superheterodyne radio receiver. The signal and local oscillating waves usually have different frequencies (ω_S and ω_L). When $\omega_S - \omega_L$, the detection scheme is called the *homodyne* detection.

Let the optical fields of the signal and the local oscillator be represented by

$$E_S = A_S \cos(\omega_S t + \phi_S) \tag{12.13}$$

and

$$E_L = A_L \cos(\omega_L t + \phi_L), \tag{12.14}$$

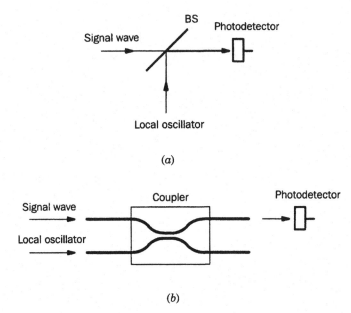

FIGURE 12.27 Optical heterodyne detection using (a) a beam splitter, and (b) an optical coupler

where the subscripts S and L represent the signal and local oscillators, respectively; and A, ω, and ϕ represent amplitude, angular frequency, and phase, respectively. The complex amplitude of the superimposed fields is then given by

$$E = A_S \cos^2(\omega_S t + \phi_S) + A_L \cos^2(\omega_L t + \phi_L). \qquad (12.15)$$

The detected intensity is the square of the amplitude averaged over the response time of the detector:

$$
\begin{aligned}
I(t) = \Big\langle & A_S^2 \cos(\omega_S t + \phi_S) + A_L^2 \cos(\omega_L t + \phi_L) \\
& + A_S A_L \cos[(\omega_S + \omega_L)t + (\phi_S + \phi_L)] \\
& + A_S A_L \cos[(\omega_S - \omega_L)t + (\phi_S - \phi_L)] \Big\rangle, \qquad (12.16)
\end{aligned}
$$

where $\langle \cdot \rangle$ denotes the time average. Note that the frequency of visible and near infrared light is of the order of 10^{14} Hz; there is no photodetector available to respond to such a high frequency. The detected intensity is, in fact, always the average over many cycles of optical waves. If the frequencies of the signal and local oscillator are very close, the detected intensity can be written as

$$
\begin{aligned}
I(t) &= \frac{1}{2} A_S^2 + \frac{1}{2} A_L^2 + A_S A_L \cos[(\omega_S - \omega_L)t + (\alpha_1 - \alpha_2)] \\
&= I_S + I_L + 2\sqrt{I_S I_L} \cos[(\omega_S - \omega_L)t + (\alpha_1 - \alpha_2)] \qquad (12.17)
\end{aligned}
$$

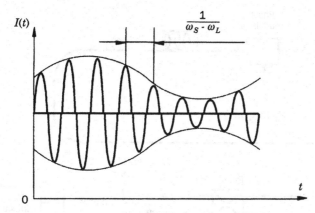

FIGURE 12.28 Intensity detected by a heterodyne detector.

where I_S and I_L are the intensities of the signal and local oscillating waves, respectively.

The time dependence of the detected intensity $I(t)$ is sketched in Figure 12.28. The third term in Eq. 12.17, which oscillates at the difference frequency $(\omega_S - \omega_L)$, carries the useful information. All information about the optical signal (i.e., amplitude, frequency, and phase) can be fully recovered from the output of the coherent detector.

An essential condition for the proper mixing of the local oscillating field and the received optical field is that they must be locked in phase and have the same polarization in order to permit interference to take place. This places stringent requirements on the two lasers and the fiber. The lasers must be working at the single-frequency mode and have minimal phase and intensity fluctuations. The local oscillator is phase locked with respect to the received optical field by means of a control system that adjusts the phase and frequency of the local oscillator adaptively (using a phase locking loop). Only single-mode fibers can be used to avoid the unwanted modal noise. Polarization-maintaining fibers are generally required; otherwise, the receiver must contain an adaptive polarization-compensation system.

A schematic diagram of a coherent fiber optic communication system using two lasers and phase modulation is shown in Figure 12.29. The local oscillating field is mixed with the received optical signal using a directional optical coupler. One branch of the coupler output contains the sum of the two fields and the other branch contains their difference. By referring to Eq. 12.17, the detected intensities can be written as

$$I_+(t) = I_S + I_L + 2\sqrt{I_S I_L} \cos[(\omega_S - \omega_L)t + (\alpha_1 - \alpha_2)] \tag{12.18}$$

and

$$I_-(t) = I_S + I_L - 2\sqrt{I_S I_L} \cos[(\omega_S - \omega_L)t + (\alpha_1 - \alpha_2)]. \tag{12.19}$$

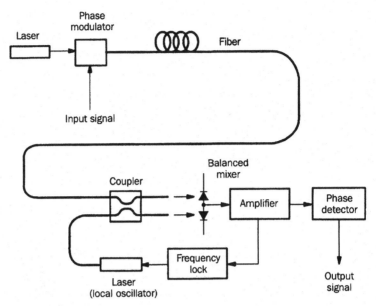

FIGURE 12.29 Coherent fiber-optic communication system.

The outputs from the two branches are subtracted electronically, yielding:

$$I(t) = 4\sqrt{I_S I_L}\cos[(\omega_S - \omega_L)t + (\alpha_1 - \alpha_2)], \qquad (12.20)$$

which is then demodulated to recover the transmitted information. This type of coherent receiver is known as a *balanced mixer*. It has the advantage of canceling out intensity fluctuations of the local oscillator. A number of coherent fiber optic communication systems have been implemented at a wavelength of 1.55 mm (where fiber attentuation is a minimum) with bit-rate–distance products matching the theoretical expectations.

12.7   Fiber Sensors

In addition to communications, in the last two decades optical fibers have been widely used to develop various fiber sensors. Owing to their merits, such as low weight, small size, low power consumption, environmental ruggedness, immunity from electromagnetic interference, and low cost, optical fiber sensors have the potential to outperform many other types of sensor in such areas as aerospace, defense, manufacturing, medicine, and construction. The convergence of fiber optic technology, which was largely driven by the telecommunications industry, in combination with low cost optoelectronic components has enabled optical fiber sensor technology to approach its ideal potential for many applications.

Both the parameters to be sensed and the mechanisms utilized by fiber sensors vary in a broad range. However, most fiber sensors can be classified into two categories: *intensity attenuation fiber sensors* and *interferometric fiber sensors*. In an intensity attenuation fiber sensor, the environmental variations (such as temperature fluctuation and acoustic vibration) change the optic characteristics of the fiber, by which the intensity of the light wave that passed through the fiber is modulated. In an interferometric fiber sensor, the phase of the light wave propagating along the fiber is modulated by environmental perturbations. Such a phase modulation is then detected interferometrically by comparing the modulated phase with the initial phase. Examples of fiber sensors in both categories are described in the following.

12.7.1 INTENSITY ATTENUATION FIBER SENSORS

In Section 12.2, we briefly discussed bending-induced losses in optical fibers. A simple fiber sensor based on microbending-induced losses can be easily constructed, as shown in Figure 12.30. Light from a light source is coupled into a multimode fiber. Any leaky mode that is inadvertently coupled to the cladding layer is removed in a mode-stripping region. This can be accomplished by coating the fiber with an index-matched oil so that the light coupled into the cladding propagates out into the oil and is absorbed. As the light wave passes beyond the mode-stripping region into the microbending portion, the amount of light coupled to the cladding region depends on how tightly the fiber is bent. When the bending radius of the fiber is varied due to the motion of the bending plate, the amount of light energy coupled to the cladding also changes. The light wave that exits the microbending region

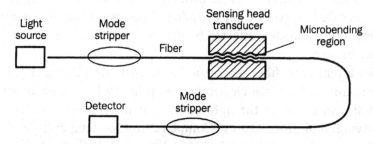

FIGURE 12.30 An intensity attenuation fiber sensor using microbending plates as the transducer.

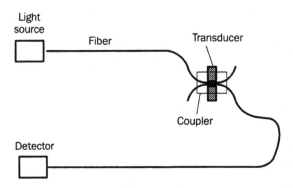

FIGURE 12.31 A fiber sensor based on evanescent coupling.

then passes through a second mode-stripping region where the radiation modes traveling in the cladding are removed. The intensity detected by the detector is amplitude modulated by an amount corresponding to the motion of the microbending plates, which may represent the amplitude of an acoustic vibration or the displacement of a mechanical part.

As discussed in Section 12.3 and shown in Figure 12.12, when a light wave travels along a single-mode fiber, it is not totally confined to the core region but extends into the surrounding glass cladding layer. The light wave that extends into the cladding layer is called an *evanescent wave*. This phenomenon, which has been exploited to fabricate single-mode fiber couplers as illustrated in Figure 12.20, can be used to develop evanescent fiber sensors. As shown in Figure 12.31, a fiber is glued into a slot in a quartz block. The fiber and the quartz block are then polished to within micrometers of the fiber core. Placing two units in contact allows the evanescent wave of one fiber to be coupled to that of another fiber. If the fibers are immersed in index-matched oil, the light will also propagate in the oil. The coplanarity of the polished surface is insured by two reference planes made in epoxy resin at the end of the polishing operation in contact with the polishing plate. Light is introduced into one of the fibers, and then propagates to a region where the second fiber core is placed in close proximity. The transfer of light between the two fibers due to evanescent coupling depends on the distance between the two fiber cores, which is affected by the environmental perturbations such as pressure or stress.

Intensity attenuation fiber sensors are inherently simple and require only a modest amount of interface electronics. A primary limitation of such fiber sensors is the variation in the light source intensity caused by temperature changes, aging, and other causes. Compensation for temperature changes is feasible, but it adds to the complexity and the cost of the system.

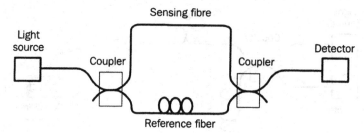

FIGURE 12.32 Schematic diagram of a Mach–Zehnder-type interferometric fiber sensor.

12.7.2 INTERFEROMETRIC FIBER SENSORS

In an interferometric fiber sensor, the external perturbation modulates the phase factor of the light wave propagating within the fiber, which is mostly caused by stretching or expanding of the sensing fiber due to the environmental perturbation to be sensed. If the length of a fiber changes by one wavelength λ, there will be a 2π radian phase shift of the propagating light wave at the exit end of the fiber. For example, if the wavelength of the propagating light is 760 nm and the refractive index of the fiber core is 1.5, a 1 rad phase shift is equivalent to a change in fiber length of only

$$\Delta l = \frac{\lambda}{2\pi\eta} = \frac{670\,\text{nm}}{2 \times 3.14 \times 1.5} = 71\,\text{nm}.$$

It is possible to measure a phase shift considerably less than 1 rad by using interferometric techniques. It is apparent then that one does not need to stretch the fiber very much in order to create an easily observable effect.

Figure 12.32 illustrates schematically an interferometric fiber sensor that is based on the configuration of the Mach–Zehnder interferometer. The light from the source is launched into a polarization preserving fiber. An optical fiber coupler splits the input light wave into two fibers: a reference fiber and a sensing fiber. The light passed through these two fibers is then combined by another coupler, and their interferometric intensity is detected by a photodetector. The optical path in the sensing fiber is sensitive to changes in the environment (e.g., changes in temperature). When the temperature increases, so does the length of the sensing fiber. Therefore, the phase of the light wave from the sensing fiber is delayed with respect to the phase of the light wave that has traveled through the reference fiber. Detection of a temperature change of the order of 0.01°C has been reported with such fiber sensors.

The Michelson-type interferometric fiber sensor, as shown in Figure 12.33, is another type of commonly used interferometric fiber sensor, in which a collimated laser beam is launched into the input fiber and is then splitted into the two arms of the Michelson interferometer by a fiber coupler. Reflection

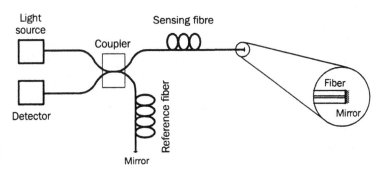

FIGURE 12.33 Schematic diagram of a Michelson-type interferometric fiber sensor.

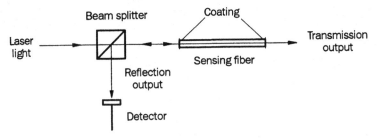

FIGURE 12.34 The Fabry–Perot interferometric fiber sensor.

mirrors are fabricated at the ends of both the reference fiber and the sensing fiber. The reflected light from the two arms is combined again by the same fiber coupler and their interference intensity is detected by the photodetector. Since the light traverses through the sensing fiber twice, the sensitivity of the Michelson interferometric fiber sensor is twice that of the Mach–Zehnder type fiber sensor.

Both the Mach–Zehnder-type and the Michelson-type fiber sensors have two arms, with the reference fiber and sensing fiber separated. This makes them vulnerable to environmental factors because minor environmental changes experienced by the reference fiber can significantly reduce the accuracy of measurements. To overcome this limitation, common-path interferometric fiber sensors have been developed.

The Fabry–Perot interferometric fiber sensor is an example of a common-path fiber sensor. As shown in Figure 12.34, a piece of optical fiber is coated with high-reflectivity, low-transmission mirror at both ends to form a Fabry–Perot cavity (see Section 6.4 for more details of the Fabry–Perot cavity). Due to the high reflectivity of the mirrors, the light wave bounces back and forth, thus experiencing phase delay within the fiber cavity many times. The intensity of the light exiting the fiber cavity is given by

$$I(\phi) = \frac{A^2 T^2}{(1 - R)^2} \frac{1}{1 + [4R/(1 - R)^2] \sin^2 \phi}, \qquad (12.21)$$

where T and R are the transmission and reflection coefficients of the mirrors, A is the amplitude of the input light, and ϕ is the total phase delay for a single transit through the fiber cavity (i.e., $k\eta L$). This output intensity is the result of multiple-beam interference and is detected by the photodetector. An additional advantage of the Fabry–Perot fiber sensor is its high sensitivity, which is much higher than that of the Mach–Zehnder or Michelson fiber sensors (in terms of phase detection) owing to the multiple beam interference. A drawback is the complex relationship between the phase delay ϕ and the output intensity I, as given in Eq. 12.21.

EXAMPLE 12.8

Assume that a number of interferometric sensors are embedded in a smart structure to monitor the stress and strain of certain materials in which sensing fibers are imbedded. Let the wavelength of the light source be 850 nm, and the refractive index of the fiber core be 1.5. If the output intensity detected by the photodetector changes from zero to a maximum value and then to half the maximum value, calculate the strain that the sensing fiber has experienced.

Since a π phase shift corresponds to the interferometric intensity change from zero to the maximum value, the phase change of the light passing through the sensing fiber is given as

$$\Delta p = 1.5\pi.$$

Thus the strain of the sensing fiber would be

$$\Delta l = \frac{1}{2} \frac{\lambda \, \Delta p}{2\pi\eta} = \frac{1}{2} \times \frac{850 \times 1.5\pi}{2\pi \times 1.5} = 212.5 \, \text{nm}.$$

Notice that the factor of 1/2 in the above equation is due to the fact that the light wave travels through the fiber twice in a Michelson interferometer. It can be seen that the sensing fiber has extended by only about 0.2 μm (approximately 213 nm).

12.7.3 FIBER SPECKLEGRAM SENSORS

All the three interferometric fiber sensors described in the preceding section utilize primarily single-mode fibers. In comparison to multimode fibers, single-mode fibers are more fragile, more expensive, and more difficult to multiplex. Therefore, multimode fiber sensors are desired in many practical applications. In this section, we present a multimode common-path fiber sensor in which the spatial information content is exploited for sensing.

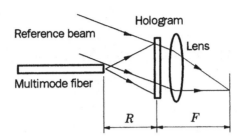

FIGURE 12.35 Formation of the fiber specklegram.

Let us assume that laser light is launched into a multimode sensing fiber, as demonstrated in Figure 12.35. At the exit end of the sensing fiber, the complex light distribution over the hologram aperture can be written as $a(x, y) \exp[i\varphi(x, y)]$, which represents a complex speckle field. If the complex speckle field is combined with an oblique reference beam, the recorded speckle hologram (called the fiber specklegram) is given by

$$h(x, y) = \left| \exp(ikx \sin \theta) + |a(x, y)| \exp[i\varphi(x, y)] \right|^2$$
$$= 1 + |a(x, y)|^2 + 2|a(x, y)| \cos[\varphi(x, y) + kx \sin \theta], \qquad (12.22)$$

where $k = 2\pi/\lambda$ and θ is the incident angle of the reference beam. If the recorded specklegram is illuminated by the same speckle field, $|a(x, y)| \exp[i\varphi(x, y)]$, the oblique reference beam can be reconstructed, by which a high intensity correlation peak can be observed at the back focal plane of the transforming lens, as shown in Figure 12.36a. On the other hand, if the specklegram is read out by a different speckle field, the oblique reference beam will not be reconstructed, and no correlation spot will appear, as shown in Figure 12.36b. However, when the specklegram is illuminated by a slightly changed speckle field, a lower intensity correlation peak is observed. Thus, by varying the speckle content of the illumination beam, a highly sensitive multimode fiber sensor can be developed. In other words, by exploiting the spatial information content of the speckle field, a sensitive fiber specklegram sensor (FSS) can be built. The advantages of the FSS include high sensitivity, less fragility, low cost (since multimode fibers are inexpensive), a single propagation path, and the ability to multiplex.

A FSS with a photorefractive fiber (as speckle hologram medium) is shown in Figure 12.37. The sensing fiber is a step-index multimode silica fiber of about 50 μm core diameter and the photorefractive hologram is a specially doped Ce:Fe LiNbO$_3$ crystal fiber about 7 mm in length and 0.7 mm in diameter. In view of the construction process, a reflection-type fiber specklegram is made, for which the reference beam can be reconstructed by the same fiber speckle field. To introduce the transverse displacement on the

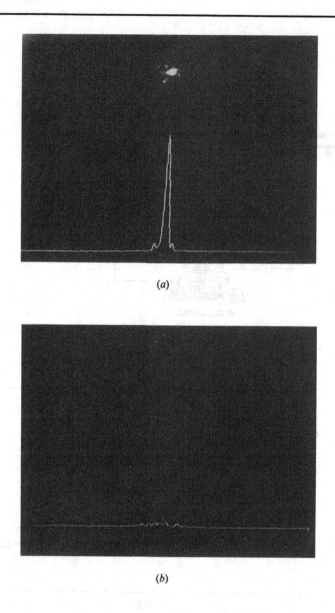

(a)

(b)

FIGURE 12.36 Reconstruction of a fiber specklegram in the focal plane with (a) unchanged fiber status and (b) changed fiber status.

sensing fiber, a cylindrical rod fixed with a piezoelectric driver is used as the bending device. A plot of the normalized correlation peak intensity as a function of transverse is shown in Figure 12.38. There is a fairly linear relationship within 1.4 µm transverse displacement, which can be used for submicrometer displacement sensing. The sensitivity of measurement can be as high as 0.05 µm and it can be improved further if a higher mode sensing fiber is used.

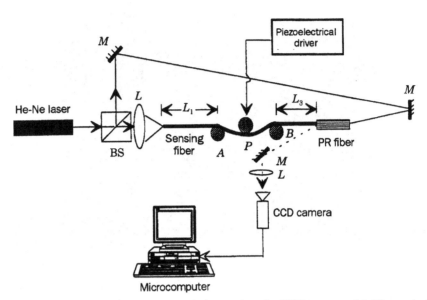

FIGURE 12.37 Experimental setup of the photorefractive FSS system. *M*, Mirror; *L*, lens; BS, beam splitter; L_1, L_3, fiber lengths; *P*, bending point.

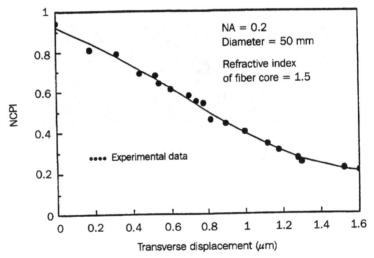

FIGURE 12.38 Normalized correlation peak intensity as a function of transversel displacement. NA, Numerical aperture.

EXAMPLE 12.9

Consider the fiber specklegram formation shown in Figure 12.35. Assume that the hologram is constructed by a fiber speckle field of $a(x,y)$,

(a) Show that the reference beam can be reconstructed by the same speckle field.

(b) Calculate the light distribution at the back focal plane of the transform lens.

Answers

(a) Referring to Eq. 12.22, we have

$$a(x,y)h(x,y) = a[1 + |a|^2 + a\exp(ikx\sin\theta) + a^*\exp(-ikx\sin\theta)],$$

in which the last term is the reconstructed reference beam, that is

$$r(x) = |a|^2 \exp(-ikx\sin\theta).$$

(b) If the preceding equation is subjected to the phase transformation of the lens, at the back focal plane we have (one-dimensional representation)

$$r(x)\tau(x) * h_f(x) = \int |a|^2 e^{-i\frac{k}{2f}(x^2 + 2fx\sin\theta)} e^{i\frac{k}{2f}(\alpha - x)^2} dx$$

$$= |a|^2 e^{i\frac{k}{2f}\alpha^2} \int e^{-i\frac{k}{f}(\alpha + f\sin\theta)} dx$$

$$= |a|^2 \delta(\alpha + f\sin\theta).$$

Thus we see that a high intensity correlation peak is diffracted at $\alpha = -f\sin\theta$.

REFERENCES

1. G. KEISER, *Optical Fiber Communcations*, second edition, McGraw-Hill, New York, 1991.
2. W. ETTEN and J. PLAATS, *Fundamentals of Optical Fiber Communications*, Prentice Hall, Englewood Cliffs, NJ, 1991.
3. H. MURATA, *Handbook of Optical Fibers and Cables*, Marcel Dekker, New York, 1988.
4. E. UDD (ed), *Fiber Optic Sensors*, Wiley, New York, 1991.
5. S. WU, S. YIN and F. T. S. YU, Sensing with fiber specklegrams, *Applied Optics*, 30 (1991) 4468.

PROBLEMS

12.1 Assume that a glass fiber has a refractive index of $\eta_1 = 1.48$ and is surrounded by air.

(a) Calculate the critical angle at the core–air boundary.

(b) Calculate the acceptance angle of the fiber.

(c) Determine the numerical aperture of the fiber.

12.2 Assume that an optical fiber has a core refractive index of 1.46 and a cladding refractive index of 1.43.

(a) Determine the critical angle at the core–cladding boundary.

(b) Find the acceptance angle if the fiber is submerged in water ($\eta_{water} = 1.33$).

(c) Calculate the numerical aperture of the fiber in water.

12.3   The refractive indices of the core and cladding of a single mode optical fiber are $\eta_1 = 1.46$ and $\eta_2 = 1.45$, respectively. If the fiber operates at a wavelength of 700 nm, determine its acceptance angle and calculate the diameter of the core.

12.4   The condenser lens shown in Figure 12.6 is required to launch a parallel laser beam into the optical fiber in Problem 12.3. If the focal length of the condenser lens is given, determine its diameter.

12.5   The typical difference in the refractive indices of the core and cladding of an optical fiber for telecommunications is about 1%. If the refractive index of the core is 1.47, estimate the numerical aperture of the fiber and the critical angle at the core-cladding interface.

12.6   Assume that the ratio of the cladding and core refractive indices of a 2 km long optical fiber is $\eta_2/\eta_1 = 0.98$. Calculate the dispersion of the fiber.

12.7   A single-mode optical fiber is used to transmit monochromatic light at a wavelength of 850 nm. If the refractive indices of the core and cladding of the fiber are 1.48 and 1.46, respectively, estimate the diameter of the core.

12.8   If the cladding refractive index of the fiber in Problem 12.7 is reduced to 1.44, estimate the new core diameter.

12.9   The refractive index difference for a fiber is defined as

$$\Delta = \frac{\eta_1^2 - \eta_2^2}{2\eta_1^2}.$$

(a) Show that Δ can be approximated by

$$\Delta \approx \frac{\eta_1 - \eta_2}{\eta_1}$$

for $\Delta \ll 1$.

(b) Show that the numerical aperture of the fiber can be approximated by

$$\text{NA} = \eta_1\sqrt{2\Delta}.$$

12.10   A step-index multimode fiber with a core diameter of 100 μm and a refractive index difference of 1% is used to transmit a light wave of 850 nm. If the refractive index of the core is 1.45, estimate the normalized frequency of the fiber and the number of guided modes. *Hint*: The normalized frequency is defined as:

$$J = \frac{2\pi}{\lambda}a\sqrt{\eta_1^2 - \eta_2^2},$$

where a is the radius of the fiber core.

12.11   A graded-index fiber has a core diameter of 80 μm with a parabolic refractive index profile. If the numerical aperture of the fiber is 0.2 and the fiber is operating at a wavelength of 700 nm, estimate the normalized frequency of the fiber and the number of guided waves. *Hint*: The number of guided modes for a graded index fiber can be shown to be [1]

$$M_g = \frac{\alpha}{\alpha + 2}(\eta_1 ka)^2 \Delta \approx \frac{\alpha}{\alpha + 2}\frac{J^2}{2},$$

where $k = 2\pi/\lambda$, α is the index profile parameter, and Δ is the core–cladding refractive index difference.

12.12   A step-index multimode fiber has a relative refractive index difference of 1.6% and operates at a wavelength of 600 nm. If the refractive index of the core is 1.5 and the number of propagation modes is 100, calculate the diameter of the core.

12.13   Assume that the core of a step-index single mode fiber is 5 μm in diameter, and that the refractive indices of the core and cladding are 1.5 and 1.45, respectively.

   (a) Calculate the relative refractive index difference of the fiber.

   (b) Determine the shortest wavelength allowed in the single-mode transmission.

12.14   If the core diameter of the fiber in Problem 12.13 is doubled, calculate the required refractive index of the cladding in order to maintain single-mode transmission.

12.15   The acceptance angle of a graded-index fiber is 7.5°. If the refractive index of the core is 1.52, determine the required refractive index of the cladding.

12.16   Assume that a graded-index fiber has a core with a diameter of 60 μm, and that the axis has a refractive index of 1.5 with a profile parameter α of 1.8. If the refractive index of the cladding is 1.495, calculate the number of guided modes when the fiber operates at a wavelength of 700 nm.

12.17   Referring to Example 12.7, if one wants to transmit 10 000 telephone channels via a single optical fiber by combining time-division and wavelength-division multiplexing, how many wavelength channels are required?

12.18   If the loss of an optical fiber is 0.5 dB/km and the input power launched into the fiber by a transmitter is 1 mW, calculate the output power after the optical signal propagates 10 km along the fiber.

12.19   Assume that an optical repeater is required in a fiber link wherever the intensity of the optical signal reduces by 50%. If the loss of the fiber is 0.3 dB/km, determine the distance between the adjacent repeaters.

12.20   Referring to the straight sleeve connector shown in Figure 12.18a, if the numerical aperture of the fiber is 0.25, the diameter of the fiber core is

50 µm, and the distance between the two fiber ends is 10 µm, calculate the coupling efficiency.

12.21 Assume that the wavelength of the light source of an embedded Mach–Zehnder-type interferometric fiber sensor is 670 nm, and that the refractive index of the fiber core is 1.5. If the sensing fiber has undergone an extension of 0.3 µm, describe the changes in the detected intensity.

12.22 State the reasons why mode-stripping regions are necessary in a micro-bending intensity-attenuation fiber sensor.

12.23 Why can only multimode fiber can be used to construct microbending attenuation fiber sensors, as shown in Figure 12.23?

12.24 Assume that the reflectivity of the mirrors in a Fabry–Perot fiber sensor is 0.95. Draw a graph to show the relationship between the output intensity I and the phase delay ϕ, as given by Eq. 12.13.

12.25 Compare the advantages and disadvantages of the Mach–Zehnder, Michelson, Fabry–Perot, and specklegram fiber sensors.

12.26 If the frequency of the local oscillator in a coherent fiber communication system equals the frequency of the signal wave, what is the detected intensity?

12.27 Design an 8×8 switch using 2×2 switches.

12.28 Design a 4×4 star coupler using 2×2 directional couplers.

INDEX

Printed in the United States
by Bookmasters

Printed in the United States
By Bookmasters